Geological Engineering: Exploration and Management

Geological Engineering: Exploration and Management

Edited by **Daniel Galea**

SYRAWOOD
PUBLISHING HOUSE

New York

Published by Syrawood Publishing House,
750 Third Avenue, 9th Floor,
New York, NY 10017, USA
www.syrawoodpublishinghouse.com

Geological Engineering: Exploration and Management
Edited by Daniel Galea

International Standard Book Number: 978-1-68286-124-0 (Hardback)

Printed in the United States of America.

Contents

Preface   VII

Chapter 1   **Analysis of Geodetic Network Established inside the Dobšinská Ice Cave Space**   1
Juraj Gašinec, Silvia Gašincová, Vladislava Zeliz̆naková, Jana Palková, Žofia Kuzevičová

Chapter 2   **Method of Rigid Overlying Strata Failure Assessment of Extracted Seams and its Practical Application**   11
Eva Jiránková

Chapter 3   **Geodetic Determining of Stockpile Volume of Mineral Excavated in Open Pit Mine**   16
Slavomír Labant, Hana Staňková, Roland Weiss

Chapter 4   **Evaluation Methods of Swot Analysis**   27
Michal Vaněk, Milan Mikoláš, Kateřina Žváková

Chapter 5   **Comparison of the Method of Least Squares and the Simplex Method for Processing Geodetic Survey Results**   36
Silvia Gašincová, Juraj Gašinec

Chapter 6   **Impacts of Measuring and Numerical Errors in LSM Adjustment of Local Geodetic Network**   52
Silvia Gašincová, Dušan Knežo, Ladislav Mixtaj, Peter Harman

Chapter 7   **Modelling the Uncertainty of Slope Estimation from a Lidar-Derived Dem: A Case Study from a Large-Scale Area in the Czech Republic**   60
Ivan Mudron, Michal Podhoranyi, Juraj Cirbus, Branislav Devečka, Ladislav Bakay

Chapter 8   **Estimation of Avalanche Hazard in the Settlement of Magurka Using Elba+ Model**   75
Martin Bartík, Matúš Hríbik, Miriam Hanzelová, Jaroslav Škvarenina

Chapter 9   **Application of GPR During Investigation Concerning Causes of Pavement Failure and Road Subgrade Quality in Granitoid Massif Near Simtany**   83
Luděk Kovář, Pavel Pospíšil

Chapter 10   **Evaluation of the Data Quality of Digital Elevation Models in the Context of Inspire**   94
Radoslav Chudý, Martin Iring, Richard Feciskanin

Chapter 11   **Assessing Relations between Water Supply and Demand in the**
             Odra and Morava River Basins   **110**
             Jan Thomas, Miroslav Kyncl, Silvie Langarová

Chapter 12   **Research on Petrophysical Properties of Chosen Samples from the**
             Point of View of Possible CO₂ Sequestration   **118**
             Martin Klempa, Michal Porzer, Petr Bujok, Ján Pavluš

Chapter 13   **Determination of Elevator Shaft Uprightness Applying the Terrestrial**
             Laser Scanning Method   **126**
             Ľudovít Kovanič

Chapter 14   **Risk Assessment in Mining-Related Project Management**   **141**
             Michal Vaněk, Yveta Tomášková, Alena Straková, Kateřina Špakovská,
             Petr Bora

Chapter 15   **Occupational Competence for Improving Industrial Enterprise**
             Competitive Standards   **148**
             Lucie Krčmarská, Igor Černý, Michal Vaněk

Chapter 16   **Homogeneous Magnetic Field Source for Attenuated Total Reflection** **159**
             Doc. Dr. Ing. Michal Lesňák, RNDr. František Staněk, Ph.D., Prof. Ing.
             Jaromír Pištora, CSc., Ing. Jan Procházka

Chapter 17   **Acidification Process in the Area of the Abandoned Ľubietová - Podlipa**
             Cu-Deposit, Slovakia   **166**
             Vojtech Dirner, Jozef Krnáč, Lenka Čmielová, Eva Lacková, Peter Andráš

Chapter 18   **Contribution of Electrical Resistivity Tomography Applied to the Slope**
             Deformation Survey in Lidečko   **176**
             Bladimir Cervantes, Aleš Poláček, Jaroslav Ryšávka

Chapter 19   **Application of Discriminate Analysis to Prediction of Company Future**
             Economic Development   **185**
             Radmila Sousedíková, Jaroslav Dvořáček, Igor Savič

Chapter 20   **Monitoring GNSS Test Base Stability**   **194**
             Marie Subiková, Rostislav Dandoš

Permissions

List of Contributors

Preface

This book was inspired by the evolution of our times; to answer the curiosity of inquisitive minds. Many developments have occurred across the globe in the recent past which has transformed the progress in the field.

Geological engineering is an interdisciplinary approach to study the applications of geological sciences and engineering. It includes management of engineering works through evaluation of geological factors involved. Geological engineering covers a broad spectrum of fields like drilling engineering, petrochemicals, civil engineering, gas processing etc. Geotechnical engineering, mineralogy and planetary geology are also studied under this discipline. This book examines various techniques, methods and practices developed in this field. Students, researchers and professionals engaged in this field will find this text beneficial.

This book was developed from a mere concept to drafts to chapters and finally compiled together as a complete text to benefit the readers across all nations. To ensure the quality of the content we instilled two significant steps in our procedure. The first was to appoint an editorial team that would verify the data and statistics provided in the book and also select the most appropriate and valuable contributions from the plentiful contributions we received from authors worldwide. The next step was to appoint an expert of the topic as the Editor-in-Chief, who would head the project and finally make the necessary amendments and modifications to make the text reader-friendly. I was then commissioned to examine all the material to present the topics in the most comprehensible and productive format.

I would like to take this opportunity to thank all the contributing authors who were supportive enough to contribute their time and knowledge to this project. I also wish to convey my regards to my family who have been extremely supportive during the entire project.

Editor

ANALYSIS OF GEODETIC NETWORK ESTABLISHED INSIDE THE DOBŠINSKÁ ICE CAVE SPACE

Juraj GAŠINEC [1], Silvia GAŠINCOVÁ [2], Vladislava ZELIZŇAKOVÁ [3], Jana PALKOVÁ [4]

Žofia KUZEVIČOVÁ[5]

[1] *Assoc. prof., Ing., PhD., Institute of Geodesy, Cartography and Geographic Information Systems, Faculty of Mining, Ecology, Process Control and Geotechnologies, Technical University of Košice, Park Komenského 19, 043 84 Košice, Slovak Republic, +421 55 602 2846*
e-mail: juraj.gasinec@tuke.sk

[2] *Assoc. prof., Ing., PhD., Institute of Geodesy, Cartography and Geographic Information Systems, Faculty of Mining, Ecology, Process Control and Geotechnologies, Technical University of Košice, Park Komenského 19, 043 84 Košice, Slovak Republic, +421 55 602 2846*
e-mail: silvia.gasincova@tuke.sk

[3] *Ing. Vladislava Zelizňáková, Institute of Geodesy, Cartography and Geographic Information Systems, Faculty of Mining, Ecology, Process Control and Geotechnologies, Technical University of Košice, Park Komenského 19, 043 84 Košice, Slovak Republic, +421 55 602 2449*
e-mail: vladislava.zeliznakova@tuke.sk

[4] *Ing. Jana Palková, Institute of Geodesy, Cartography and Geographic Information Systems, Faculty of Mining, Ecology, Process Control and Geotechnologies, Technical University of Košice, Park Komenského 19, 043 84 Košice, Slovak Republic, +421 55 602 2449*
e-mail: jana.palkova@tuke.sk

[5] *Assoc. prof., Ing., PhD., Institute of Geodesy, Cartography and Geographic Information Systems, Faculty of Mining, Ecology, Process Control and Geotechnologies, Technical University of Košice Park Komenského 19, 043 84 Košice, Slovak Republic, +421 55 602 2916*
e-mail: zofia.kuzevicova@tuke.sk

Abstract

The present article summarizes the progress and results of geodetic works during the construction of a geodetic network inside the Dobšinská Ice Cave underground space to monitor temporal and spatial changes in its ice filling. In order to objectively evaluate the changes, parameter estimations of the first- and second-order of the geodetic network from the set of field geodetic measurements were provided, and a robust analysis of the network was applied in terms of the assessment of impacts of potential outlier measurements on the network geometry.

Abstract

Predložený príspevok sumarizuje priebeh a výsledky geodetických prác počas budovania polohovej geodetickej siete založenej v podzemných priestoroch Dobšinskej ľadovej jaskyne za účelom monitorovania časových a priestorových zmien jej ľadovej výplne. V snahe objektívneho vyhodnotenia týchto zmien boli zo súborov terénnych geodetických meraní stanovené odhady I. a II. rádu geodetickej siete a z hľadiska posúdenia vplyvu potenciálnych odľahlých meraní na geometriu siete bola aplikovaná robustná analýza tejto siete.

Key words: positional geodetic network, robust analysis, Dobšinská Ice Cave

1   INTRODUCTION

The Dobšinská Ice Cave ranks among the most important world's caves. Its magnificent ice filling has remained the same for thousands of years at an altitude of only 920 to 950 meters. The ice filling is not static; it changes depending on climatic conditions and gravitational distortions. On the surface of the ice filling and on the walls of tunnels along a scenic route cut into the ice, various large and small morphological shapes, created by running and dripping water as well as air flow and sublimation of ice, are produced. In addition to a geoscience and environmental point of view, this issue is particularly important namely in terms of safety and maintaining footpaths of the scenic route for visitors. In 2010 and 2011, the mentioned issue started to be addressed in an innovative way by the project VEGA No 1/0786/10 based on the cooperation of the Institute of Geodesy, Cartography and Geographic Information Systems at the BERG Faculty, Technical University in Košice with the State Nature Conservancy of the Slovak Republic and the Slovak Caves Administration in Liptovský Mikuláš. Its main objective was the exact recording and digital modelling of changes in ice filling in caves by means of contactless measuring methods, due to their protection and operation needs. From the perspective of the objective assessment of glaciation development, it is, of course, very important to measure changes in ice filling at the highest level of precision which would be impossible without any well-built horizontal and vertical networks.

2   DOBŠINSKÁ ICE CAVE

The Dobšinská Ice Cave is situated on the southwestern border of the Slovak Paradise National Park in the Spiš-Gemer Karst which ranks among the most important karst areas in Central Europe. The cave is located in the cadastral area of the town of Dobšiná, Rožňava district, at a distance of about 18 km from the town. The cave and its surroundings belong to the National Nature Reserve Stratená [10]. The entrance to the cave, which is on the northern slope of the Duča Hill at an altitude of 969 m, has long been known as the *"ice hole"*. It takes about 25 minutes to walk from the road up to the cave. As regards natural conditions, the Dobšinská Ice Cave is part of the genetic system Stratenská Cave, consisting of 6 separate caves: *Dobšinská Ice Cave, Duča Cave, Stratenská Cave and Dog Holes Cave, Military Cave, Green Cave and Sinter Cave* (Fig. 1). In this system, five genetic levels and two horizons, of which the fourth genetic level is the most developed and the most important, were classified. The entire system was created by two underground streams – the Tiesňava brook and the Hnilec River [1].

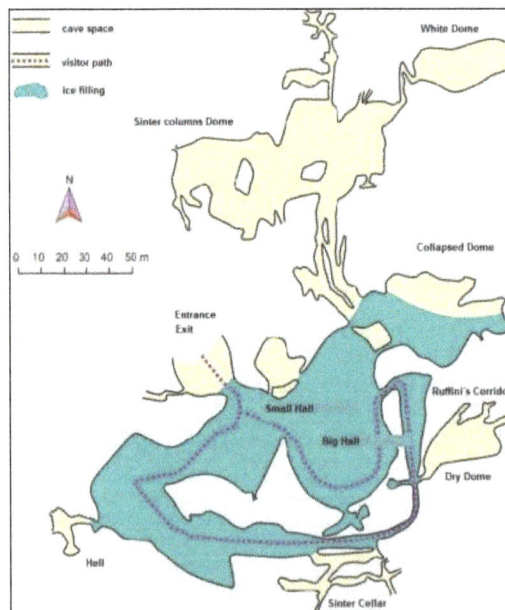

Fig. 1 Dobšinská Ice Cave [4]

Currently, the Dobšinská Ice Cave is largely filled with ice extending here and there up to the ceiling and dividing its upper part into two separate sections – the Small Hall and the Great Hall (Fig. 2).

Fig. 2 Forms of ice filling in the Small and Great Halls

3   ESTABLISHMENT OF GEODETIC NETWORK IN DOBŠINSKÁ ICE CAVE SPACES

Surveying measurements took place in collaboration with personnel of the Slovak Caves Administration based in Liptovský Mikuláš. With regard to the short, two-year duration of the VEGA project, two stages of measurements have been carried out so far. In the first stage of surveying works which took place in March 2011, spatial measurements of the Great Hall and the Small Hall of the Dobšinská Ice Cave were made by the terrestrial laser scanner Leica ScanStation C10 and through the motorized universal measuring Trimble ® VX ™ Spatial Station; in the second stage, the universal measuring station Leica Viva TS15 and the same laser scanner Leica ScanStation C10 were used. In the first stage of surveying works, positional and vertical connections to the preserved points of underground positional and vertical control was realized in the coordinate system of the Datum of Uniform Trigonometric Cadastral Network (S-JTSK) and the Baltic Vertical Datum – After Adjustment (Bpv). The vertical connection was made due to the subsequent creation of a spatial model of the Dobšinská Ice Cave. Since a considerable number of the points were damaged, it was necessary to monument a new minor geodetic control in the second phase of surveying works within solving the project. The point monumentation was performed so that any damage or destruction of the ice filling occurs, the monumented points cannot be damaged or possibly destroyed by natural processes ongoing in the cave or by visitors themselves, and good visibility of as greatest number as possible of other points in the geodetic network are preserved. The structure of the built geodetic network consisting of 11 points is demonstrated in Fig. 3.

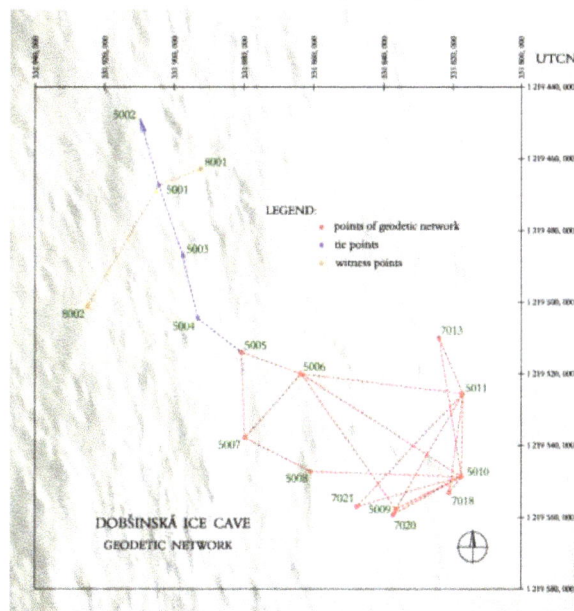

Fig. 3 The clear outline of the network of surface and underground surveying points

Surface points of the new created network were attached to the National Spatial Network; the connection was implemented via the Slovak Spatial Observation Service (SKPOS) using signals of the Global Navigation Satellite Systems (GNSS). For the static GNSS measurement lasted for three hours, two two-frequency Leica GPS1200 receivers and Leica GPS900 were used which identify the points of the orientation line 5001-5002 located approximately in a distance of 1047 m from each other. From the orientation line 5001-5002, the surface surveying witness marks No 8001 and No 8002, the points of underground control of the cave No 5004-5011 monumented in solid, unweathered parts of the rock cave ceiling with surveying pins (Fig. 4) as well as the points No 7013, 7018, 7020 and 7021 monumented with reflecting labels, were determined. With regard to the fact that the point No 5001 is monumented before entering the cave, and could be damaged e.g. during reconstruction works on the pavement, the part of the built geodetic network are also the mentioned witness marks No 8001 and 8002 which can be used in this case in subsequent geodetic measurements. The witness marks were not subject to the adjustment and the robustness analysis of the geodetic network.

Fig. 4 The monumentation of a surveying point in the Small Hall of the Dobšinská Ice Cave

The connection of the geodetic network to the S-JTSK was implemented unilaterally by the connected and oriented traverse consisting of the points 5001, 5003, 5004 and 5005 with an orientation to the point 5002 (Fig. 3).

4 ADJUSTMENT AND ROBUSTNESS ANALYSIS OF POSITIONAL GEODETIC NETWORK

The measurement carried out in situ by using a static method was processed through the use of the Leica Geo Office software. Cartesian coordinates Y, X and Z of the measured points 5001 and 5002 in the European Terrestrial Reference System ETRS 89 (B, L, $H_{elips.}$) are the results of the processing; such as they are transformed to the coordinates Y, X and H in the S-JTSK and the Bpv systems through a transformation service provided by the Office of Geodesy, Cartography and Cadastre of the Slovak Republic (Tab. 1).

Tab. 1 The coordinates of the points 5001 and 5002 determined in the S-JTSK and Bpv systems

Point	Y [m]	X [m]	H [m]
5001	331,904.430	1,219,467.427	969.349
5002	332,194.234	1,218,472.894	871.125

The estimation of parameters of the first order of the geodetic network was implemented by the method of least squares applied to the model of adjustment of intermediary measurements with the following conditions for the unknowns:

$$\mathbf{v} = \mathbf{A}.\mathbf{d\hat{C}} - \mathbf{dl},$$
$$\mathbf{0} = \mathbf{G^T}.\mathbf{d\hat{C}},$$
$$\Sigma_l = \sigma_0^2.\mathbf{Q}_l,$$

(1)

Where:

$\mathbf{v}$ – vector of measured quantities,

$\mathbf{A}$ – design matrix,

$\mathbf{d\hat{C}}$ – vector of complements of adjusted coordinates,

$\mathbf{dl}$ – vector of complements of measured values ($\mathbf{dl=l-l^0}$),

$\mathbf{G}$ – datum matrix,

Σ_l – covariance matrix of measured values,

σ_0^2 – a priori unit weight variance factor,

$\mathbf{Q}_l$ – cofactor matrix of measured quantities [7].

Whereas the number of measured values (n = 44, 19 measured lengths, 25 measured horizontal directions) is greater than the number of the determined parameters (k = 22, 11 points of the network for each point of the network, two coordinates are determined for each point of the network), there are redundant measurements for adjustment (LSM) in the network (Tab. 2). In order to avoid undue weighting of individual measured values due to the influence of the specific cave environment on the surveying process, the appropriate weighting coefficients of the cofactor matrix Ql (estimation of parameters of II order) were calculated by means of the method MINQUE (Minimum Norm Quadratic Unbiased Estimation) [5],[8] and are represented by an estimated standard deviation of the measured lengths $m_d = 1,4\ mm$ and the directions m $_a = 1,49\ mgon$ for the motorized universal measuring station Leica Viva TS15. For testing the residues (corrections) of the measured values in terms of the detection of outlier measurements which could contaminate the set of measured values due to the influence of the cave complex physical environment on the surveying process, the Pope τ–test [3] was used:

$$T_i = \frac{|v_i|}{s_0 \cdot \sqrt{q_{v_i}}} \approx \tau_{f,(1-\alpha 2/2)}. \qquad (2)$$

Where:

T – test statistics,

v_i – corrections,

s_0 – posteriori variance factor,

q_{vi} – cofactor of corrections,

$\tau_{f,(1-\alpha 2/2)}$ – The Pope test critical value for the chosen level of significance α

Tab. 2 Coordinates and accuracy characteristics of points of the positional geodetic network

Point	Y [m]	X [m]	s_y [mm]	s_x [mm]	s_{yx} [mm]	s_P [mm]	a_s [mm]	b_s [mm]	σ [^{g}]	a_c [mm]	b_c [mm]
5005	331880.840	1219513.959	0.0	0.0	0.0	0.0	0.0	0.0	0.00	0.0	0.0
5006	331863.605	1219520.031	0.6	0.2	0.5	0.7	0.7	0.0	321.56	2.3	0.0
5007	331879.790	1219537.898	1.0	0.6	0.8	1.1	1.0	0.5	327.97	3.6	1.6
5008	331860.951	1219547.233	1.3	0.9	1.2	1.6	1.4	0.9	82.76	4.7	3.0
5009	331836.520	1219557.762	1.9	1.6	1.8	2.5	2.1	1.3	60.58	7.3	4.5
5010	331817.615	1219548.690	1.4	2.2	1.8	2.6	2.5	0.8	32.13	8.5	2.9
5011	331817.182	1219525.537	0.8	2.2	1.7	2.4	2.2	0.8	3.11	7.7	2.7
7013	331823.952	1219509.954	1.4	2.1	1.8	2.5	2.2	1.2	375.10	7.5	4.3
7018	331821.131	1219553.159	1.6	2.1	1.9	2.7	2.5	0.9	40.09	8.6	3.1
7020	331837.318	1219559.308	2.0	1.6	1.8	2.6	2.2	1.4	64.59	7.6	4.6
7021	331847.769	1219557.039	1.9	1.6	1.8	2.5	1.9	1.6	83.98	6.6	5.3

Legend:

Y, X – coordinates of points in S-JTSK,

s_y, s_x – mean coordinate errors of point in direction of axes Y and X,

s_{yx} – mean coordinate error of point,

s_P – mean point position error,

a_s, b_s – major and minor semi-axes of standard error ellipse,

a_c, b_c – major and minor semi-axes of absolute 95% confidence level error ellipse

σ   – convolution of major semi-axis of error ellipse.

The error ellipses (Fig. **5**) indicate the error propagation in the direction from the fixating point 5005 to the network. The network as a whole can be characterized by the mean average coordinate error $m_{xy} = 1,5\ mm$ and the mean positional error $m_p = 2,1\ mm$.

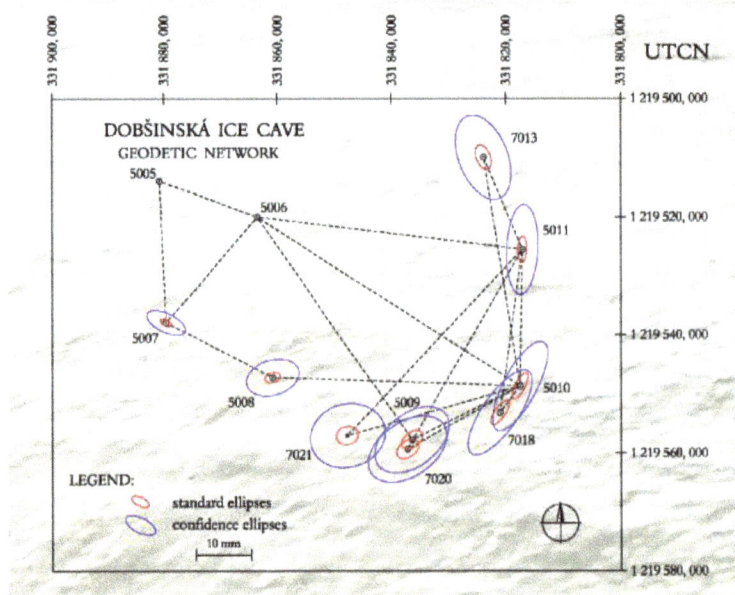

Fig. 5 Standard and confidence error ellipses

In many cases, the terrestrial measurements are tested only in a statistical sense (testing blunders in the set of measured data, testing posteriori variance factor value, testing absolute and relative confidence ellipses, testing posterior estimates of residues) [7]. In the event that a blunder in the statistical tests of estimated residues is revealed, the incorrect measurement can be corrected (in practice most excluded). A problem occurs when a blunder is not revealed during testing, or the test does not recognize any blunders. The aim of the so-called robustness analysis is to determine the degree of the network robustness – to determine the effect of undetected blunders [6]. The degree of robustness of the network is determined by its degree of deformation. The easiest way to describe the deformation of the network consists in the displacement of individual points of the network. The shifts cause a problem as their estimates are datum-dependent, i.e. their estimates depend not only on the network geometry and the accuracy of measurements, but also on the choice of the method of adjustment which has nothing to do with the deformation of the network. If the deformation is to be used to quantify the robustness of the network, then the deformation characteristics must reflect only the network geometry, the type and accuracy of measurements.

Let us denote the shift of a point P_i of the network as follows:

$$\Delta \mathbf{x_i} = \begin{bmatrix} \Delta x_i \\ \Delta y_i \end{bmatrix} = \begin{bmatrix} u_i \\ v_i \end{bmatrix},$$

(3)

in the shift vector $\Delta \hat{\mathbf{x}} = \left(\mathbf{A^T.Q_l^{-1}.A}\right)^{-1}.\mathbf{A^T.Q_l^{-1}}.\sqrt{\lambda_0}.\dfrac{q_{l_i}}{\sqrt{r_i}}$ where each coordinate differences are replaced due to the

simplification of notation by the symbols u_i and v_i, the symbol r_i reflects the redundancy of the network.

Then the gradient tensor in respect to the position is given by:

$$\mathbf{E_i} = grad(\Delta \mathbf{x_i}) = \begin{bmatrix} \dfrac{\partial u_i}{\partial x} & \dfrac{\partial u_i}{\partial y} \\ \dfrac{\partial v_i}{\partial x} & \dfrac{\partial v_i}{\partial y} \end{bmatrix},$$

(4)

where $\Delta \mathbf{x_i}$ is the shift vector of the point P_i. The matrix $\mathbf{E}$ is a so-called matrix of deformation [2], [3] or strain matrix (of the point P_i) and is independent on the method of adjustment – datum. The strain matrix elements of each network point can be determined in several ways. The simplest of them is to obtain partial derivatives right from the shifts. Let us take, for example the point $P_i = P_0$ with the position vector $\mathbf{r}_i = (x_i, y_i) = \mathbf{r_0}$ and the adjacent points P_j with the position vectors $\mathbf{r_j}$. For the point P_i and each point P_j, two equations for two planes corresponding to the shift components u_j and v_j can be then written as follows:

$$\forall j = \ldots: \quad a_i + \left(\frac{\partial u_i}{\partial x}\right)(x_j - x_i) + \left(\frac{\partial u_i}{\partial y}\right)(y_j - y_i) = u_j,$$

$$\forall j = \ldots: \quad b_i + \left(\frac{\partial v_i}{\partial x}\right)(x_j - x_i) + \left(\frac{\partial v_i}{\partial y}\right)(y_j - y_i) = v_j,$$

$$(5)$$

where all the partial derivatives, the absolute members a_i and b_i coordinates x_i and y_i relate to the point P_i. To determine the strain matrix of any point P_i of the network, all points of the geodetic network were used (Fig. **6**). The arrows at the connecting line of two points do not indicate the direction of measurement, but the correlation between the points.

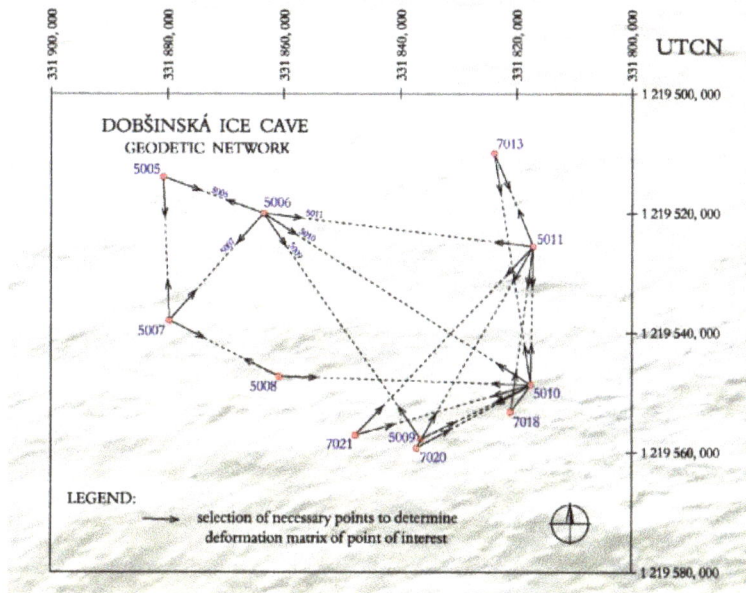

Fig. 6 Relationships between points in the analysis of robustness

Any potential change in the measurement causes a potential deformation of the entire network. To study the degree of deformation caused by potential gross errors in the measurements, only the greatest deformation of each point is to be taken into account. This greatest potential deformation corresponds to the weakest point in the network – a network can be just as strong (robust) as its weakest point.

To describe the dimension of deformation, the following is used [6]:

principal strain σ (strain in scale):

$$\sigma = \frac{1}{2}\left(\frac{\partial u}{\partial x} + \frac{\partial v}{\partial y}\right),$$

$$(6)$$

full shear γ (strain in configuration):

$$\gamma = \frac{1}{4}\cdot\sqrt{\left(\frac{\partial u}{\partial x} - \frac{\partial v}{\partial y}\right)^2 + \left(\frac{\partial u}{\partial y} + \frac{\partial v}{\partial x}\right)^2},$$

$$(7)$$

local rotation ω (strain in orientation):

$$\omega = \frac{1}{2}\left(\frac{\partial v}{\partial x} - \frac{\partial u}{\partial y}\right).$$

$$(8)$$

The values of the principal strain σ, full shear γ, and the rotation ω of each point are illustrated on the relevant maps (Fig. **7**, Fig. **8** and Fig. **9**) by means of circles whose radius corresponds to the numerical value of the appropriate deformation primitive. The analysis results clearly show that the point 7018 is the weakest point in the network in terms of robustness in scale, in configuration and in orientation.

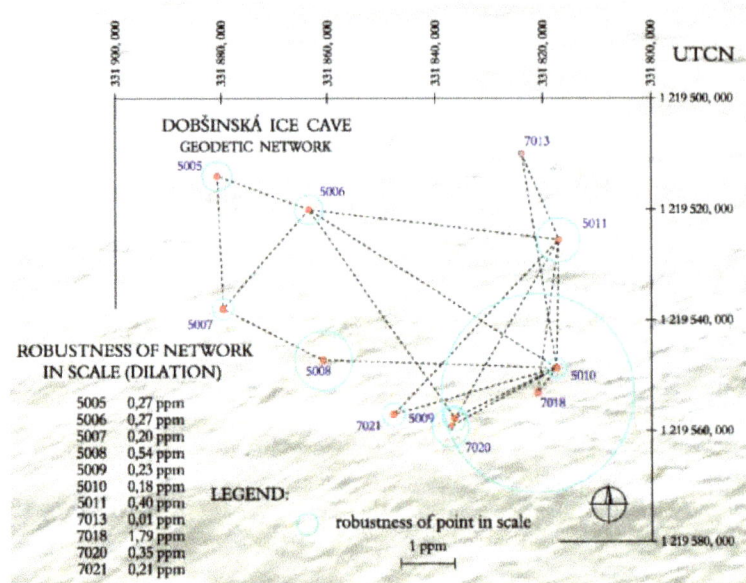

Fig. 7 Robustness of the positional geodetic network in scale

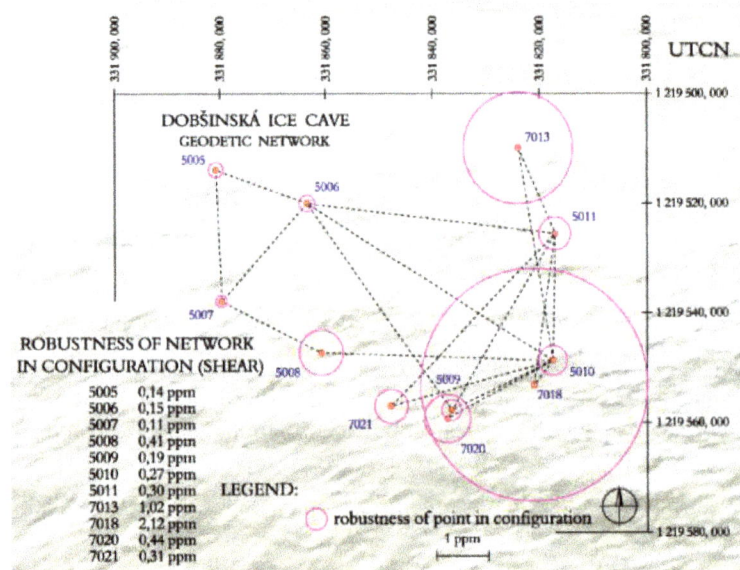

Fig. 8 Robustness of the positional geodetic network in configuration

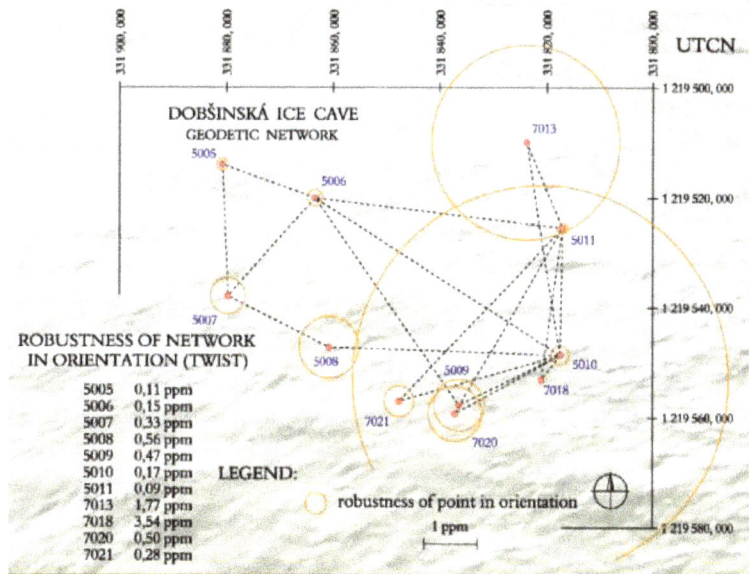

Fig. 9 Robustness of the positional geodetic network in orientation

5   CONCLUSIONS

The robustness analysis highlights the weak points in the network, the use of which may lead to biased monitoring results of ice filling. For its use, it would be appropriate to make further supporting measurements binding to the points 7013 and 7018 and effectively increase the robustness and homogeneity of the network in its weak points.

REFERENCES

[1]   BELLA, P.: Glaciálne ablačné formy v Dobšinskej ľadovej jaskyni. In: *Aragonit*, ISSN 1335-213X, 2003, vol. 8, p. 3-7.

[2]   BERBER, M., DARE, P. J., VANÍCEK, P., CRAYMER, M. R.: On the Application of Robustness Analysis to Geodetic Networks. *Proceedings of Canadian Society for Civil Engineering*, 31st Annual Conference, June 4-7, Moncton, NB, Canada, 2003.

[3]   CASPARY, W.F.: Concepts of network and deformation analysis. First edition. Kensingthon: School of surveying The University of New South Wales, 1987, p.187. ISBN 0-85839-044-2.

[4]   ZELIZŇAKOVÁ, V., GAŠINCOVÁ, S.: Analýza robustnosti polohovej geodetickej siete zriadenej v podzemných priestoroch Dobšinskej ľadovej jaskyne. In: Quaere 2013 : interdisciplinární konference doktorandů a odborných asistentů : mezinárodní vědecká konference : recenzovaný sborník příspěvků : vol. 3 : 20. - 24.května 2013, Hradec Králové. - Hradec Králové. Magnanimitas, 2013 p. 2835-2844, ISBN 978-80-905243-7-8

[5]   IŽVOLTOVÁ, J. : Teória chýb a vyrovnávací počet I. Príklady ku cvičeniam. Žilinská univerzita, 2004, s. 121,. ISBN 9788080702397.

[6]   VANÍČEK, P., CRAYMER, M. R., KRAKIWSKY, E. J.: Robustness analysis of geodetic horizontal networks. In: *Journal of Geodesy*, 2001, vol. 75, p. 199 – 209.

[7]   WEISS, G., ŠÜTTI, J.: Geodetické lokálne siete I. 1.vydanie. Košice, Vydavateľstvo Štroffek, 1997, s.130. ISBN 80-967636-2-8 .

[8]   FABIÁN, M., KOŽÁR, J., SOKOL, Š.: Varianty odhadu parametrov 1. a 2. rádu v geodetickej sieti budovanej po etapách. Geodetický a kartografický obzor, 37 (79), 1991, č. 5, s. 89-93.

[9]   VYHLÁŠKA č. 87/2013 Úradu geodézie, kartografie a katastra Slovenskej republiky z 8. apríla 2013, ktorou sa mení a dopĺňa vyhláška Úradu geodézie, kartografie a katastra Slovenskej republiky č. 461/2009 Z. z., ktorou sa vykonáva zákon Národnej rady Slovenskej republiky č. 162/1995 Z. z. o katastri nehnuteľností a o zápise vlastníckych a iných práv k nehnuteľnostiam (katastrálny zákon) v znení neskorších predpisov v znení vyhlášky Úradu geodézie, kartografie a katastra Slovenskej republiky č. 74/2011 Z. z.

[10]   www.ssj.sk/jaskyne/spristupnene/dobsinska-ladova

RESUMÉ

Vytvorená meračská sieť po svojom dobudovaní aj v spodnej časti Dobšinskej ľadovej jaskyne umožní vytvorenie jej presného digitálneho modelu, umožňujúceho na základe opakovaných expedičných meraní popísať exaktnými aj empirickými závislosťami fyzikálne procesy zmien jej ľadovej výplne. Kvantitatívne nové údaje o sezónnych, cyklických a trendových zmenách ľadovej výplne v Dobšinskej ľadovej jaskyni vo vzťahu k zmenám a sezónnemu režimu klimatických procesov, doplnia doterajšie poznatky o tejto významnej, z celosvetového hľadiska unikátnej zaľadnenej jaskyne pre jej bezpečné a trvalo udržateľné využívanie.

2

METHOD OF RIGID OVERLYING STRATA FAILURE ASSESSMENT OF EXTRACTED SEAMS AND ITS PRACTICAL APPLICATION

Eva JIRÁNKOVÁ

Ing., Ph.D., Institute of Geodesy and Mine Surveying, Faculty of Mining and Geology VŠB - Technical University of Ostrava, 17. listopadu 15, Ostrava, tel. (+420) 59 699 54 29
e-mail: eva.jirankova@vsb.cz

Abstract

The method of rigid overlying strata failure assessment of extracted seams is based on the simultaneous assessment of surface subsidence and seismic activity considering spatio-temporal progress of mining depending on the rock mass character and previous mine activity. If no complete failure of the firm overlaying layers occurs, the surroundings of the worked-out area is considerably supercharged and a risk of anomalous geomechanical phenomena occurrence substantially increases. The paper explains the mechanism of a rigid overlying strata failure under specific condition.

Abstrakt

Metoda hodnocení porušování pevného nadloží exploatovaných slojí je založena na současném hodnocení poklesů povrchu a seismické aktivity vzhledem k časoprostorovému postupu dobývání v závislosti na charakteru horského masivu a předchozí hornické činnosti. Pokud nedojde k prolomení pevných nadložních vrstev, je okolí vydobytého prostoru značně přitíženo a podstatně se zvyšuje nebezpečí vzniku anomálních geomechanických jevů. V článku je vysvětlen mechanismus porušování pevných nadložních vrstev v konkrétních podmínkách.

Key words: strutting arch, surface subsidence, extent of breakthrough, rigid overlying strata breakthrough, subsidence trough

1 INTRODUCTION

Due to the extraction of coal seams the original balance of rock mass is affected and a redistribution of stress occurs, i.e. changes in directions and size of applied main stresses appear. Around the mined out area, always a stress increase occurs, resulting in the compression of the goaf surroundings that reveals itself by a certain measurable surface subsidence. Determining the surface subsidence value with respect to the extent and thickness of the mined out coal faces is important to recognize the conditions under which a deformation of rigid overlying strata occurred. In many cases, a strutting arch is formed over the mined out area and no breach through the entire thickness of the unfaulted rigid overlying strata appears. During the formation of the strutting arch a high concentration load of rocks can occur and anomalous geo-mechanical phenomena can appear. However, also in cases when a breakthrough of an unfaulted roof takes place, the breakthrough extent does not require to be further extended by subsequent mining. Occurred overhangs of the unfaulted firm layers tailed into the non-undermined roof participate in the considerable surcharge of the affected area. The method of the roof failure assessment based on the measurements of surface subsidence conduces to a better overview of main roof failures of extracted seams.

2 DESCRIPTION OF THE ACTUAL STATE OF THE SOLVED ISSUES

Whether a complete failure of rigid overlying strata over the coal face has occurred or not, it is considerably affected by the size of stress applied around the coal face. Provided that any complete failure of the rigid overlying strata has not occur yet, then in the surrounding of the coal face a high stress is applied. This stress rises with increasing mined-out width. Theoretically, it is therefore possible to deduce a rigid overlying strata failure from determining the above applied stress. To determine the stress in a rock mass the following methods are used:

- - Deformometric method by means of discharging the drill core (overcoring),

- - Direct stress measurement by hydrofrac.

Both these methods are considerably expensive and provide the information on stress in a certain place and at a certain time. For continuous findings of stress they are not relevant in terms of plant-scale.

Another method of the indirect assessment of the stress condition of rock mass is drill tests. This method, however, assesses the stress state at a small distance from the workings and does not provide the information on the total stress state of the rock mass.

The next method used for the evaluation of stress and deformation states of a rock mass is mathematical and physical modelling. The mathematical modelling of the rock mass, in which a process of longwall mining is in progress, is not so far at such level to be possible to find out, at which stage of mining a rigid overlying strata failure will occur. The physical modelling by the method of equivalent materials gives indeed a theoretical chance to determine the breakthrough of firm layers, but the model cannot take into account all considerable variety of affecting factors existing in the extracted rock mass. Moreover, it is a very expensive method, which was the reason, why it is not used under the given conditions in the Czech Republic any longer.

3   EXPLANATION OF THE PROPOSED ASSESSMENT METHOD GIST

In assessing rigid overlying strata under specific conditions we proceed as follows:

1. Data collection. The required information involves the results of periodic surface height measurements and the mine-engineering information on mining.

2. Data processing. The mine-engineering information must be spatio-temporally classified.

3. Data assessment. In assessing rigid overlying strata, it is appropriate to compare the measured values of subsidence with theoretical calculations, determine the mining factor and determine the time of a breakthrough that can be specified by a simultaneous assessment of seismic activity records. For the determined time of the complete failure, the coal face width and the characteristics of affecting factors are then specified for further processing.

The roof failure assessment according to the above procedure must be performed under various mining-geological conditions in order to be possible to process the database of the cases being assessed and carry out quantitative and qualitative assessments according to the following schema, Fig. 1.

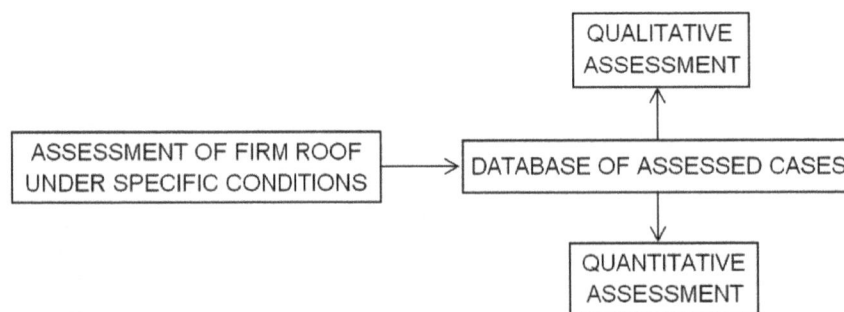

Fig. 1 Schema of the procedure of processing the assessment results for rigid overlying strata

As mentioned above, the method of rigid overlying strata failure assessment of extracted seams is based on the simultaneous assessment of surface subsidence and seismic activity considering spatio-temporal progress of mining depending on the rock mass character and previous mine activity. However, it is important, whether it is the retroactive assessment of earlier mining or the assessment of the actual state of rigid overlying strata failure of current mining.

Provided that the earlier mining is being assessed, when a coherent series of results of periodical altimetry, mine-engineering information and records of registered seismic phenomena are available, the assessment results are demonstrative and depending on the quality of input data precise. Such assessments provide the valuable information on the roof failure condition of the assessed seams for the whole period of mining, i.e. when the breakthrough of the rigid overlying strata occurred related to the width of goaf and when the overrun of overhangs of unfaulted carboniferous rocks occurred with respect to the spatio-temporal progress of subsequent mining.

A usable result of the earlier assessments is the determination of the coal face width at the time of breakthrough. Results from more localities then serve to the quantitative assessment depending on the mechanical character of roof of the seam being assessed and the mining depth. Another usable result of the earlier assessments is the qualitative processing of the gained data from the assessed localities, i.e. the determination of all affecting factors that can be characterized by an appropriate parameter. The qualitative assessment is processed individually, namely for the cases when a periodical roof failure of extracted seams occurred.

Unlike the assessments of the results of earlier extraction, the actual condition of the rigid overlying strata failure of actual extraction is assessed. Although resulting again from the measured values of surface subsidence and registered seismic phenomena, it is necessary to have also a good overview of the roof failure of the formerly extracted seams and assessment experience from similar localities. From this point of view, the assessment of the current failure condition is always a prediction, because the correctness of the actual assessment result proves itself only from the assessment at the time of the following measurement.

The assessment success of the actual state at the same time depends especially on relevant location of the surface points with respect to mining, frequency of surface measurements and on sufficient knowledge of natural conditions and mine-technical information on mining.

4   INPUT PARAMETERS FOR THE PROPOSED ASSESSMENT METHOD

The data collection for the proposed assessment method consists of the results of surface altimetry (Jiránková et al. 2009), geological information, geophysical measurements and mine-engineering information on mining (Jiránková 2008).

The mine-engineering information represents the data gained from:

- Basic mine maps
- Structural, mine, survey boreholes, e.g. geological profiles and determination of compressive strength, thickness and compactness coefficient of individual layers of unbroken roof
- Carbon contour maps
- Detailed tectonic maps
- Records of seismic activity in given areas
- Overviews of performed non-breaking large-scale blasting operations in given areas, or other methods of rigid overlying strata weakening

From the basic mine map, the information is obtained on thicknesses, dimensions and shape of mined-out areas, altitude of extracted parts of the seam, information on tectonic faults, state of monthly progress of the coal head. The geological information is gained from core boreholes; it is especially a macro-petrographic description of rocks and their thickness and lumpiness. The carbon contour map including the results of surface measurements is used for determining the cover thickness and thickness of overlaying carboniferous layers. The detailed tectonic maps provide the information on mutual position of tectonic faults with respect to mining. From the seismic records, the information is obtained on the location, time and amount of released energy of the registered phenomena. From the overviews performed by non-breaking large-scale blasting operations, the information is gained on the place and time of the shot of boreholes at the individual stages of non-breaking large-scale blasting operations, quantity of used explosive, borehole parameters (charging plan) and seismic effect of non-breaking large-scale blasting operations.

5   QUANTITATIVE ASSESSMENT

The result of the quantitative assessment of the achieved results is the determination of functional dependence between the goaf width (at which the breakthrough occurred) and the natural conditions characterized by the inflexibility coefficient (kn) of the rigid overlying strata and the depth of extraction.

The subject of further research is to process, by the proposed method of the roof failure assessment a sufficient number of localities, and to perform the quantitative assessment by the above described method. The assumed behaviour of the main lines is plotted in Fig. 2 in connection with the expected functional dependence between the goaf width and the inflexibility coefficient.

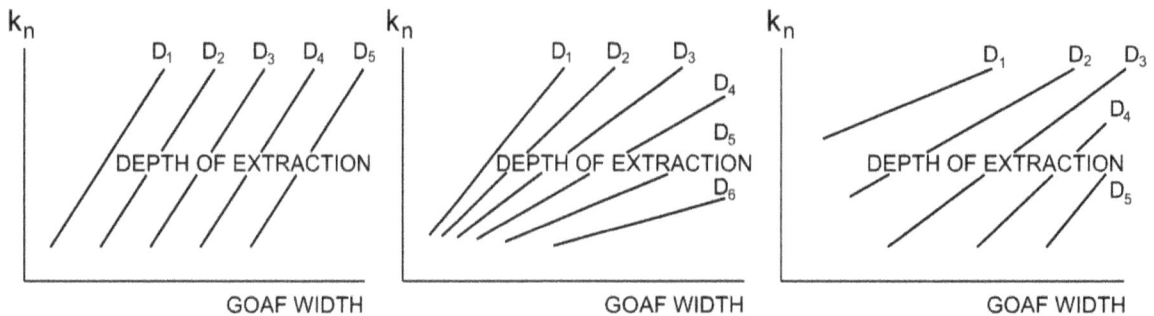

Fig. 2 The expected behaviour of main lines

6 QUALITATIVE ASSESSMENT

The qualitative assessment creates a knowledge database of the assessed cases. As already presented, by the quantitative assessment the functional dependence can be found only between the goaf width at the time of breakthrough, inflexibility coefficient characterizing natural conditions of the unbroken roof and depth of extraction. However, there is much more factors affecting during the mechanisms of failure. The qualitative assessment is based on these affecting factors that can be in some way classified. These are:

- Thickness of unbroken roof
- Number of interlayer interfaces of the unbroken roof
- Number of significant tectonic faults situated nearby the mining
- Depth of extraction
- Extracted thickness
- Number of boreholes of performed non-breaking large-scale blasting operations
- Mass of the used explosive during the large-scale non-breaking blasting operations
- Velocity of coal face progress
- Inflexibility coefficient

Further affecting factors involve the ruggedness and the shape of goaf that can be assessed only by comparison with the formerly assessed localities. The qualitative assessment must be also amended by the information on the strata members locating in the roof (Suchá, Saddle etc.) with respect to their different properties. The Saddle Member is formed predominantly from thick competent rocks and is characterized by a low number of bedding anisometry. These properties of the Saddle Member enable considerable stresses to be concentrated. Therefore, also the number of registered significant seismic phenomena in the Saddle Member is substantially higher than in the Suchá Member.

7 CONCLUSION

The method of the roof failure assessment is based on the simultaneous assessment of surface subsidence and seismic activity with respect to the progress of mining considering the mine-geological conditions of extraction. The result of the method is an interpretation of tensile deformations, which could cause a breaktrough of firm overlaying layers or vice versa such that a strutting arch would be created over the goaf, where a quasi-equilibrium stress state occurred. Another usable result of the assessment method is an interpretation of mined-out area dimensions, at which the breakthrough of firm overlaying layers occurred under specific conditions.

The method is applicable in such areas, where the mining is performed by the method of roof-controlled longwall working with an extracted thickness greater than 1 m. It is possible to use practically the found out functional dependence only in such areas, for which it was determined, e.g. for mines of OKD.

Currently the active or passive means of rock burst prevention are applied, in all successfully, in the area of the extracted seam and effective roof (seams rock burst are not practically recorded). The possibilities of prediction and prevention by active means for phenomena of a regional character with the place of occurrence in a wider surroundings and in a main roof are limited with respect to the complexity of this problem solution under OKD rock burst conditions. The method results serve in the area of geo-mechanics for completing the actual methods of rock burst prevention, in particular (with respect to the utilization of surface measurements for the assessment) in the area of a main roof. The result of the quantitative assessment should serve to plan and project mining activities.

ACKNOWLEDGEMENTS

The research presented in this paper was supported by the Grant Project no. SP2012/148.

REFERENCES

[1] JIRÁNKOVÁ, E. Assessment of Failure Condition of the Extracted Seams of the Firm Roof in Dependence on the Rock Mass Character. GeoScience Engineering, http://gse.vsb.cz, Volume LIV(2008), No.1 p. 1-10, ISSN 1802-5420

[2] JIRÁNKOVÁ, E., MUČKOVÁ, J. Data Collection for Development of Assessment Methods of Rigid overlying strata Failure Based on Mine Surveying Observations. GeoScience Engineering, http://gse.vsb.cz, Volume LV(2009), Issue No.4, ISSN 1802-5420

RESUMÉ

Metoda hodnocení porušování pevného nadloží je využitelná při hlubinném dobývání mocných slojí (s mocností větší než 1m) metodou směrného stěnování na řízený zával. Z výsledků hodnocení porušování pevného nadloží dobývaných slojí je možné posoudit, zda byly horniny pevného nadloží nad výrubem deformovány nebo se nad výrubem vytvořila vzpěrná klenba. Výhodou využití povrchových měření je možnost interpretace změn ve vyšším nadloží.

3

GEODETIC DETERMINING OF STOCKPILE VOLUME OF MINERAL EXCAVATED IN OPEN PIT MINE

Slavomír LABANT [1], *Hana STAŇKOVÁ* [2], *Roland WEISS* [3]

[1] *MSc., PhD., Institute of Geodesy, Cartography and Geographic Information Systems, Faculty of Mining, Ecology, Process Control and Geotechnologies, Technical University of Košice, Park Komenského 19, 043 84 Košice, Slovak Republic, +421 55 602 2859*
e-mail: slavomír.labant@tuke.sk

[2] *MSc., PhD., Institute of Geodesy and Mine Surveying, Faculty of Mining and Geology, VSB-Technical University of Ostrava, 17. listopadu 15, 708 33 Ostrava - Poruba, Czech Republic, +420 59 732 1234*
e-mail: hana.stankova@vsb.cz

[3] *MSc., PhD., Institute of Geotourism, Faculty of Mining, Ecology, Process Control and Geotechnologies, Technical University of Košice, Letná 9, 042 00 Košice,Slovak Republic, +421 55 602 2332*
e-mail: roland.weiss@tuke.sk

Abstract

In the contemporary geodetic practice it is practically a must to use modern geodetic apparatuses and a variety of the CAD (Computer Aided Design) software for processing and visualising spatial data. The present paper deals with geodetic surveying of Kecerovce open pit mine to determine, for the purpose of mine reopening and commencing with mining of andesite, the volume of non-extracted volumes of andesite. The open pit mine is situated on the foot of Slanské vrchy mountain range. Determining of the auxiliary survey control points and the quarry vicinity was performed by GNSS technology and RTK method. Detailed surveying of the open pit mine was performed through an electronic total station. By measurements attained spatial data were processed by pertinent proprietary software. Subsequently, the determined spatial coordinated were imported into the graphic-calculating softwares for further processing and visualisation. These graphical-calculating applications make possible not only 3D modelling and visualising of surfaces but also their analysing, especially then determining the volumetric data that represent various aspects necessary to assess as activities within the related branches so possible future development.

Abstrakt

V súčasnej geodetickej praxi je nevyhnutnosťou používať moderné geodetické prístroje a rôzne CAD (Computer Aided Design) softvéry pre proces spracovania a vizualizácie priestorových údajov. Tento príspevok sa zaoberá geodetickým zameraním povrchového lomu Kecerovce za účelom určenia objemu nevyťažených zásob andezitu pre znovu otvorenie lomu a začatie ťažby andezitu. Predmetný lom je situovaný na upätí Slanských vrchov. Určenie pomocných geodetických bodov a okolia lomu sa vykonalo technológiou GNSS RTK metódou. Podrobné zameranie lomu bolo realizované univerzálnou meracou stanicou. Priestorové údaje získané z meraní sa spracovali v príslušných firemných softvéroch. Následne získané priestorové súradnice boli importované do graficko-vypočtových softvérov pre ďalšie spracovanie a vizualizáciu. Tieto graficko-výpočtové softvéry ponúkajú okrem 3D modelovania povrchov a vizualizácie aj ich analýzu, najmä určenie objemových údajov reprezentujúcich rôzne aspekty pre posudzovanie činností v daných odvetviach s možným ďalším rozvojom.

Key words: calculation cubature, open pit, modelling, visualising, terrain surface

1 INTRODUCTION

In the geodetic practice, determining of the volume presents a highly frequent task. Geodetic measurements present the basis for volumes determining, and using them realised are precision surveying of the terrain shape. In question are determining of various volumes of stockpile, mineral deposit resources or recoverable reserves, potential volume of waste dump, possibly of non-extracted stockpiles remaining in the extractable open pit mine area. According to [1] the quarry is both a workplace and plant for extracting minerals shallowly situated under the earth surface. Within the quarry, most frequently determined are being volumes for the purpose of determining volumes of already mined minerals or possibly for determining extractable volumes of minerals. For the reason, performed within the quarry are geodetic surveys so that, e.g. by a combination of GNSS and a terrestrial technology, precisely surveyed was the quarry terrain shape. These activities stand for field works. The office work is to process the attained data to a format suitable for further processing to determine the volume out of acquired data.

2   DETERMINING STOCKPILE VOLUMES OF SURFACE DEPOSITS

Calculation of the stockpile volume presents a purposeful and self-contained procedure of collecting, processing and assessing the data on geological position and quantity of industrial minerals in their natural or anthropogenic embedding, which allows coming to decisions on their further utilisation.

At determining volumes, a frequently applied step is to determine volume of certain areas, which is performed by multiple ways:

- by calculation equations for elementary formations,

- by planimetric means,

- using a grid,

- by integrating – by integral calculus.

Determining volumes and overburden rocks in open pit mines falls amongst important and integral parts of the mine surveying operations. Core of determining volumes of bodies that are delineated by irregular (topographic) areas dwells in that that irregular bodies are appropriately broken down to smaller geometrical bodies, volumes of which are determined by pertinent solid geometry equations. Even though calculations of volumes are approximate only, they prove to be sufficient for a raft of practical tasks. The calculation of volumes issue is extensively described in [2], which is performed by multiple ways:

- disassembling to regular geometric formations,

- by use of cross-sections,

- from surface/grid levelling results,

- from the form of contour lines,

- by use of digital triangular 3D model.

a.   Determining volumes by use of cad systems

Software surface modelling and subsequent determining of the volume presents the most frequently used way of arriving to information on the stockpile (quantity) of a mineral (industrial mineral) in given locality (storages of the mineral, waste dump, etc.). Determining the stockpile volume of minerals in open pit mines is predominantly utilised to plan the quarry extraction and operation. Modelling a body surface is an integral part of almost all infrastructural projects, from designing slopes at line structures through modelling the landscape and the terrain up to planning of mining operations.

Best the issue can be resolved in software environments e.g. AutoCad, MicroStation etc., which contain effective tools for the body surface modelling. When designing models of bodies calculated is volume of the material. Yet, modelling of a measured real surface in a software environment presents just an intermediate step to determining the volume (see also [3]).

Used for determining the volume of irregular objects most frequently are three methods:

- use of defined surfaces – the method is based on defining two surfaces with subsequent assigning of points one of which performs the function of the real surface (measured surface – upper surface) and the other one performs the function of reference surface (vertical alignment – lower surface). Both the upper and bottom surface are determined by triangulation based on spatial data. The solid circumscribed by these surfaces is divided into n triangular prisms (Fig. 1) and volume of a prism is determined using the equation:

$$V_i = P \frac{\sum\limits_{i=1}^{3} H_i}{3}.$$

(1)

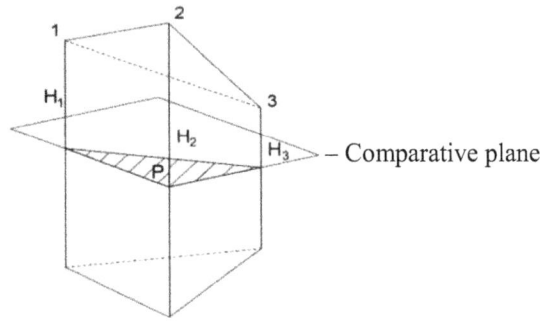

Fig. 1 Illustration of a triangular prism

- use of horizontal sections – contour lines – non-symmetrical object can be divided into layers applying a system of horizontal planes crossing the body. The surveyed terrain will be graphically represented by a map elevation-related features in which can be represented by contour lines. Equidistance of contour lines, i.e. thickness of a layer is to be determined at designing the plant see [4]. To more precisely determine volume of a layer, it is appropriate to use the Simpson's equation, specifically for volume of the layer between P_{570} and P_{571} (Fig. 2):

$$V_{570,571} = \frac{h}{6}\left(P_{570} + 4\left(\frac{\sqrt{P_{570}} + \sqrt{P_{571}}}{2} \right)^2 + P_{571} \right). \tag{2}$$

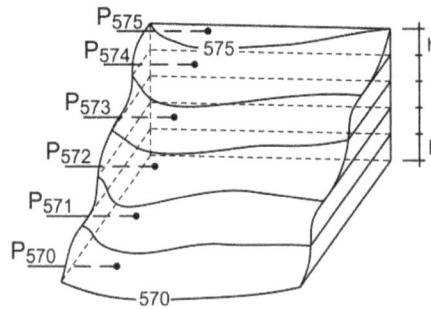

Fig. 2 Dividing the formation by horizontal sectioning

The overall formation volume is calculated by integrating partial volumes of individual layers $V_C = \sum V_i$.

- use of parallel vertical cross sections – this way of calculating is much more demanding than the one utilising surface defining. At this way of volume calculations is the volume determined on the basis of mathematical difference between the terrain surface and the developed surface of the corridor. To determine the volume of excavation or possibly embankments producing earthworks developed has to be surface of the corridor, terrain and traces of cross sections. Cross sections are as a rule selected as running parallel in intervals of 5, 10, 20, 50, possibly 100 m depending on segmentation of shape. Volume of the formation limited by two sections (Fig. 3) is determined analogically with the preceding case:

$$V_i = \frac{h}{6}\left(P_1 + 4\left(\frac{\sqrt{P_1} + \sqrt{P_2}}{2} \right)^2 + P_2 \right). \tag{3}$$

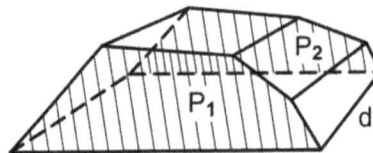

Fig. 3 Cross sections of the formation

b. Factors that influence precision of determining volumes

The overall precision of determining volumes of bodies is influenced by errors occurring during performing measurement and processing activities (errors in the terrain or landscape measurement, erroneously set up graphical supporting materials, errors in planimetry, etc.), as well as approximations errors. The latter errors are the most materially influencing overall imprecision of the volume being determined. Hence, it follows that a body of real dimensions and shape is replaced by a geometrised shape that most closely reflects the real shape, which inevitably follows from the fact that used to determine a volume is mathematical expression.

According to [5], precision of determining volumes by software means is predominantly influenced by the below factors:

- the number of geometric survey control points on the ground plan contour and on surface of the pile by which is the extent and shape of the pile as of an irregular body determined,

- ignorance of the form of the pile subsoil, i.e. height form of the subsoil surface, which is during calculations approximated by a planar area,

- manner of the mathematical expression of the pile surface, segmentation and irregularities in the surface area of the pile,

- way of the pile geometrisation – its disassembling into a set of specific types of elementary geometrical bodies, as well as utilisation of certain relations for determining their volumes.

3   CREATION OF THE DIGITAL TERRAIN MODEL

Digital terrain model is being, in the field of geo-informatics, used approximately since 1950. It presents an integral part of digital processing in the geographic information systems (GIS) environment. Further, it provides the space for modelling, analysing and depicting with topography and the terrain relief phenomena [6].

The relief digital model (DMT), according to [6], describes surface of the Earth void of vegetation and human-built structures such as buildings, bridges, etc. DMT, according to [7], works exclusively with elevations (or heights above sea level) of points. Digital models include a series of cartography techniques by help of which expressed can be the form and shape of the terrain using the [8]:

- contoured model,

- colour hypsometry,

- terrain hachures

- slope

- 3D model.

According to [7], the terrain digital model (DMT) – is digital representation of the Earth´s surface in the memory of a computer composed of data and an interpolating algorithm that allows deriving heights of the intermediate points.

By the surface types, DMT can be divided in:

- Polyedric model of the terrain – planes are represented by irregular triangles. The grid of triangles is made up using triangulation algorithms. Used at construing is the Constrained Delaunay Triangulation. The most significant presentation is TIN. TIN represents the vector-based description of the polyedric model at introducing typological relations among individual triangles.

- Raster model of the terrain – consists of regular planes with common edges = grid. According to [8], this can be developed by deterministic interpolation method IDW and a stochastic method, e.g. trend, kriging.

- Plate model of the terrain – according to [9] utilises dividing the surface into irregular plates of varying sizes, usually of triangular shapes. Made use of at partial planes is nonlinear interpolation that takes into account form of the area of the neighbouring plates.

DMT precision is, as detailed in [10] influenced by the below factors:

- density and arrangement of input points,

- data collection methods (spatial coordinates) – the geodetic methods used fall amongst the most exact ones. Measurements are performed by use of universal measuring stations, the GNSS technology, photogrammetry along with distance Earth surveying (DES),

- interpolation methods – are used to estimate values at places where measured data are unavailable (distribution and density of elevation points varies).

Modelling of DMT using various interpolation methods

Points of known values can be arranged both regularly and irregularly. The serious of methods that include calculations of unknown values occurring between these regularly or irregularly arranged points are called interpolation procedures – methods. Interpolation serves in the process of terrain modelling calculation of values in areas where the values were not measured. In question at interpolating contour lines usually is determining of height (of coordinate z) or calculation of position (coordinates x, y). *Interpolation* stems, according to [11], in that that an unknown function that characterises real form of the investigated plane is replaced, based on the given real values of selected points, by a function of simpler type. The function is labelled interpolating function and the points of discrete geodetic control are called the interpolation nodes. Based on the values of heights of the interpolation nodes the functions allow calculating, with a required precision, detailed form of height values among input points, and possibly form of partial derivations for determining morphometric quantities of the geo-relief. The most commonly used are methods based on weighted linear average of surrounding values. In general, these can be expressed, according to [12], [13] using the equation:

$$z^* = \Sigma(w_i \cdot z_i),$$ (4)

where: z^* is the value estimate,

w_i is the weight (spaceless number from 0 to 1),

z_i are known values.

Classification of interpolation methods

Existent are various points of view according to which can be interpolating methods divided (according to [14]):

a) Deterministic methods

b) Stochastic methods

Deterministic methods perform interpolation directly from the input point measured values. Non-utilised is the probability theory that calculated after each point will be identical final estimate.

Stochastic methods include a randomness element. The resulting spatial prediction is understood to be one of the many that could have been developed. The method is based on the statistic model that assumes spatial dependence among input points.

4   THE INVESTIGATED OPEN PIT MINE

The quarry is located in the Košice region, in cadastral territory of Kecerovský Lipovec laying on the western side of Slanské vrchy (Fig. 4 left). Andesite has been extracted from the discussed open pit mine since 60s of the past century, and though the quarry is at present not in use it is being prepared for commencement. Andesite is extrusive igneous rock of grey to dark grey colour and when it is effloresced it can attain yellowish, ochre or red shades. Andesites, according to [15], occur in the form of superficial deposits or small shallow intrusions. Andesite is most commonly intended for constructional purposes.

The geodetic surveying was performed for the purpose of determining the stockpile volume of andesite found in the quarry. Results of geodetic surveying will be used as a supporting fact for re-starting mining of the mineral, whereas the quarry remains up to the day inactive. Geodetic surveying preceded the open pit mine terrain reconnaissance (Fig. 4). Its purpose was to get acquainted with the terrain, to propose appropriate surveying methods and arrange temporarily stabilised surveying points. Combination of the GNSS a terrestrial technology (in more detail discussed in [15] [17]) was used to profile the andesite open pit mine. The very measurement was performed on March 6, 2012 in bright sunny weather conditions. Dual frequency GNSS tracker Leica GPS900CS (RTK method) was used to locate the points of measurement as well as detailed points in the grassy and bushy terrain outside of the quarry. Based on the reconnaissance results decided was on arrangement of five profiling points for terrestrial surveying. The principle of surveying in geodetic local nets is dealt with in, for example [16].

View from Google Earth   **View from above**   **North view**

Fig. 4 Pictures of the quarry

Decided that used for detailed surveying of the open pit mine would be a selective method based on using universal imaging total station (UMS) Leica Viva TS 15. Position of the device – points 5001 to 5005 were determined by use of GNSS receiver Leica GPS900CS. Points of measurement were temporarily stabilised by crosses engraved in stationary rocks, whereas the subsoil did not allow using another way of stabilising (e.g. reinforcing bars).

Measurements of accessible planimetric and vertical elements of the open pit mine were performed by UMS on a surveying prism. Inaccessible detailed points, along with upper and lower edges of the open pit rock face, were determined via prism-free measuring mode that uses passive reflection of rays. Within the terrain, profiled were some 700 detailed planimetric and vertical points.

Measurement of the grassy and bushy terrain surrounding the open pit mine was performed by use of GNSS receiver and the RTK method, which resulted in localisation of approximately 150 points.

The measurement data were exported from the UMS in the form of index of coordinates XYZ in the *.txt format. Data obtained from the GNSS receiver were processed by software Leica Geo Office 7.0, and the output of the software presented also a list of coordinates XYZ in the *.txt format. The data were subsequently processed by CAD software AutoCAD CIVIL 3D 2008 (*.dwg) and by MicroStation V8 complemented by TerraModeler (*.dgn) superstructure/extension.

To secure higher quality visualisation of specific points these are in TIN, where elevation relations are depicted in colours (Fig. 5)

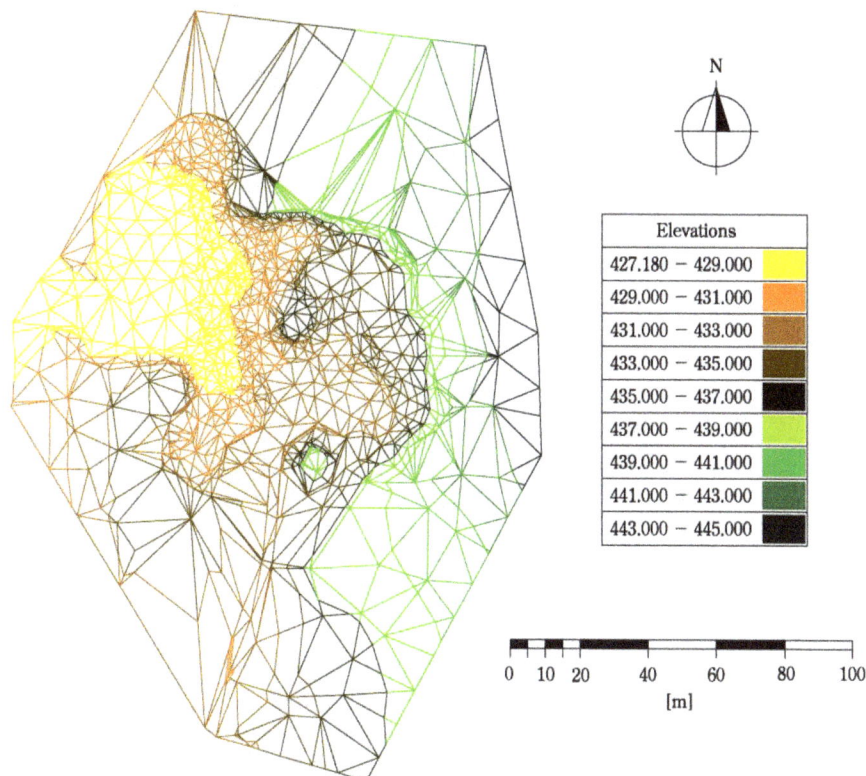

Fig. 5 Visualisation of detailed points through TIN

a. Precision of determining volumes

The process of creation of a 3D model and of determining its shape-wise properties (distance, area, volume) does not often presents the goal of task-solving; it rather presents a means to its resolving. The terrain model presents the basic element of all projects and is used at creating longitudinal profiles, models of the corridor, at calculating volumes or creating terrestrial objects.

The most recent CAD software provide continuous interconnection of all objects with the surface, and there is the guarantee that results will be at all times up-to-date, whilst at modelling of a terrain establishing of volumes is performed automatically (for more information kindly see [18]).

Precision of determining the 3D model volume depends also on consistent work performed during modelling of the surface. If a body surface is created unrealistically, realistically solved will not be the solution goal either (e.g. the volume being determined). 3D model of the discussed open pit mine was processed in three types of the terrain digital model.

DMT modelling is most often performed using the TIN model (Fig. 5). The grid of triangles is made up by use of triangulation algorithms (Delaunay triangulation). Entering the process of creation are break lines so that the quarry was represented in as much detail as possible. Polyedric TIN model approximates real terrain more appropriately than the raster model provided that the terrain points are also appropriately arranged. To develop real terrain, the measured points were located on singularities.

Contour model (Fig. 6) consists of contours with pertinent elevation (height) data. Contour lines break down to basic contour lines (with contour interval of 1 m), reinforced contours (with contour interval of 5 m) and supplementary contour lines (auxiliary – with interval of 0.5 m).

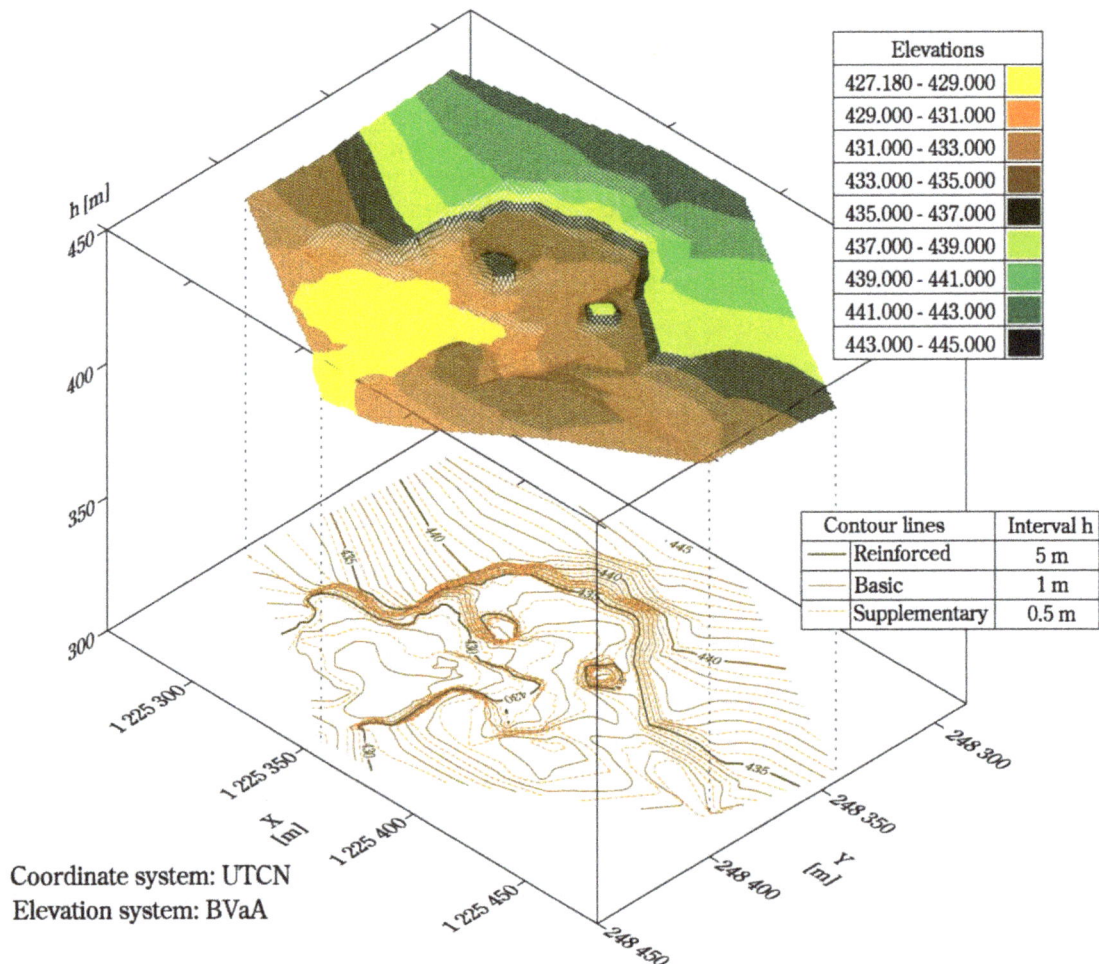

Fig. 6 DTM visualisation by the combination of GRID and the contour line model

GRID falls amongst regular raster structures. The GRID model is made up of regular grid of regular shapes, most frequently of squares (Fig. 6), when the surface is divided into cells. Each cell is bearer of the height above sea level relating to the centre of the square (GRID) or to that of the grid node. [10]

b. Determining the stockpile volume by software

The modelled open pit mine terrain was further processed to determine the stockpile volume of the mineral. All of the three ways of determining the volume differ in the quantity of stockpile determining procedure.

Calculation of the volume using two terrains is, compared with the other two suggested procedures, relatively fast. Currently modelled terrain had to be supplemented by another terrain that represented reference plane delimiting the reserves from the bottom. The software sets volume of the body delineated by the two surfaces. The lower terrain modelling was performed by using the open pit mine boundary points assigned to which was the selected elevation for reference plane on the level of 420 m above sea level (Fig. 7). The plane was selected for the need to establish the lowest exploiting level. Calculated based on comparison of two surfaces was the numeric value of the stockpile volume. Presented in Fig. 8 is an illustration of both surfaces in the GRID structure.

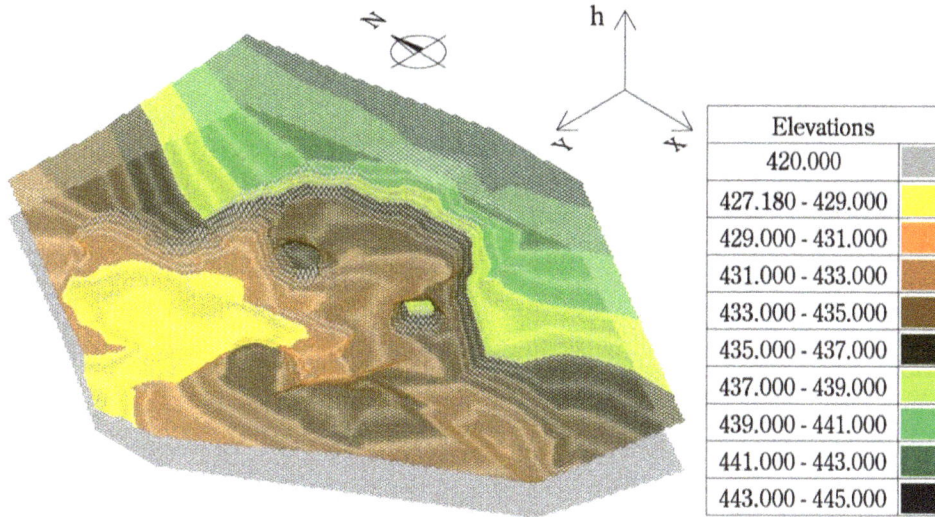

Fig. 7 Iso view of visualisation of two terrains using GRID

Calculating the volume by use of horizontal sections lied in generating the contour lines (Fig. 8) and in determining the volume of areas of surfaces circumscribed by selected contour lines having the interval of $h = 2m$. Partial volumes were determined independently for each horizontal layer, see Tab. 1 accoridng to (2).

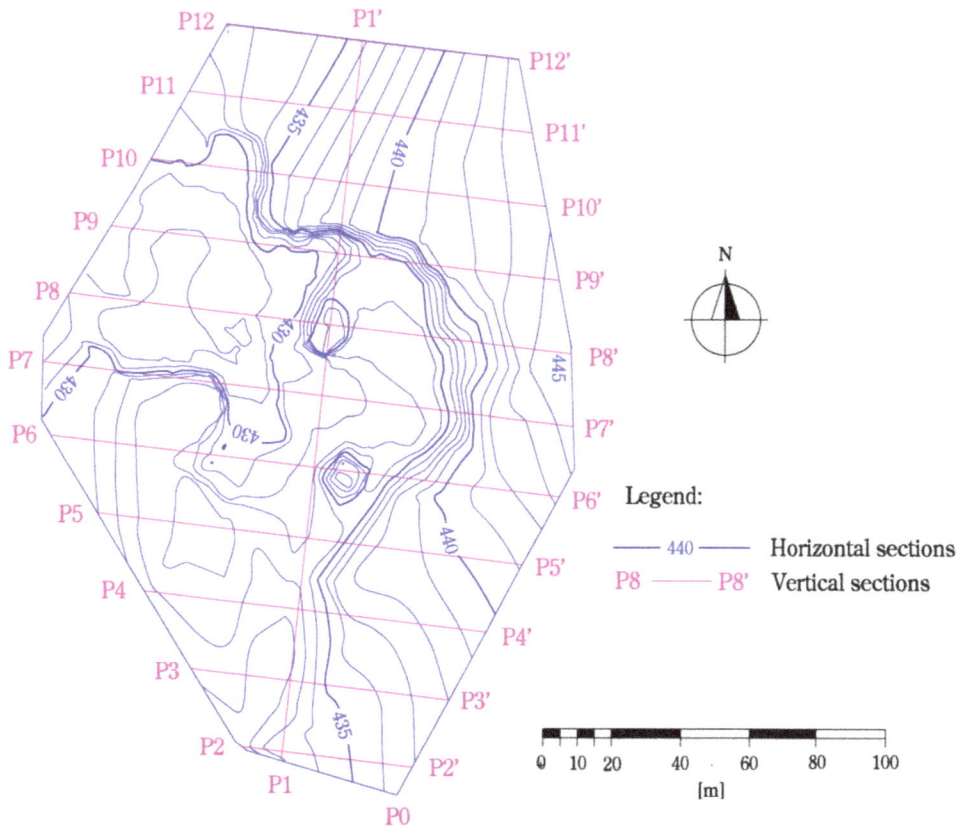

Fig. 8 Illustrations of vertical and horizontal sections

Calculating the volume by use of parallel vertical cross sections was performed employing 11 created cross sections. Further to the request of the principal and based on ratios of precision and size of the quarry, the section interval was selected to be per 20 m (Fig. 8, P0, P1-P12). Calculated, based on by software generated parallel vertical cross sections and on determining their planar areas, was the extractable mineral's volume calculated. Volumes of irregular spatial formations bounded by vertical sections were determined based on the relations known from stereometry (Tab. 2) according to (3).

Tab. 1 Calculation of the volume using horizontal sections

Elevation of the contour line	Contour delineated area	Volumes between contour lines	Total volume
[m]	[m^2]	[m^3]	[m^3]
420	23339.70		
		46659.4	
422	23339.70		
		46659.4	
424	23339.70		
		46659.4	
426	23339.70		
		45806.9	
428	22477.20		
		41783.7	
430	19306.50		
		36224.8	
432	16918.26		
		29579.4	352942.2
434	12661.10		
		21838.2	
436	9177.10		
		15993.7	
438	6816.58		
		11546.2	
440	4729.60		
		7193.8	
442	2464.20		
		2864.1	
444	399.90		
		133.3	
445	0.00		

Tab. 2 Calculation of the volume using cross sections

Section	Section area [m^2]	Spacing of sections [m]	Partial volume [m^3]	Total volume [m^3]
P0	0.00			
		10.04	2375.2	
P2-P2'	709.71			
		20.00	17720.2	
P3-P3'	1062.31			
		20.00	25676.7	
P4-P4'	1505.36			
		20.00	34483.7	
P5-P5'	1943.01			
		20.00	42209.3	
P6-P6'	2277.92			
		20.00	43589.7	351757.1
P7-P7'	2081.05			
		20.00	40416.3	
P8-P8'	1960.58			
		20.00	38087.9	
P9-P9'	1848.21			
		20.00	38011.3	
P10-P10'	1952.92			
		20.00	36870.8	
P11-P11'	1734.16			
		20.00	31716.0	

c.   Comparison of the volume calculations

The surface between individual cross sections does not have to present a linear area (planar structure) at all times. This uncertainty influences precision of the volume calculation. The volume calculated based on two surfaces and Microstation V8 and AutoCAD environments renders relevant results. Selected for the reference volume when comparing final volumes was the one determined within Microstation V8 environment. Volumes determined using two defined surfaces in the CAD software are, at such high quantities of non-extracted stockpiles, differing by minute per-cent value (0.03 %). The difference between volume determined using horizontal and vertical cross-sections demonstrates with both software products larger but statistically insignificant deviations falling within − 0,21 % and + 0,18 %. Imprecisions of these determined volumes are especially due to segmentation of the terrain, morphological shape of areas, size of interval of horizontal and vertical cross-sections, and hence due to generalisation of bordering areas between individual cross-sections **(Chyba! Chybný odkaz na záložku.).**

Tab. 3 Final evaluation and analysis of determining the stockpile volume

Software environment	Volume determined using	Volume	Difference	
			[m³]	[%]
MicroStation V8 with Terramodeler extension	Two surfaces	352 217.8	-	-
	Vertical cross sections	351 757.1	+ 460.7	+ 0.13
	Horizontal sections	352 942.2	- 724.4	- 0.21
AutoCAD Civil 3D 2008 CZ	Two surfaces	352 311.5	- 93.7	- 0.03
	Vertical cross sections	351 586.3	+ 631.5	+ 0.18
	Horizontal sections	352 825.3	- 607.5	- 0.17

5 CONCLUSIONS

The geodetic surveying was performed to determine the volume of exploitable stockpiles of andesite in the mentioned quarry. The arrived at spatial information were used for further processing by CAD systems. The basic step of software processing was modelling of the terrain based on by measurements attained data that would be most closely reflect the terrain real shape. In total considered during processing were 6 ways of determining the stockpile volume within two selected CAD software suites (MicroStation V8, AutoCAD). By mutual comparing of six volumes it was determined that the results attained showed statistically insignificant differences in the interval of <–0.21 %, + 0.18 %>. Used for the modelling process were during field measurements attained spatial data in combination with modern terrestrial and GNSS technologies. The arrived at results are presented in both graphical and tabular forms. Upon processing the considered issue we have concluded that the relatively most precise method of determining the volume seems to be utilisation of the method of two surfaces. Utilisation of the method is not influenced by selection of the software to be used. Presently, in the geodesy practice is the processed issue significantly applicable. Ensuring attainment of maximum precision is contingent to quality of the geodetic gear used and to the experience of the surveyor not only at performing field work but at processing as well. The contribution of authors is in the field of the issue under consideration a comparison of presently used procedures at determining volumes of irregular formations.

REFERENCES

[1]   Cehlár, M., Engel, J., Rybár, R., Mihok, J.: Surface mining. BERG Faculty, TU Košice, 2005. 328 pp. ISBN: 80-8073-271-X.

[2]   MICHALČÁK, O. et al.: Engineering Surveying II. Alfa, Bratislava, 1990. ISBN 80-05-00678-0.

[3]   FILIP, J.: AutoCAD Civil 3D 2008 CZ. Educational materials - updated version, 2007.

[4]   ŠÜTTI, J.: Geodesy. Alfa, Bratislava, 1987.

[5]   ČERNOTA, P., LABANT, S., WEISS, G., HARMAN, P.: Determine of the inert waste volume and landscaping proposal landfill. Coal - Ores - Geological Survey. No. special (2012), p. 6-11. - ISSN 1210-7697.

[6]   KLIMÁNEK, M: Digital terrain models (1). (online - cited 2013-04-23).

[7]   Dictionary of terms of land surveying a cadastre of real estates, (online - cited 2013-04-23) http://www.vugtk.cz/slovnik/1050_digitalni-model-reliefu--digitalni-model-terenu-(dmr--dmt).

[8]   BAYER, T.: Digital terrain models. (online - cited 2013-04-23).

[9]   TUČEK, J: GIS principles and practice Computer Press, 1997.

[10] VOŽENÍLEK, V. et al.: Integration of GPS / GIS in geomorphological research. Palacky University, Faculty of Science, Olomouc 2001. 185 pp.

[11] SABOLOVÁ, J., KUZEVIČOVÁ, Ž., GERGEĽOVÁ, M., KUZEVIČ, Š., PALKOVÁ, J.: Surface modelling in 2D scalar field in GIS environment - 2012. Trends and innovative approaches to business processes, 15 International scientific conference proceedings. 10.-11.12.2012, Košice TU, 2012 S. 1-6. - ISBN 978-80-553-1126-5.

[12] HORÁK, J. Introduction to geostatistics and interpolation of spatial data: Syllabus for seminar participants geostatistics. Ostrava: VSB - Technical University of Ostrava. 2002. p. 3-5.

[13] BRUNČÁK, P: Creation of DEM by means of the various interpolation methods. In: JUNIORSTAV 2011.

[14] LI, J., HEAP, A, D.: Areview of Spatial Interpolation Methods for Environmental Scientists. Geoscience Australia, Record 2008/23, 137 pp. ISBN 978 1 921498 30 5.

[15] MUČKOVÁ, J.; ČERNOTA, P.; BARTÁK, P., MIKOLÁŠ, M.: Mining Subsidence Monitoring of Highway between Ostrava and Fradek-Místek, CZE, Inzynieria Mineralna 1(29): 31–40. ISSN 1640-4920

[16] WEISS, G., ŠÜTTI, J.: Geodetic local networks I. 1st edition. Štroffek, Košice 1997. 88 pp. ISBN 80-967-636-2-8.

[17] WEISS, G., JAKUB, V., WEISS, E.: Compatibility of geodetic points and their verification. 1st edition. TU, Košice, F BERG, 2004. 139 pp. ISBN 80-8073-149-7.

[18] http://www.autodeskclub.cz/sewer-autocad-civil-3d (online - cited 2013.01.25).

RESUMÉ

Článok je venovaný určovaniu objemov nepravidelných telies, konkrétne rieši určenie objemu ložiskových zásob andezitu v povrchovom lome. Modelovanie povrchov a následné určenie objemu je najpoužívanejší spôsob získania informácií o stave zásob resp. množstva nerastu príp. inej suroviny v predmetnej lokalite čí v úložisku nerastu, na skládke a pod. Jedným z dôležitých krokov je modelovanie digitálneho modelu terénu, kde sa využívajú štruktúry TIN a GRID. Avšak modelovanie reálne meraného povrchu v softvérovom prostredí je iba medzikrokom k určovaniu objemov. Na určenie objemu nepravidelných telies sa využívajú najčastejšie tri spôsoby a to pomocou dvoch zadefinovaných povrchov, rovnobežných zvislých priečnych rezov a horizontálnych rezov. Vykonala sa analýza výsledných objemov z jednotlivých softvérov a spôsobov určenia objemu ložiskových zásob. Rozdiel objemov určených pomocou dvoch nadefinovaných povrchov v dvoch rôznych CAD softvéroch je zanedbateľný. Pri porovnaní s ostatnými spôsobmi sa rozdiel v percentuálnom vyjadrení rozdielu objemu pohyboval v rozmedzí od –0,21 % po +0,18 %.

4

EVALUATION METHODS OF SWOT ANALYSIS

Michal VANĚK [1], *Milan MIKOLÁŠ*[2], *Kateřina ŽVÁKOVÁ* [3]

[1] *doc. Ing. Ph.D., Institute of Economics and Control Systems, Faculty of Mining and Geology, VŠB –
Technical University of Ostrava 17. listopadu 15/2172, Ostrava, tel. (+420) 59 732 3336
e-mail: michal.vanek @ vsb.cz*

[2] *doc. Ing. Ph.D., Institute of Mining Engineering and Safety, Faculty of Mining and Geology, VŠB –
Technical University of Ostrava 17. listopadu 15/2172, Ostrava, tel. (+420) 59 732 3179,
e-mail: milan.mikolas@vsb.cz*

[3] *Ing., Hyundai Motor Manufacturing Czech.
Průmyslová zóna Nošovice, 739 51 Dobrá, tel. (+420) 59 614 1417,
e-mail: Katerina.Zvakova@hyundai-motor.cz*

Abstract

Strategic management is an integral part of top management. By formulating the right strategy and its subsequent implementation, a managed organization can attract and retain a comparative advantage. In order to fulfil this expectation, the strategy also has to be supported with relevant findings of performed strategic analyses. The best known and probably the most common of these is a SWOT analysis. In practice, however, the analysis is reduced to mere presentation of influence factors, which does not allow more precise determination of a strategic concept. The content of the article tries to remove this drawback, submitting for public discussion two possible approaches to evaluate the SWOT analysis providing important information for the selection process of an organization's strategic orientation.

Abstrakt

Strategické řízení je neoddělitelnou součástí vrcholového managementu. Formulací správné strategie a její následnou realizací může řízená organizace získat a udržet komparativní výhodu. Aby strategie naplnila toto očekávání, je mimo jiné nezbytné, aby byla opřena o relevantní závěry provedených strategických analýz. Nejznámější a zřejmě i nejpoužívanější z nich je SWOT analýza. V praxi je však analýza redukována na pouhé uvedení vlivových faktorů, což neumožňuje exaktnější určení strategické koncepce. A právě tento nedostatek se snaží odstranit obsah článku, kterým jsou oborné veřejnosti předkládány k diskusi dva možné přístupy k vyhodnocení SWOT analýzy poskytující významné informace pro process volby strategické orientace organizace.

Key words: management, strategy, SWOT analysis, municipality.

1 INTRODUCTION

One key area, which the top management of companies, institutions, municipalities and non-profit organizations (hereinafter referred to as an "organization") deals with, is strategic management which results in formulating their own strategies. Veber's strategy means determining the direction the organization should develop. (Veber, 2007) However, the question is whether such strategic management has a sense under the existing conditions of dynamic development of globalized economy and society. In the publication *Strategic Management* (Keřkovský, 2006), the following arguments can be found: "Conditions are changing so fast that managers cannot plan anything, especially on a long-term basis." Also, Peter Drucker asked questions based on a sceptical view of strategic management like these: "What may a strategy result from in phase of rapid changes and vast uncertainty, which is the period to which the world faces at the threshold of the 21st century?", "Are there any premises which strategies of organizations, especially business organizations, could result from?" "Are there any certainties?". (Drucker, 2000) Drucker sees the way-out in five phenomena he identified as present certainties: the collapsing birth rate in the developed world, shifts in the distribution of disposable income, definition of performance, global competitiveness, growing incongruence between economic globalization and political splintering. As we can see, the phenomena are more social and political rather than economic in their nature. (Drucker, 2000) There is no doubt that it depends primarily on the life philosophy and experience of managers, on their approach to the strategic management. Mainly in practice, but also in theoretical considerations we can see some scepticism. However, the authors of the article are convinced that it is strategic management which can help to attract and retain a comparative advantage of an organization. Veber states in his book *Management*: "In terms of social and economic instability strategic management should be an anchor, which is intended to allow analysing current and expected future situation and suggest the most appropriate direction of development of an organization." (Veber, 2007)

Under current turbulent conditions, the search for fixed points is certainly very challenging, but in our opinion necessary. The basis for this finding can be seen especially in the area of strategic analyses, since in accordance with Keřkovský: "every strategy should be formulated on the basis of real facts detected during analyses that are focused on significant events affecting strategic decisions." (Keřkovský, 2006)

Management theory and practice in particular, know and apply a number of strategic analyses that focus on identifying internal and external environmental factors affecting an organization and prediction of their future development. The best-known ones include a PEST analysis, Porter's model, analysis of sources, BSG model and especially then a SWOT analysis.

Our article is devoted just to a SWOT analysis. The SWOT analysis is a widely used methodology which can be met not only in business practice, but also in the area of municipalities. The analysis frequently ends with the initial definition of various factors like strengths and weaknesses, opportunities and threats of an organization. This is an inadequate and misleading understanding of the SWOT analysis. The SWOT analysis should direct strategic management to a correct strategy or to its starting points based on the mutual comparison and evaluation of relationships between strengths and weaknesses, opportunities and threats. A mere listing of influence factors is insufficient to determine critical strategic positions of an organization. Therefore, we sought a way to establish a credible organization's strategic position on the basis of SWOT. We think we found an interesting way of evaluating SWOT analyses, on which we would like to inform the professional public.

2 BASIC MATRIX OF SWOT ANALYSIS

Apparently, there is no professional publication dedicated to strategic management and planning, which would not mention the SWOT analysis. Keřkovský indicates its exact characteristics like this: "SWOT analysis is a valuable source of information when formulating strategy. The thing is that the basic logic of a strategy proposal is based on its very nature." (Keřkovský, 2006) The analysis is usually described as a tool for the diagnosis of strengths and weaknesses, opportunities and threats of an organization. The description is often supplemented with an illustrative scheme of a relational SWOT matrix. We, in our opinion, present the most cogent view of the SWOT matrix, which appears in the publication *Management* by the authors Koontz and Weihrich (Koontz), see Tab. 1 and Fig. 1. The matrix both incites to the choice of an appropriate strategic direction and draws attention to the dynamics of internal and external environment, which is very valuable with regard to the use of the analysis to strategic purposes.

Tab. 1 : SWOT matrix for strategy formulation

Internal factors / External factors	Internal strengths (S) — Strengths of management, operations, finance, marketing, research and development, engineering.	Internal weaknesses (W) — Weaknesses in the spheres mentioned in the cell "internal strengths"
External opportunities (O) — *(Consider also a risk) such as current and future economic conditions, political and social changes, new products, services and technology.*	**SO strategy: Maxi – Maxi** — *Potentially the most successful strategy which uses strengths of an organization for utilizing opportunities.*	**WO strategy: Mini – Maxi** — *Such as a development strategy to overcome weaknesses in order to take advantage of opportunities.*
External threats (T) — *Such as lack of energy, competitiveness and shortcomings in the areas mentioned in the previous cell "external opportunities"*	**ST strategy: Maxi - Mini** — *Such as the use of internal strengths to overcome the threats or to circumvent them.*	**WT strategy: Mini - Mini** — *Such as reduction, winding-up or joint venture.*

Source: Koontz [3]

However, scientific literature (e.g., Koontz, 1993; Keřkovský, 2002; Veber, 2007) does not usually describe how to evaluate the strengths and weaknesses, opportunities and threats. Certain guidance is provided by Kotler in his book *Marketing Management*: "Marketing opportunities can be classified according to attractiveness and success probability. Threats should be classified according to their seriousness and probability of occurrence. Strengths and weaknesses should be classified according to performance and importance." Nevertheless, the guidance is more marketing-oriented and it is questionable whether it is applicable even outside the business sphere.

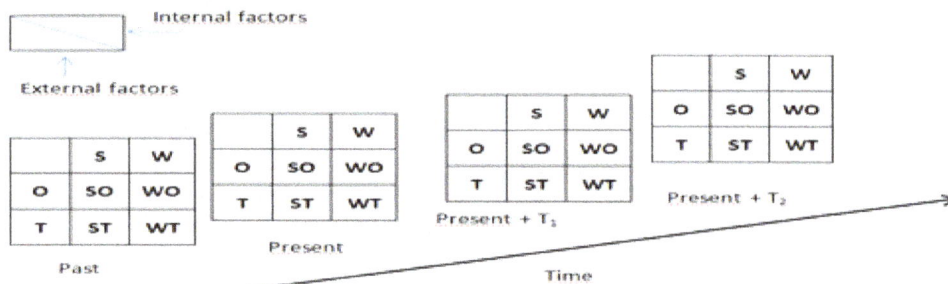

Fig. 1: Dynamics of a SWOT analysis

Source: Koontz [3]

If the assessment of a SWOT analysis is not standardized, an own approach is offered to analysts how to evaluate the information and pinpoint the direction of other strategic considerations. Often, however, as mentioned in the text above, a SWOT analysis ends with mere enumeration of influence factors, see Fig. 2.

Fig. 2: Illustration of a SWOT analysis

Source: The development program of the Moravian-Silesian Region [8]

3   PROPOSALS OF SWOT ANALYSIS EVALUATION FOR STRATEGIC PLANS

The initial, if possible comprehensive identification of factors representing S (strengths), W (weaknesses), W (opportunities), T (threats) is insufficient to determine a strategic orientation. Therefore, we offer to expert discussion two relatively exact approaches for evaluating the intensity of their mutual relations and the intensity of influence on the strategic orientation of an organization. We acknowledge the approaches are not very original, but their application in assessing a SWOT analysis is original.

The **first approach** is based on the evaluation of mutual relationships between strengths and weaknesses, opportunities and threats using the symbols +, - and 0, as follows:

- positive relationship +

- strong positive relationship + +

- negative relationship -

- strong negative relationship - -

- no relationship 0

The evaluation itself is carried out in the following steps:

1) Depending on the intensity and direction of a relationship we assign the symbols (+ +, +, 0, -, --) the interrelationships between strengths and weaknesses, opportunities and threats.
2) The symbols (+ +, +, 0, -, --) will be substituted by numbers (2, 1, 0, -1, 2).
3) For the individual quadrants of the SWOT matrix, we set the strategic potential as a sum of values assigned to the relevant relationships.
4) The highest absolute value of the strategic potential defines the strategic orientation in the SWOT matrix.

The **second approach** is essentially based on the determination of weights for individual influence factors of the relational SWOT matrix. We proceed as follows:

1) The evaluation of the SWOT matrix factors with values in the interval of <-10, 10>. The evaluation is performed by an expert estimate. It results from the logic that weaknesses and threats are assigned a negative sign.
2) The entire SWOT matrix (strengths and weaknesses, opportunities and threats) is evaluated by the sum of the weights of items.
3) According to the evaluation numbers of the appropriate quadrants of the relational SWOT matrix the coordinates of the point which determines the resulting strategic quadrant, are defined. Zero of the coordinate system is located at the intersection of all four quadrants. In each semi-axis, the value of the relevant SWOT matrix area is highlighted. The larger absolute value of the pair of strengths and weaknesses, opportunities and threats determines the required coordinates of the point and the relevant strategic quadrant of the SWOT matrix.

For greater clarity, both approaches are applied to the SWOT analysis of the town of Vratimov.

4   SWOT ANALYSIS OF THE TOWN OF VRATIMOV

The town of Vratimov is a municipality located in the Moravian-Silesian Region, particularly in the northwest tip of the Frýdek-Místek District. It consists of two boroughs: Vratimov and Horní Datyně. Officially, it belongs to the Ostrava-City District and according to its cadastre it borders with the towns of Ostrava, Šenov and Paskov. Together with other 7 towns it belongs to the Silesian Gate Region and thus forms a gateway to the Beskydy Mountains Euroregion. [6] It performs tasks of separate competency to the extent provided by law and delegated competency in the District of Administration – Vratimov, Horní Datyně and Řepiště. [6]

According to the Czech Statistical Office there lived 6 850 inhabitants in Vratimov on 1st January 2010. Residents of the town of Vratimov mostly commute for work to conurbation of four large towns in surroundings – Ostrava, Frýdek Místek, Karviná and Havířov. The registered unemployment rate in Vratimov was 8,9 % on 31 December 2009. The total number of enterprises in Vratimov was 1,199 on 31 December 2010. [7]

Based on the examination of available information and results of own empirical investigation strengths and weaknesses, opportunities and threats of the town were identified, see Tab. 2.

Tab. 2: Overview of strengths and weaknesses, opportunities and threats of the town of Vratimov and weights of influence factors

STRENGTHS	WEAKNESSES
S1 – A favourable location of the town, public transport (10)	W1 – Sewerage system not completed, in some parts of the town gas distribution (-7)
S2 – Plenty of space for relaxation, recreation and sport (7)	W2 – A very poor condition of both local and state roads and pavements (-10)
S3 – Very good conditions for cultural and social activities of citizens (7)	W3 – Problems with aberrant citizens of the town (-5)
S4 – A sufficient network of schools and school facilities (7)	W4 – Lack of parking space (-6)
S5 – Cleanliness and plenty of green (8)	W5 – Excessive traffic (-9)
S6 – A quality environment for living (8)	W6 – Industrial load, home heating with solid fuels – poor quality of air and environment (-10)
S7 – Suitable areas for construction of new family houses (5)	W7 – Missing facilities for seniors, playgrounds and places for young people (-8)
S8 – Supply and level of food shops (8)	
average: 6; sum: 60	average: -7.86; sum:-55
OPPORTUNITIES	THREATS
O1 – New housing construction (6)	T1 – Traffic load (-10)
O2 – Bring people to more active interest in the municipality development (9)	T2 – Lack of finance (-8)
O3 – Building up playgrounds (10)	T3 – Air pollution – local heating and large enterprises (-10)
O4 - Strengthening missing services (competing pharmacy, grocery store in the outskirts of the town, confectionery, fast food, local TV, ...) (7)	T4 - Increasing crime (-8)
O5 - Improving care for seniors (retirement home, ...) (7)	T5 - Failure to receive grant funding from the EU (-8)
O6 - Building a new multi-purpose hall (10)	T6 - Industrial zone in the town centre, landfill, incineration of waste (-9)
O7 - Improvement draw-downs of grants from the EU (9)	
O8 - Development of small and medium businesses (5)	
average: 7.86; sum: 63	average: -8.83; sum: -53

Source: inherent processing

In order to perform the evaluation of the SWOT analysis, we applied the above two approaches. The results of the assessment of strengths and weaknesses, opportunities and threats by the application of the ***first approach*** are given in Annexes 1 to 3. Annex 1 shows the evaluation of the relational SWOT matrix factors using symbols. Annex 2 then illustrates their numerical representation. Annex 3 is derived from Annex 2 and shows the strategic potential of individual quadrants of the SWOT matrix.

The annexes, especially then Annex 3, show that the highest absolute value of strategic potential is 34, which corresponds to the strategic quadrant S – O. Therefore, the strengths and opportunities dominate in Vratimov. The second highest absolute value of 30 represents the strategic quadrant S – T that is followed by the quadrant W – T (19) and the last quadrant is W – O (5). The management should, therefore, formulate a strategy based on the strengths of Vratimov, while leveraging opportunities that occur for the town.

The SWOT analysis can be evaluated using the **second approach we proposed** as well. Its application means that the influence factors of the SWOT analysis are assigned weights, see Tab. 2. Once weights are assigned to the factors, they are add, giving us the evaluation numbers of the relevant SWOT matrix quadrants, see Tab. 2. Subsequently, these numbers are plotted in a coordinate system, see Fig. 3. The last step is to find a point that defines the strategic target quadrant. The searched coordinates of the point are given by a higher absolute value of numbers on the horizontal and vertical axes, see Fig. 3.

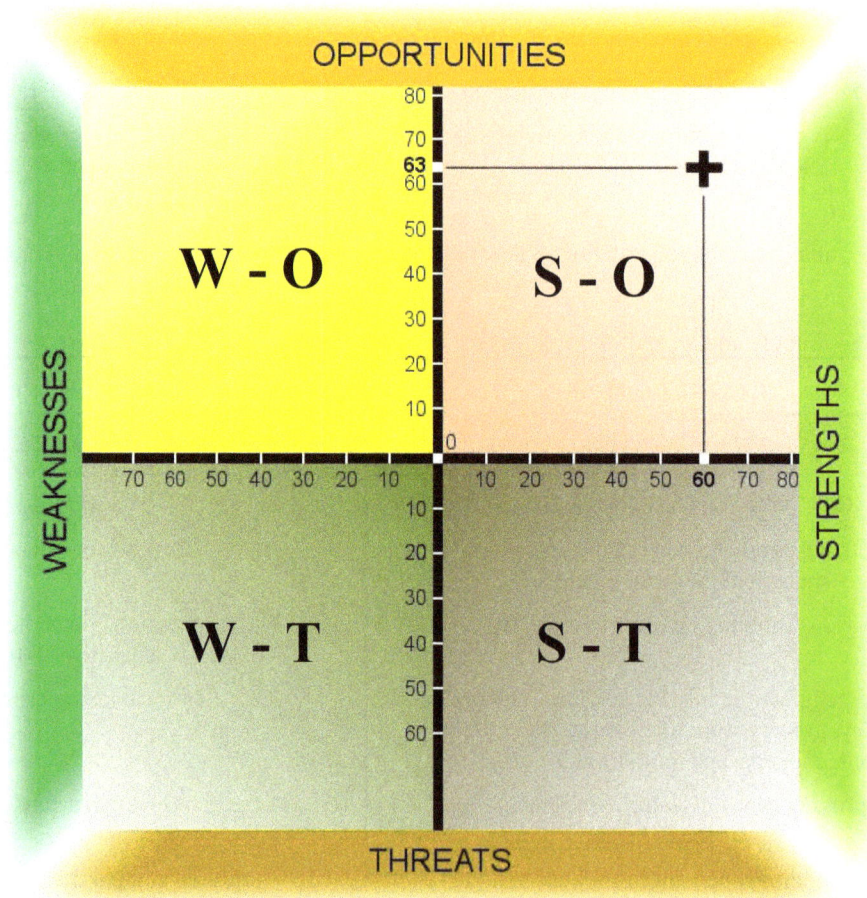

Fig. 3: Determining the strategic quadrant

Source: inherent processing

The figure shows that the target strategic quadrant is the quadrant S – O.

5 CONCLUSIONS

Both our proposed approaches for the SWOT analysis evaluation in case of the analysis of the town of Vratimov resulted in the target S – O strategic quadrant. To generalize this conclusion, the management of an organization can receive with this quadrant some guidance for the formulation of a specific strategy.

As indicated in Fig. 2, the analyst should not forget about the dynamics and thus the possibility of changes or development of the items under consideration in time. It can therefore be recommended to conduct the SWOT analysis in a time perspective to obtain more stereoscopic vision for strategic considerations.

The aim of this paper was not to carry out a SWOT analysis of the town of Vratimov, but to inform the professional public about our proposed procedures for its evaluation. We believe that the advantage of the proposed approaches is the efforts to extract exact information from SWOT analyses for identifying and verifying strategic concepts of organizations. We also believe that this-way-set conclusion of strategic analysis allows building a future strategy of a managed organization much harder to prevent frequent changes in strategic plans. It is understandable that so-called strategic gaps cannot be completely eliminated, but a clearer assessment of SWOT analysis can help to reduce their size and frequency.

Whether our proposed approaches for the evaluation of a SWOT analysis will fulfil their ambitions, it depends mainly on taking our suggestions into practice.

REFERENCES

[1]   DRUCKER, P. F., Výzvy managementu pro 21. století. 1. vyd. Praha: Management Press, 2000. 187 s. ISBN 80-7261-021-X.

[2]   KEŘKOVSKÝ, Miroslav; VYKYPĚL, Oldřich. Strategické řízení : teorie pro praxi. 1. vydání. Praha : C. H. Beck, 2002. 172 s. ISBN 80-7179-578-X.

[3]   KOONTZ, H. WEIHRICH, H. Management. 10. vydání. Praha : Victoria Publishing, 1993. 659 s. ISBN 80-85605-45-7.

[4]   KOTLER, Philip. Marketing management. 10. rozšířené vydání. Praha : Grada Publishing, 2001. 719 s. ISBN 80-247-0016-6.

[5]   VEBER, Jaromír a kol. Management. Základy, prosperita, globalizace. 1. vydání (dotisk) Praha : Management Press, 2007. 700 s. ISBN 978-80-7261-029-7

[6]   *Oficiální   stránky   města   Vratimova*.   [online].   [cit.   2012-01-12].   Dostupný   z   WWW < http://www.vratimov.cz>

[7]   *Veřejná   databáze   ČSÚ*.   [online].   [cit.   2012-01-12].   Dostupný   z   WWW < http://vdb.czso.cz/vdbvo/hledej.jsp?vo=null&q_text=vratimov&q_rezim=3>.

[8]   *Program   rozvoje   Moravskoslezského   kraje*.   [online].   [cit.   2012-01-12].   Dostupný   z   WWW <http://www.bestpractices.cz/praktiky/obecne_dokumenty/prk.pdf >.

RESUMÉ

Jednou z klíčových oblastí, které se věnuje vrcholový management podniku, instituce, municipality i neziskové organizace (dále jen organizace) je strategické řízení, které vyúsťuje ve formulování vlastní strategie. Manažerská teorie a zejména praxe zná a užívá řadu strategických analýz, které se zaměřují na identifikaci ovlivňujících faktorů vnějšího a vnitřního prostředí zájmové organizace a predikci jejich dalšího vývoje. Mezi nejznámější patří PEST analýza, Porterův model, analýza zdrojů, BSG model a SWOT analýza.

A právě SWOT analýze se věnuje i náš článek. SWOT analýza je široce používanou analýzou, se kterou se můžeme setkat nejen v podnikové praxi, ale i v oblasti municipalit. Odborná literatura však obvykle nepopisuje, jak vyhodnotit silné a slabé stránky, příležitosti a hrozby.

Samotná identifikace položek hodnocených oblastí je nesporně přínosná, avšak dle našeho názoru nedostatečná. Proto k odborné diskuzi nabízíme dva naše přístupy.

První přístup je založen na ohodnocení vzájemného vztahu silných a slabých stránek příležitostí a hrozeb pomocí symbolů +, - a 0. Druhý přístup je v podstatě založen na určení vah jednotlivým položkám SWOT. Pro větší názornost byly oba přístupy aplikovaný na SWOT analýze města Vratimov.

Annex 1: Evaluation of the relational SWOT matrix factors using symbols

	Opportunities								Threats					
	O1	O2	O3	O4	O5	O6	O7	O8	T1	T2	T3	T4	T5	T6
S1	++	0	0	+	0	0	+	++	-	0	0	0	0	-
S2	0	+	++	0	0	++	++	0	-	-	--	0	-	-
S3	0	+	0	0	0	0	+	0	0	--	0	0	-	0
S4	0	0	0	0	0	0	+	0	0	-	0	0	-	0
S5	++	+	0	0	0	0	+	0	--	0	--	0	0	--
S6	++	+	0	0	0	0	+	0	--	0	--	-	0	--
S7	++	+	0	+	0	0	+	0	-	0	--	0	0	-
S8	+	0	0	++	0	0	0	++	0	0	0	0	0	0
W1	--	0	0	0	0	0	--	0	0	0	0	0	--	0
W2	0	0	0	0	0	0	0	0	--	-	0	0	0	0
W3	0	0	0	0	0	0	0	0	0	0	0	--	0	0
W4	-	0	0	0	0	0	0	0	-	0	0	0	0	0
W5	-	0	0	0	0	0	0	0	--	0	--	0	0	-
W6	-	0	0	0	0	0	0	0	-	0	--	0	0	--
W7	-	-	+	0	+	+	+	0	0	-	0	0	0	0

Annex 2: Numerical representation of the relational SWOT matrix factors

	Opportunities								Threats					
	O1	O2	O3	O4	O5	O6	O7	O8	T1	T2	T3	T4	T5	T6
S1	2	0	0	1	0	0	1	2	-1	0	0	0	0	-1
S2	0	1	2	0	0	2	2	0	-1	-1	-2	0	-1	-1
S3	0	1	0	0	0	0	1	0	0	-2	0	0	-1	0
S4	0	0	0	0	0	0	1	0	0	-1	0	0	-1	0
S5	2	1	0	0	0	0	1	0	-2	0	-2	0	0	-2
S6	2	1	0	0	0	0	1	0	-2	0	-2	-1	0	-2
S7	2	1	0	1	0	0	1	0	-1	0	-2	0	0	-1
S8	1	0	0	2	0	0	0	2	0	0	0	0	0	0
W1	-2	0	0	0	0	0	-2	0	0	0	0	0	-2	0
W2	0	0	0	0	0	0	0	0	-2	-1	0	0	0	0
W3	0	0	0	0	0	0	0	0	0	0	0	-2	0	0
W4	-1	0	0	0	0	0	0	0	-1	0	0	0	0	0
W5	-1	0	0	0	0	0	0	0	-2	0	-2	0	0	-1
W6	-1	0	0	0	0	0	0	0	-1	0	-2	0	0	-2
W7	-1	-1	1	0	1	1	1	0	0	-1	0	0	0	0

Annex 3: Strategic potential of the relational SWOT matrix quadrants

	Opportunities								Threats					
	O1	O2	O3	O4	O5	O6	O7	O8	T1	T2	T3	T4	T5	T6
S1														
S2														
S3														
S4				34							-30			
S5														
S6														
S7														
S8														
W1														
W2														
W3														
W4				-5							-19			
W5														
W6														
W7														

5

COMPARISON OF THE METHOD OF LEAST SQUARES AND THE SIMPLEX METHOD FOR PROCESSING GEODETIC SURVEY RESULTS

Silvia GAŠINCOVÁ [1], Juraj GAŠINEC [2]

[1] *Assoc. prof.,Ing., PhD., Institute of Geodesy, Cartography and Geographic Information Systems, Faculty of Mining, Ecology, Process Control and Geotechnologies, Technical University of Košice Park Komenského 19, 043 84 Košice, Slovak Republic, +421 55 602 2846 e-mail: silvia.gasincova@tuke.sk*

[2] *Assoc. prof., Ing., PhD., Institute of Geodesy, Cartography and Geographic Information Systems, Faculty of Mining, Ecology, Process Control and Geotechnologies, Technical University of Košice, Park Komenského 19, 043 84 Košice, Slovak Republic, +421 55 602 2846 e-mail: juraj.gasinec@tuke.sk*

Abstract

The present paper is devoted to the use of the simplex method in the processing of results from geodetic measurements as compared with the standard used method of least squares. Using the simplex method, a minimization problem is usually solved in a standard tabular form by rearranging lines and columns in order to find an optimal solution. The paper points out the simpler, more stable and more efficient way to solve a problem of linear programming through a matrix of relations.

Abstrakt

Predložený príspevok je venovaný použitiu simplexovej metódy pri spracovaní výsledkov geodetických meraní v porovnaní so štandardne používanou metódou najmenších štvorcov (MNŠ). Minimalizačný problém je pri simplexovej metóde zvyčajne riešený tabuľkovou formou, preskupovaním stĺpcov a riadkových operácií, ktorých cieľom je nájdenie optimálneho riešenia. V príspevku je poukázané na numericky nenáročnejšiu, stabilnejšiu a efektívnejšiu cestu riešenia problému lineárneho programovania pomocou maticových vzťahov.

Key words: method of least squares, simplex method, L1 norm, L2 norm, outlier detection

1 INTRODUCTION

In geodetic practice, the least square method (LSM) ranks among standard used processing methods. This method is based on the vector of corrections of the L2 norm which together with the L1 norm is most often applied to the processing of results of geodetic measurements. However, a prerequisite for the proper functioning of the LSM is a normal distribution of errors; otherwise the created probabilistic model is not correct. For this reason, the geodetic practice started to use variant processing methods as well. Such methods include, for example, robust estimation procedures that preserve their function in a certain neighbourhood of a normal distribution, i.e. not fail in case of a moderate failure to comply with this requirement. The more the method is resistant, the more it is robust. Several types of these methods are known, from the robust M-estimates [5],[6],[7],[8] to linear programming methods to which undoubtedly the simplex method belongs. The LSM and the simplex method will be demonstrated on two examples; on the examples of a regression line and a geodetic network. In the first case, the models of a regression line and a geodetic network are loaded by the normal distribution of measurements, in the other the line and the geodetic network are loaded by the value of an experimental outlier.

2 METHOD OF LEAST SQUARES

The essence of this method lies in minimizing the sum of squares of deviations in measurements of the behaviour of any quantity or physical phenomenon (3). The least square method is based on the condition of so-called *L2-norm*; the norm is the number assigned to each n-dimensional vector of residual deviations

$v = (v_1, v_{2,\cdots}, v_n)$ that in some sense characterizes its size [3] , [13] . In geodesy, the most commonly used types of objective functions are as follows:

$$\rho(v) = \left(\sum_{i=1}^{n} |v_i|^p \right)^{\frac{1}{p}} = \min. \qquad i \in \langle 1, n \rangle \qquad (1)$$

Where:

p - a parameter defining a special type of an objective function,

v_i – a vector of residual deviations (vector of corrections).

Assuming that $p = 2$ *(L2 norm)*, the objective function is as follows:

$$\rho(v) = \left(\sum_{i=1}^{n} |v_i|^2 \right)^{\frac{1}{2}} = \min \qquad (2)$$

which leads to the least square method that leads under certain conditions to the most reliable estimates of unknown quantities, and hence it is the most commonly used method in geodetic practice. The least square method will be explained on a one-dimensional linear model, while all the estimation methods will be demonstrated on the example of a regression line.

Let us assume that the following linear relationship exists between a variable y, variables X (Fig. 1) and a random component u_i:

$$y_i = \beta_0 + \beta_1 X_{i1} + \beta_2 X_{i2} + \cdots + \beta_k X_{ik} + u_i \qquad \text{for } i = 1, 2 \cdots n$$
$$y = \beta_0 + \beta X + u \qquad (3)$$

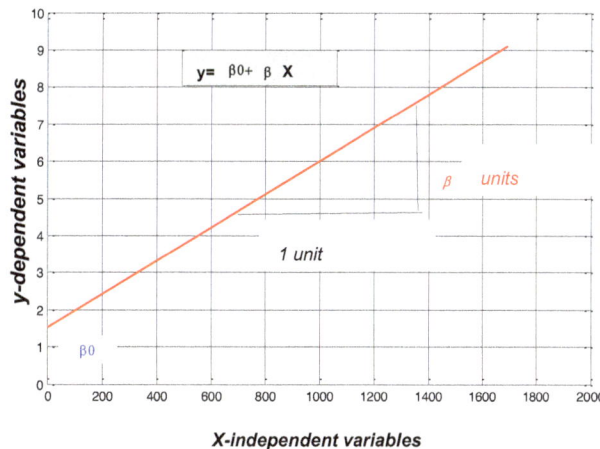

Fig. 1 One-dimensional linear model

Where:

y_i – dependent variables,

X_1 , X_2, ... X_k – independent variables,

β_0 – the parameter indicating a value of the variable y provided that the variable X is equal to 0, the so-called intercept

β - the regression coefficient defining the slope of the regression line,

u_i – the random component.

Provided that we have n pairs of the independent variable X and the dependent variable Y $[x_1, y_1], [x_2, y_2], \cdots [x_n, y_n]$ and are sufficiently convinced of the linear dependence, we can construct the line that best describes this fact:

$$\hat{y}_i = b_0 + b_1 x_i \qquad \text{kde i} = 1, 2 \cdots n \ldots \ldots \text{vyrovnávajúca regresná priamka} \qquad (4)$$

adjusting regression line

Where:

$\hat{y}_i$ – the adjusted one (theoretical value) of the dependent variable,

x_i – the dependent variable for the i-th observation,

b_0 – the point estimate of the parameter β_0,

b – the point estimate of the parameter β.

For residual deviations (in the geodesy vector of corrections) $\mathbf{v} = \mathbf{v}_1, \mathbf{v}_2, \cdots, \mathbf{v}_n$, the relation $v_i = y_i - \hat{y}_i$ applies.

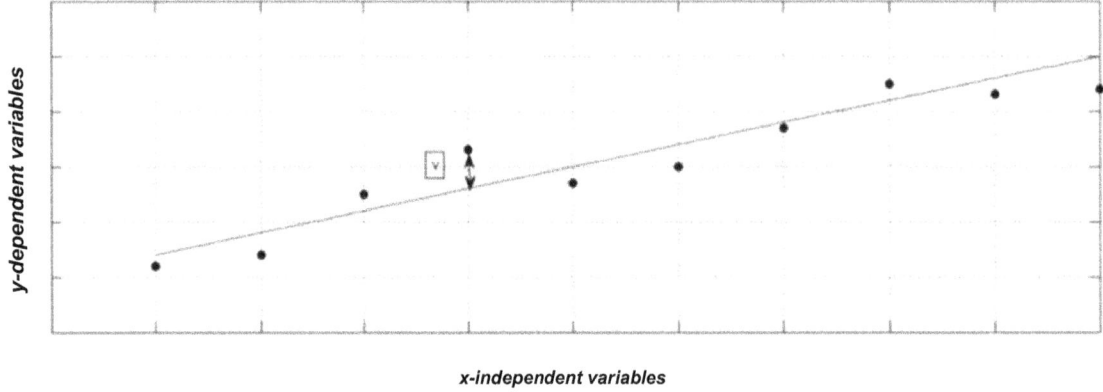

Fig. 2 Residual deviations

Since the residual deviations may take a positive as well as negative value, they null each other. The nullifying problem can be solved just by applying the LSM, the principle of which consists in the sum of squares of deviations and not of the deviations themselves. The LSM formulation is then as follows:

$$\sum_{i=1}^{n}(y_i - \hat{y}_i)^2 = \sum_{i=1}^{n}(y_i - b_0 - b_1 x_i)^2 = \sum_{i=1}^{n}(v_i)^2 = \min. \tag{5}$$

The minimum of a function of two variables can be found by placing its partial derivatives under both variables (coefficients b_0 a b_1) equal to zero:

$$\frac{\partial\left(\sum_{i=1}^{n}(y_i - b_0 - b_1 x_i)^2\right)}{\partial(b_1)} = 2\left(\sum_{i=1}^{n}(y_i - b_0 - b_1 x_i)(-x_i)\right) = 0,$$

$$\frac{\partial\left(\sum_{i=1}^{n}(y_i - b_0 - b_1 x_i)^2\right)}{\partial(b_0)} = 2\left(\sum_{i=1}^{n}(y_i - b_0 - b_1 x_i)(-1)\right) = 0. \tag{6}$$

By an appropriate algebraic transformation, we get a system of normal equations of two variables b_0 and b_1:

$$\sum_{i=1}^{n} y_i = n b_0 + b_1 \sum_{i=1}^{n} x_i,$$

$$\sum_{i=1}^{n} y_i x_i = b_0 \sum_{i=1}^{n} x_i + b_1 \sum_{i=1}^{n} x_i^2. \tag{7}$$

For the variables b_0 and b_1 it is true:

$$b_0 = \frac{\sum_{i=1}^{n} y_i \sum_{i=1}^{n} x_i^2 - \sum_{i=1}^{n} x_i y_i \sum_{i=1}^{n} x_i}{n\sum_{i=1}^{n} x_i^2 - (\sum_{i=1}^{n} x_i)^2} \quad b_1 = \frac{n\sum_{i=1}^{n} x_i y_i - \sum_{i=1}^{n} x_i \sum_{i=1}^{n} y_i}{n\sum_{i=1}^{n} x_i^2 - (\sum_{i=1}^{n} x_i)^2} \tag{8}$$

In the processing of geodetic measurements, methods of adjustment are used, by which the most probable value of a quantity and its accuracy characteristics are determined. In determining the parameters of the regression line, a mediating adjustment was used [1],[10],[9],[12], where the relationship between the measured and unknown variables is expressed by the intermediating function of the searched unknown parameters, the so-called estimates:

$$y_i = l_i + v_i = f(\widehat{\theta_1}, \hat{\theta}_2, \cdots \hat{\theta}_k) \tag{9}$$

For the vector of residues, the following is valid:

$$\mathbf{v} = \mathbf{A}\hat{\Theta} - \mathbf{l}, \tag{10}$$

leading to the Guss-Markov model:

$$\mathbf{v} = \mathbf{A}\hat{\Theta} - \mathbf{l}$$
$$\textstyle\sum_l = \sigma_0^2 \mathbf{Q}_l \tag{11}$$

It applies for LSM:

$$\rho = \sum_{i=1}^{n}(p_i v_i)^2 = \sum_{i=1}^{n} v^T \mathbf{P} v = \min., \quad \mathbf{P}\dots\dots \text{matica váhovýchkoeficientov}. \tag{12}$$

matrix of weight coefficients

The minimum of the function of two variables can be found again by placing the partial derivatives of the function (12) equal to zero $\dfrac{\partial(v^T \mathbf{P} v)}{\partial\hat{\Theta}}$ and so getting the system of two equations, from which the searched variables (estimates) can be determined:

$$(\mathbf{A}^T\mathbf{PA}\hat{\Theta}) - (\mathbf{A}^T\mathbf{Pl}) = 0 \rightarrow (\mathbf{N}\hat{\Theta}) - (\mathbf{A}^T\mathbf{Pl}) = 0, \mathbf{N} \text{ - matica koeficientov normálnychrovníc} \tag{13}$$

matrix of coefficients of normal equations

When using the LSM for searching the estimates, (indices), the following applies after adjusting the parameters:

$$\hat{\Theta} = (\mathbf{A}^T\mathbf{PA})^{-1}(\mathbf{A}^T\mathbf{Pl}) = \mathbf{N}^{-1}(\mathbf{A}^T\mathbf{Pl}). \tag{14}$$

3   SIMPLEX METHOD

The simplex method is an iterative computational procedure that is used to find optimal solutions whereas the objective (minimized) function must be in a canonical form. The minimization problem is typically solved by means of methods of linear programming in a tabular form by rearranging columns and line unless the objective function optimization is reached. The paper presents a simpler procedure of processing based on the principle of matrix solution [2],[4]. This method will again be presented on the example of a regression line. The model line was chosen not only because of its simplicity, but especially considering the fact that in the processing of results from geodetic measurements there is a need, quite often, to determine the accuracy of the measured length which can be expressed by the following relationship:

$$STD = appm + b \dots \text{modified form of regression line} \tag{15}$$

Where:

STD - the standard deviation of a measured length,

a - the parameter reflecting the impact of errors dependent on the measured length which take into account the influence of the physical environment,

b - the parameter reflecting the impact of the errors independent of the measured length.

The functional relationship for the correction of the intermediate variable can be expressed as follows:

$$\underset{(n,1)}{\mathbf{v}} = \underset{n,k}{\mathbf{A}} \cdot \underset{k,1}{\hat{\Theta}} - \underset{n,1}{\mathbf{f}}, \tag{16}$$

where *v* is *the vector of corrections*, *A the matrix of coefficients*, $\hat{\Theta}$ is *the vector of unknown parameters (estimates)*, *f* is *the vector of observations*, *n* is *a number of measurements*, and *k* is *a number of necessary measurements (determined parameters)*. Within the search of an optimal solution, first the measured values *f* are divided to the **basic variables** (*needed measurements*) and **non-basic variables** (*r = n-k*, r- *a number of redundant measurements*).

$$\begin{bmatrix} \mathbf{v}_{(1)} \\ \mathbf{v}_{(2)} \end{bmatrix} + \begin{bmatrix} \mathbf{A}_{(1)} \\ \mathbf{A}_{(2)} \end{bmatrix} \cdot \hat{\Theta} = \begin{bmatrix} \mathbf{f}_{(1)} \\ \mathbf{f}_{(2)} \end{bmatrix}. \tag{17}$$

The vector *f* can be broken down as follows:

$$\begin{bmatrix} \mathbf{f}_{(1)} \\ \mathbf{f}_{(2)} \end{bmatrix} = \begin{bmatrix} \mathbf{l}_{(1)}^{\circ} \\ \mathbf{l}_{(2)}^{\circ} \end{bmatrix} - \begin{bmatrix} \mathbf{l}_{(1)} \\ \mathbf{l}_{(2)} \end{bmatrix}. \tag{18}$$

In the case of observations with varying accuracy, the L1 norm can be derived from the objective function L2 norm:

$$\rho_{L2} = \mathbf{v}^T \mathbf{P} \mathbf{v} = \sum_{i=1}^{n} \mathbf{v}_i^2 p_{ii} = \sum_{i=1}^{n} (\mathbf{v}_i \sqrt{p_{ii}})(\mathbf{v}_i \sqrt{p_{ii}}) = \min. \tag{19}$$

To determine the individual elements of the weighting matrix $\mathbf{P}$, the Cholesky decomposition is used:

$$\mathbf{P} = \mathbf{S}.\mathbf{S}^T. \tag{20}$$

For the functional relationships of measured values, it applies:

$$\mathbf{S}\mathbf{A}\hat{\Theta} \approx \mathbf{S}.\mathbf{f}. \tag{21}$$

The relationship for the objective function for the L1 norm after the adjustment takes the form:

$$\rho_{L1} = (vecd(\mathbf{S})^T |\mathbf{v}| \to \min. \tag{22}$$

Where $vectd\,()$ is a column vector and the matrix $|\mathbf{v}|$ has the following form:

$$|v| = [|v|]_{i=1}^{n} \text{ with elements on diagonal.} \tag{23}$$

For calculating the unknown parameters, it applies:

$$\hat{\Theta} = (\mathbf{S}_{(1)}.\mathbf{A}_{(1)})^{-1}\mathbf{S}_{(1)}.\mathbf{f}_{(1)} = \mathbf{A}_{(1)}.\mathbf{f}_{(1)}, \tag{24}$$

Where $\mathbf{S}_{(1)}$ is the diagonal matrix with dimensions (k x k).

The vector of corrections is divided into two parts:

$$\mathbf{v}_{(1)} = \mathbf{f}_{(1)} - \mathbf{A}_{(1)}\hat{\Theta} = \mathbf{f}_{(1)} - \mathbf{A}_{(1)}\mathbf{A}_{(1)}^{-1}\mathbf{f}_{(1)} = \mathbf{0}, \dots.. \text{ residues are always equal 0} \tag{25}$$

For the non-zero vector of corrections $\mathbf{v}_{(2)}$, it applies:

$$\mathbf{v}_{(2)} = \mathbf{f}_{(2)} - \mathbf{A}_{(2)}\hat{\Theta} = \mathbf{f}_{(2)} - \mathbf{A}_{(2)}\mathbf{A}_{(1)}^{-1}\mathbf{f}_{(1)}. \tag{26}$$

The assignment is solved by linear programming methods and is defined as follows:

$$\begin{aligned} r^T \mathbf{x} &= \rho_{L1}, \\ \mathbf{A}\mathbf{x} &= g. \end{aligned} \tag{27}$$

where $\mathbf{x} \geq 0$, r is a *minimizing funkcion* also called „*cost vector*" of size of t x 1, $t=2k+2n$, x is *the vector of non-negative variables*, A is *the matrix of coefficients*.

In order to meet the condition $\mathbf{x} \geq 0$ for linear programming, the calculated estimates and the vector of corrections is divided into two non-negative components:

$$\begin{aligned} \hat{\Theta} &= \delta - \gamma \quad \delta, \gamma \geq 0, \\ \mathbf{v} &= \mathbf{y} - r \quad \mathbf{y}, r \geq 0 \end{aligned}$$

$$\begin{aligned} \mathbf{v_1} &= y_i \text{ ak } v_i > 0, r_i = 0, -\mathbf{v_i} = r_i \text{ ak } v_i < 0, y_i = 0, \\ \mathbf{v}_i &= 0 \text{ ak } y_i = 0, r_i = 0, |v_i| = y_i + r_i \end{aligned} \tag{28}$$

The target function (27) can be written in a matrix form as well:

$$\begin{bmatrix} \mathbf{S}_{(1)}\mathbf{A}_{(1)} & -\mathbf{S}_{(1)}\mathbf{A}_{(1)} & \mathbf{I} & 0 & -\mathbf{I} & 0 \\ \mathbf{S}_{(2)}\mathbf{A}_{(2)} & -\mathbf{S}_{(2)}\mathbf{A}_{(2)} & 0 & \mathbf{I} & 0 & -\mathbf{I} \end{bmatrix} \begin{bmatrix} \delta \\ \gamma \\ \mathbf{S}_{(1)}y_{(1)} \\ \mathbf{S}_{(2)}y_{(2)} \\ \mathbf{S}_{(1)}z_{(1)} \\ \mathbf{S}_{(2)}z_{(2)} \end{bmatrix} = \begin{bmatrix} \mathbf{S}_{(1)}f_{(1)} \\ \mathbf{S}_{(2)}f_{(2)} \end{bmatrix} \tag{29}$$

$$\begin{bmatrix} 0^T & 0^T & 1^T & 1^T & 1^T & 1^T \end{bmatrix} \begin{bmatrix} \delta \\ \gamma \\ \mathbf{S}_{(1)}y_{(1)} \\ \mathbf{S}_{(2)}y_{(2)} \\ \mathbf{S}_{(1)}z_{(1)} \\ \mathbf{S}_{(2)}z_{(2)} \end{bmatrix} = \phi_{L1}$$

where $\mathbf{1}$ is the all-ones vector ($1^T = [1, 1 \cdots 1]$). The default simplex table has the following structure:

$$T = \begin{bmatrix} \rho_{L1} & \mathbf{z}^T \\ \mathbf{g} & \mathbf{A} \end{bmatrix} =$$

$$= \begin{matrix} 0 & 0^T & 0^T & 1^T & 1^T & 1^T & 1^T \\ \mathbf{S}_{(1)}.\mathbf{f}_{(1)} & \mathbf{S}_{(1)}.\mathbf{A}_{(1)} & -\mathbf{S}_{(1)}.\mathbf{A}_{(1)} & \mathbf{I} & 0 & -\mathbf{I} & 0 \\ \mathbf{S}_{(2)}.\mathbf{f}_{(2)} & \mathbf{S}_{(2)}.\mathbf{A}_{(2)} & -\mathbf{S}_{(2)}.\mathbf{A}_{(2)} & 0 & I & 0 & -\mathbf{I} \end{matrix} \quad (30)$$

The calculation can be tabulated using the rearrangement operations of columns and lines; in the paper, the procedure of transforming the simplex table using a *"pivot"* matrix $\mathbf{K}$ was used:

$$\mathbf{K} = \begin{bmatrix} I & 0^T & 1^T \\ 0 & \mathbf{S}_{(1)}\mathbf{A}_{(1)} & 0 \\ 0 & \mathbf{S}_{(2)}\mathbf{A}_{(2)} & \mathbf{I}^* \end{bmatrix}, T^* = \mathbf{K}^{-1}.T , \quad (31)$$

$\mathbf{I}^*$ - the variable pivot matrix; its diagonal elements have a value +1, or -1 to meet the condition met$v_{(2)} \geq 0$.

The iterative process is carried out by multiplying the initial simplex table (30) with the pivot matrix in inverse form. The structure of the transformed simplex table after the adjustment is as follows:

$$T^* = \begin{matrix} \mathbf{u}_1 & 0^T & 0^T & 1^T + 1^T\mathbf{U}_2 & 1^T - 1^T\mathbf{I}^* & 1^T - 1^T\mathbf{U}_2 & 1^T + 1^T\mathbf{I}^* \\ \mathbf{A}_{(1)}^{-1}.\mathbf{f}_{(1)} & \mathbf{I} & -\mathbf{I} & \mathbf{A}_{(1)}^{-1} & 0 & -\mathbf{A}_{(1)}^{-1} & 0 \\ \mathbf{u}_3 & 0 & 0 & \mathbf{U}_2 & \mathbf{I}^* & \mathbf{I}^* & -\mathbf{I}^* \end{matrix} \quad (32)$$

Where the elements of the table can be determined according to the following relationship:

$$\mathbf{u}_1 = 1^T \mathbf{I}\mathbf{S}_{(2)}\mathbf{A}_{(2)}\mathbf{A}_{(1)}^{-1}f_{(1)} - 1^T\mathbf{I}^*\mathbf{S}_{(2)}f_{(2)},$$
$$\mathbf{U}_2 = \mathbf{I}^*\mathbf{S}_2\mathbf{A}_{(2)}(\mathbf{S}_{(1)}\mathbf{A}_{(1)})^{-1}, \quad (33)$$
$$\mathbf{u}_3 = -\mathbf{I}^*\mathbf{S}_{(2)}\mathbf{A}_{(2)}\mathbf{A}_{(1)}^{-1}f_{(1)} + \mathbf{I}^*\mathbf{S}_{(2)}f_{(2)}.$$

The objective function is transformed to minimize the function $\mathbf{r}$ so-called reduced „*cost*" vector (34) which can be simplified in comparison with the original form, because its elements $1^T - 1^T\mathbf{I}^*$ and $1^T - 1^T\mathbf{U}_2$ have a value of 0 or 2.

$$\mathbf{r}^T = [0^T, 0^T, 1^T + 1^T\mathbf{U}_2, 1^T - 1^T\mathbf{I}^*, 1^T - 1^T\mathbf{U}_2, 1^T + 1^T\mathbf{I}^*] \quad (34)$$

The optimal solution is found when for all the elements of the vector $\mathbf{r} \geq 0$ applies which can be transformed to the following condition:

$$-1^T \leq \mathbf{S}_{(2)}\mathbf{A}_{(2)}\mathbf{S}_{(1)}\mathbf{A}_{(1)} \leq 1^T \quad (35)$$

4 EMPIRICAL DEMONSTRATION

In the following chapter, the method of least squares and the simplex method are presented on the example of a regression line and a geodetic network. The regression line and the geodetic network in the first example are loaded by a normal distribution of measurements; due to the investigation of the properties of the used estimation methods, the line and the geodetic network are loaded by an experimental outlier before the adjustment.

4.1 Application of estimation methods using the example of a regression line loaded by a normal distribution of measurements

The regression line in this case is expressed by the equation of a rangefinder as follows:

$$STD = \sigma_D = 2.ppm d + 3[mm] \cdots rovnica\ regresnej priamky, \quad (36)$$

Where:
STD - a standard deviation of the measured length,
d   - the measured length,
ppm - parts per million (10^{-6}).

The deterministic model of the regression line loaded by a normal distribution of measurements is presented in a graphic form in **Chyba! Nenalezen zdroj odkazů.**. The experimental values shown in blue are simulated in the MATLAB environment by the function *normrnd (normal random numbers)*.

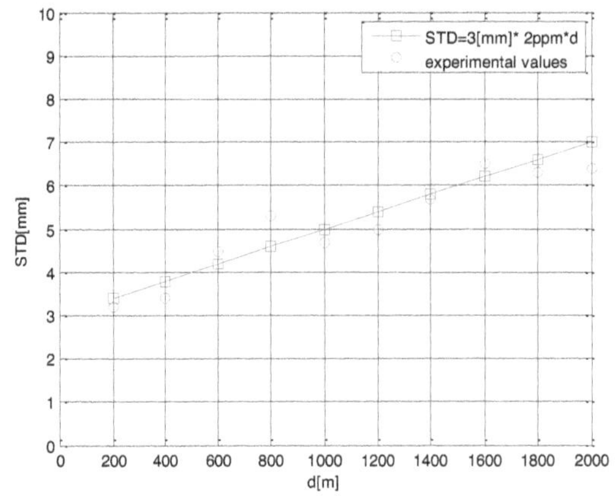

Fig. 3 Deterministic model of the regression line loaded by normal distribution

Tab. 1 Final simplex table for the regression line loaded by a normal distribution of measurements

a	1	0	-1	0	-0.625	0.625	0	0	0	0	0	0	0	0	0.625	-0.625	0	0	0	0	0	0	0	0	1.9375
b	0	1	0	-1	1.125	-0.125	0	0	0	0	0	0	0	0	-1.125	0.125	0	0	0	0	0	0	0	0	2.8125
v_2	0	0	0	0	0.875	0.125	-1	0	0	0	0	0	0	0	-0.875	-0.125	1	0	0	0	0	0	0	0	0.1875
v_3	0	0	0	0	-0.75	-0.25	0	1	0	0	0	0	0	0	0.75	0.250	0	-1	0	0	0	0	0	0	-0.5250
v_4	0	0	0	0	-0.625	-0.375	0	0	1	0	0	0	0	0	0.625	0.375	0	0	-1	0	0	0	0	0	-0.9375
v_5	0	0	0	0	0.500	0.500	0	0	0	-1	0	0	0	0	-0.500	-0.500	0	0	0	1	0	0	0	0	0.0500
v_6	0	0	0	0	0.375	0.625	0	0	0	0	-1	0	0	0	-0.375	-0.625	0	0	0	0	1	0	0	0	0.1375
v_7	0	0	0	0	-0.25	-0.750	0	0	0	0	0	1	0	0	0.250	0.750	0	0	0	0	0	-1	0	0	-0.1750
v_8	0	0	0	0	-0.125	-0.875	0	0	0	0	0	0	1	0	0.125	0.875	0	0	0	0	0	0	-1	0	-0.5875
v_{10}	0	0	0	0	-0.125	1.125	0	0	0	0	0	0	0	-1	0.125	-1.125	0	0	0	0	0	0	0	1	0.2875
r	0	0	0	0	1.125	0.875	2	0	0	2	2	0	0	2	0.875	1.125	0	2	2	0	0	2	2	0	-2.8875

Tab. 2 Results of adjustment of the regression line loaded by a normal distribution of the LSM and L1 norm

M. No.	Theoretical line			LSM			L1
	d	σ_d exp.	σ_d determ.	ε	v	Δ	v
	[mm]	[mm]	[mm]	[mm]	[mm]	[mm]	[mm]
1.	200.00	3.20	3.40	0.20	0.2473	-0.05	0.0000
2.	400.00	3.40	3.80	0.40	0.4145	-0.01	-0.1875
3.	600.00	4.50	4.20	-0.30	-0.3182	0.02	0.5250
4.	800.00	5.30	4.60	-0.70	-0.7509	0.05	0.9375
5.	1000.00	4.70	5.00	0.30	0.2164	0.08	-0.0500
6.	1200.00	5.00	5.40	0.40	0.2836	0.12	-0.1375
7.	1400.00	5.70	5.80	0.10	-0.0491	0.15	0.1750
8.	1600.00	6.50	6.20	-0.30	-0.4818	0.18	0.5875
9.	1800.00	6.30	6.60	0.30	0.0855	0.21	0.0000
10.	2000.00	6.40	7.00	0.60	0.3527	0.25	-0.2875

$$\varepsilon = \sigma_{d_{\text{determ.}}} - \sigma_{d_{\text{exp.}}}, \quad \Delta = \varepsilon - v \tag{37}$$

Parameters of regression line:

Deterministic form of line	y= 3.0[mm]+ 2.0*ppm*d
Parameters of line estimated by LSM	y= 3.1[mm]+ 1.8*ppm*d
Parameters of line estimated by Hampel method	y= 2.8[mm]+ 1.9*ppm*d

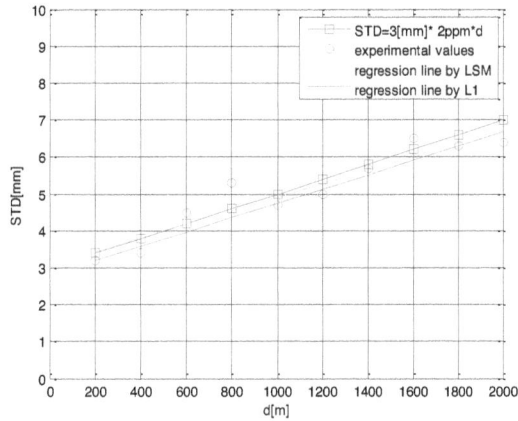

Fig. 4 Graphical interpretation of LSM and L1 norms on the example of the regression line loaded by normal distribution

From the results of the adjustment (Tab. 2) of such proposed regression line, it is evident that in case of the load of the line by normal distribution of measurements neither the LSM, nor the simplex method in the set of measured data revealed any outlier ("which go beyond") value and the results of the adjustment are very similar. The graph of the regression line behaviour after the adjustment by individual estimate procedures is presented in **Chyba! Nenalezen zdroj odkazů.**.

4.2 Application of estimation methods using the example of the regression line loaded by an experimental outlier

With regard to the fact that the present contribution is devoted to the issue of estimation methods which allow to reveal so-called outlier measurements in the set of measured data, and for which the arithmetic mean was not selected as the centrality parameter, the regression line was deliberately loaded with the only outlier (Fig. 5) in order to track the performance of such methods. In the figure, the experimental values are shown in blue; the line behaviour in deterministic form is shown in red.

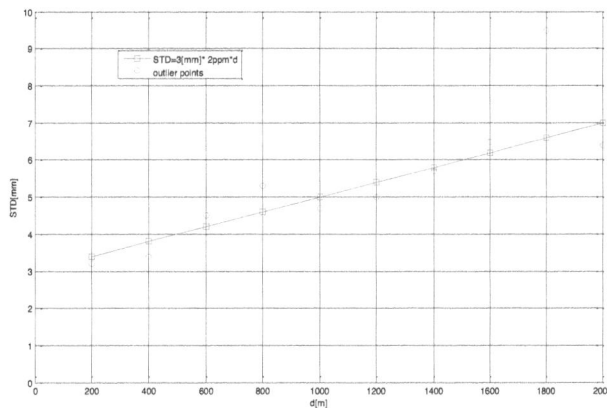

Fig. 5 Deterministic form of the regression line loaded by an experimental outlier

Such modified regression line was adjusted first by the method of least squares, and consequently by the simplex method. The results of adjustments of the regression line loaded by an experimental outlier after the adjustment by the LSM and the simplex method are interpreted in Tab. 4 and in Fig. 6.

Tab. 3 Resulting simplex table for the example of the regression line loaded by an experimental outlier

a	1	0	-1	0	-0.8333	0.8333	0	0	0	0	0	0	0	0	0.8333	-0.8333	0	0	0	0	0	0	0	0	2,0833
b	0	1	0	-1	1.1667	-0.1667	0	0	0	0	0	0	0	0	-1.1667	0.1667	0	0	0	0	0	0	0	0	2,7833
v_2	0	0	0	0	0.8333	0.1667	-1	0	0	0	0	0	0	0	-0.8333	-0.1667	1	0	0	0	0	0	0	0	0.2167
v_3	0	0	0	0	-0.6667	-0.3333	0	1	0	0	0	0	0	0	0.6667	0.3333	0	-1	0	0	0	0	0	0	0.4667
v_4	0	0	0	0	-0.5000	-0.5000	0	0	1	0	0	0	0	0	0.5000	0.5000	0	0	-1	0	0	0	0	0	0.8500
v_5	0	0	0	0	0.3333	0.6667	0	0	0	-1	0	0	0	0	-0.3333	-0.6667	0	0	0	1	0	0	0	0	0.1667
v_6	0	0	0	0	0.1667	0.8333	0	0	0	0	-1	0	0	0	-0.1667	-0.8333	0	0	0	0	1	0	0	0	0.2833
v_8	0	0	0	0	0.1667	-1.1667	0	0	0	0	0	1	0	0	-0.1667	1.1667	0	0	0	0	0	-1	0	0	0.3833
v_9	0	0	0	0	0.3333	-1.3333	0	0	0	0	0	0	1	0	-0.3333	1.3333	0	0	0	0	0	0	-1	0	2,9667
v_{10}	0	0	0	0	-0.5000	1.5000	0	0	0	0	0	0	0	-1	0.5000	-1.5000	0	0	0	0	0	0	0	1	0.5500
r	0	0	0	0	0.8333	1.1667	2	0	0	2	2	0	0	2	1.1667	0.8333	0	2	2	0	0	2	2	0	-5,8833

Tab. 4 Comparison of the LSM and the L1 norm on the example of the regression line loaded by an experimental outlier

	Theoretical line			LSM			L1
no. m	**d**	σ_d exp.	σ_d deter.	**ε**	**v**	**Δ**	**v**
	[mm]	[mm]	[mm]	[mm]	[mm]	[mm]	[mm]
1.	200.00	3.20	3.40	0.20	-0.0436	0.24	0.0000
2.	400.00	3.40	3.80	0.40	0.2594	0.14	0.2167
3.	600.00	4.50	4.20	-0.30	-0.3376	0.04	-0.4667
4.	800.00	5.30	4.60	-0.70	-0.6345	-0.07	-0.8500
5.	1000.00	4.70	5.00	0.30	0.4685	-0.17	0.1667
6.	1200.00	5.00	5.40	0.40	0.6715	-0.27	0.2833
7.	1400.00	5.70	5.80	0.10	0.4745	-0.37	0.0000
8.	1600.00	6.50	6.20	-0.30	0.1776	-0.48	-0.3833
9.	1800.00	9.50	6.60	-2.90	-2.3194	-0.58	-2.9667
10.	2000.00	6.40	7.00	0.60	1.2836	-0.68	0.5500

Parameters of regression line:

Deterministic form of line	y= 3.0[mm]+ 2.0*ppm*d
Parameters of line estimated by LSM	y= 2.7[mm]+ 2.5*ppm*d
Parameters of line estimated by L1	y= 2.1[mm]+ 2.8*ppm*d

Fig. 6 10 Graphic interpretation of the LSM and the L1 norm (simplex method) on the example of the regression line loaded by an experimental outlier

From the results of the adjustment through the LSM, it is evident that in this case the LSM found an outlier measurement at a measured length of 1800 m and just this measurement was assigned the greatest value of correction after the adjustment. Comparing the two estimation procedures, we concluded that both methods arrive at a mutually similar results and the simplex method found an outlier measurement at a measured length of 1800 m.

4.3 Adjustment of a geodetic network loaded by a normal distribution of measurements

Characteristics of the presented estimation methods will also be investigated on the example of a geodetic network. This is a simulated geodetic network where the point No. 6 is the point being determined whose position is defined by triangulation (angular) measurements (Fig. 7) . Such a network model was proposed on the grounds that the geodetic practice is often encountered with the task when it is necessary to determine the location of an inaccessible point which can be done just through triangulation measurements. The geodetic network will be processed as a binding network by both network estimative procedures, first by a normal distribution of measurements, and consequently after it is loaded by outlier experimental values.

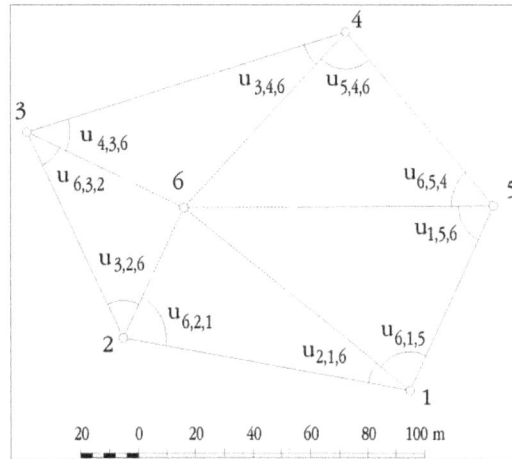

Fig. 7 Structure of the geodetic network loaded by a normal distribution of measurements

Tab. 1 Results of adjustment of the geodetic network loaded by a normal distribution through the LSM

L-S-P	l~	l	l^	v~	v	vC^	p	T.Baarda	T.Pope	s(v)	s(l^)	r *
angle	[g]	[g]	[g]	[cc]	[cc]	[cc]		–B–	–P–	[cc]		
2-1-6	31.04550	31.04500	31.04559	4.95	5.94	5.94	1.0000	1.58	1.43	3.76	4.15	0.88
6-1-5	86.04331	86.04290	86.04321	4.08	3.09	3.09	1.0000	0.82	0.74	3.76	4.15	0.88
3-2-6	57.35031	57.35090	57.35053	-5.90	-3.72	-3.72	1.0000	1.17	1.06	3.18	3.51	0.63
6-2-1	82.30387	82.30390	82.30366	-0.26	-2.45	-2.45	1.0000	0.77	0.70	3.18	3.51	0.63
4-3-6	46.63242	46.63190	46.63229	5.21	3.95	3.95	1.0000	1.18	1.07	3.33	3.68	0.69
6-3-2	43.61620	43.61630	43.61633	-0.98	0.29	0.29	1.0000	0.09	0.08	3.33	3.68	0.69
5-4-6	94.62331	94.62290	94.62322	4.11	3.15	3.15	1.0000	0.84	0.76	3.74	4.13	0.88
6-4-3	32.88348	32.88290	32.88358	5.81	6.76	6.76	1.0000	1.81	1.64	3.74	4.13	0.88
1-5-6	71.37301	71.37290	71.37303	1.07	1.28	1.28	1.0000	0.33	0.30	3.83	4.23	0.92
6-5-4	54.12859	54.12900	54.12857	-4.08	-4.29	-4.29	1.0000	1.12	1.01	3.83	4.23	0.92

Legend:
Significance level alpha   = 0.050   B - Baarda Data snooping (N(0,1))   = 2.800
Basic standard deviation of angle mesur.   = 4.000 [cc]   Number of critical measure. (Baarda data-snooping) = 0
Estimate of the accuracy of angle m. (MINQUE) = 4.417 [cc]   P - Pope Tau-test (Tau(r, 1-alfa0/2)   = 2.361
Posterior standard deviation   = 4.417 [cc]   Number of critical measure. Pope Tau-test)   = 0
Critical limit s0_posterior   = 5.569 [cc]   Adjustment efficiency   = 0.800
s0_poster^2/s0_aprior^2   = 1.219   Redundancy   = 8.000

The results of adjustment of thus proposed geodetic network by the method of least squares are presented in

Tab. 1. In the case of the network loaded by a normal distribution of measurements, the standard deviation of the measured angle takes a value of 4.000 cc. The accuracy of the angular measurement 4.417 cc was estimated by the method MINQUE [10],[12]. The fact whether the set of measured data was

infiltrated by an outlier measurement was studied using two statistical tests Baarda Data snooping test and the Pope Tau test. From the results of processing

(Tab. 1), it is clear that not only the LSM did not find outlier measurements, but none of the measurements exceeded the critical value of both test statistics. Due to the investigation of the properties of alternative estimation methods, the network was adjusted again by the simplex method (Tab. 2).

Tab. 2 Rresulting simplex table in processing the geodetic network loaded by a normal distribution

dX$_6$^	1	0	-1	0	0.032	-0.085	0	0	0	0	0	0	0	0	-0.032	0.085	0	0	0	0	0	0	0	0	0.1078
dY$_6$^	0	1	0	-1	-0.069	-0.042	0	0	0	0	0	0	0	0	0.069	0.042	0	0	0	0	0	0	0	0	-0.4497
V$_{u(2,1,6)}$	0	0	0	0	0.109	0.591	-1	0	0	0	0	0	0	0	-0.109	-0.591	1	0	0	0	0	0	0	0	6.1698
V$_{u(6,1,5)}$	0	0	0	0	-0.109	-0.591	0	-1	0	0	0	0	0	0	0.109	0.591	0	1	0	0	0	0	0	0	2.8602
V$_{u(6,2,1)}$	0	0	0	0	1	0	0	0	1	0	0	0	0	0	-1	0	0	0	-1	0	0	0	0	0	6.1600
V$_{u(4,3,6)}$	0	0	0	0	0	-1	0	0	0	-1	0	0	0	0	0	1	0	0	0	1	0	0	0	0	4.2300
V$_{u(5,4,6)}$	0	0	0	0	-0.566	0.223	0	0	0	0	-1	0	0	0	0.566	-0.223	0	0	0	0	1	0	0	0	0.9912
V$_{u(6,4,3)}$	0	0	0	0	0.566	-0.223	0	0	0	0	0	-1	0	0	-0.566	0.223	0	0	0	0	0	1	0	0	8.9288
V$_{u(1,5,6)}$	0	0	0	0	-0.193	0.499	0	0	0	0	0	0	-1	0	0.193	-0.499	0	0	0	0	0	0	1	0	0.4197
V$_{u(6,5,4)}$	0	0	0	0	-0.193	0.499	0	0	0	0	0	0	0	1	0.193	-0.499	0	0	0	0	0	0	0	-1	3.4297
r	0	0	0	0	0.386	1.001	2	2	0	2	2	2	2	0	1.614	0.999	0	0	2	0	0	0	0	2	-33.1894

Tab. 3 Estimates of the coordinates of the point No. 6 from measurements loaded by a normal distribution

Method	X° [m]	Y° [m]	dX^ [mm]	dY^ [mm]	X^ [m]	Y^ [m]
LSM	1167044.502	438102.861	-0.0374	-0.2039	1167044.50196	438102.86080
L1			0.1078	-0.4497	1167044.50211	438102.86055

Tab. 4 Comparing corrections in the geodetic network non-loaded by outlier measurements

Angle	l~ *Theoretical value of angle*	l *Measured value of angle*	v~ *Actual measurement error*	v LSM	v Simplex
L-S-P	[g]	[g]	[cc]	[cc]	[cc]
2-1-6	31.04550	31.04500	4.95	5.94	6.17
6-1-5	86.04331	86.04290	4.08	3.09	2.86
3-2-6	57.35031	57.35090	-5.90	-3.72	0.00
6-2-1	82.30387	82.30390	-0.26	-2.45	-6.16
4-3-6	46.63242	46.63190	5.21	3.95	4.23
6-3-2	43.61620	43.61630	-0.98	0.29	0.00
5-4-6	94.62331	94.62290	4.11	3.15	0.99
6-4-3	32.88348	32.88290	5.81	6.76	8.93
1-5-6	71.37301	71.37290	1.07	1.28	0.42
6-5-4	54.12859	54.12900	-4.08	-4.29	-3.43

The graphical interpretation of the results of adjustment (Fig. 8) is presented through the confidence ellipses constructed at 99% and 95% probability.

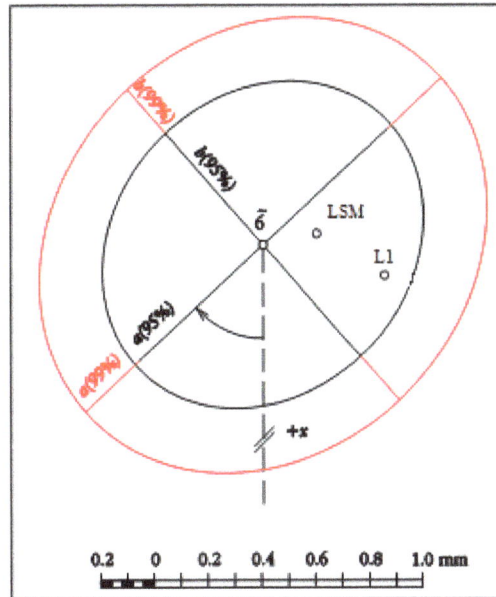

Fig. 8 Graphical interpretation of the estimation methods for adjusting the geodetic network loaded by a normal distribution of measurements

4.4 The adjustment of the geodetic network loaded by a normal distribution of measurements

To investigate the properties of alternative estimation methods enabling to detect outlier measurements that can infiltrate such a set of measured data, e.g. by a supraliminal influence of the physical environment, two measured angles were loaded in the proposed geodetic network prior to the processing (Fig. 9), once by five times the mean error of the measured angle, once by six times the estimated error. Such proposed geodetic network was again adjusted as the binding network by the (MNS) (Tab. 5) and by the simplex method (Tab. 7).

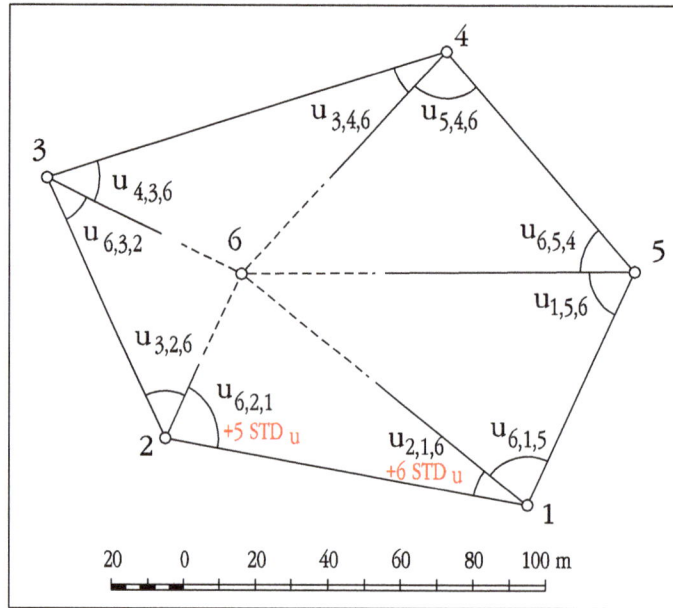

Fig. 9 Structure of the geodetic network loaded by two outlier measurements

Tab. 5 Results of adjustment of the geodetic network loaded by outlier measurements through the LSM

L-S-P	l~	l	l^	v~	v	vC^	p	T.Baarda	T.Pope	s(v)	s(l^)	r *	f
angle	[g]	[g]	[g]	[cc]	[cc]	[cc]		–B–	–P–	[cc]			
2-1-6	31.04550	31.04790	31.04581	-24.00	-20.86	-20.86	1.0000	5.55 B	2.02	3.76	10.33	0.88	66.0
6-1-5	86.04331	86.04290	86.04299	4.08	0.94	0.94	1.0000	0.25	0.09	3.76	10.33	0.88	66.0
3-2-6	57.35031	57.35090	57.34998	-5.90	-9.24	-9.24	1.0000	2.91 B	1.06	3.18	8.72	0.63	39.2
6-2-1	82.30387	82.30587	82.30421	-20.00	-16.66	-16.66	1.0000	5.25 B	1.91	3.18	8.72	0.63	39.2
4-3-6	46.63242	46.63190	46.63183	5.21	-0.71	-0.71	1.0000	0.21	0.08	3.33	9.14	0.69	44.6
6-3-2	43.61620	43.61630	43.61679	-0.98	4.94	4.94	1.0000	1.48	0.54	3.33	9.14	0.69	44.6
5-4-6	94.62331	94.62290	94.62363	4.11	7.32	7.32	1.0000	1.95	0.71	3.74	10.28	0.88	64.7
6-4-3	32.88348	32.88290	32.88316	5.81	2.60	2.60	1.0000	0.70	0.25	3.74	10.28	0.88	64.7
1-5-6	71.37301	71.37290	71.37337	1.07	4.67	4.67	1.0000	1.22	0.44	3.83	10.52	0.92	71.1
6-5-4	54.12859	54.12900	54.12823	-4.08	-7.68	-7.68	1.0000	2.01	0.73	3.83	10.52	0.92	71.1

Significance level alpha   = 0.050   B - Baarda Data snooping (N(0,1))   = 2.800
Basic standard deviation of angular meas.   = 4.000 [cc]   Number of critical meas. (Baarda)   = 3
Estimate of precision of angular meas. (MINQUE)   = 10.984 [cc]   P - Pope Tau-test (Tau(r, 1-alfa0/2)   = 2.361
Posterior standard deviation   = 10.984 [cc] Number of critical meas. (Pope Tau-test) = 0
Critical limit s0_posterior   = 5.569 [cc]   Redundancy   = 0.800
s0_poster^2/s0_aprior^2   = 7.541
Crit. ratio s0_poster^2/s0_aprior^2   = 1.938

Results of the adjustment of the geodetic network show that the LSM found two outliers just on the measured angles loaded by multiples of the median error of the measured angle. Infiltrating outlier measurements to the set of the processed data, however, was also studied through the Baarda Snooping test and the Pope test. From

Tab. 5, it is evident that the Baard test found three critical measurements, while the Pope test did not find any critical test measurement. For this reason, the simplex method (Tab. 7) was used for the processing of the geodetic network.

Tab. 6 Estimates of the point No. 6 coordinates from measurements loaded by a normal distribution

Method	X° [m]	Y° [m]	dX^ [mm]	dY^ [mm]	X^ [m]	Y^ [m]
LSM	1167044.502	438102.861	-0.6136	-0.0144	1167044.50139	438102.86099
L1			-0.2529	-0.6252	1167044.50175	438102.86037

Tab. 7 Resulting simplex table in processing the geodetic network loaded by outlier measurements

dX6^	1	0	-1	0	0.032	0.085	0	0	0	0	0	0	0	0	-0.032	-0.085	0	0	0	0	0	0	0	0	-0.2529
dY6^	0	1	0	-1	-0.069	0.042	0	0	0	0	0	0	0	0	0.069	-0.042	0	0	0	0	0	0	0	0	-0.6252
$V_{u(2,1,6)}$	0	0	0	0	-0.109	0.591	1	0	0	0	0	0	0	0	0.109	-0.591	-1	0	0	0	0	0	0	0	20.3326
$V_{u(6,1,5)}$	0	0	0	0	-0.109	0.591	0	-1	0	0	0	0	0	0	0.109	-0.591	0	1	0	0	0	0	0	0	0.3626
$V_{u(6,2,1)}$	0	0	0	0	1	0	0	0	1	0	0	0	0	0	-1	0	0	0	-1	0	0	0	0	0	25.8600
$V_{u(6,3,2)}$	0	0	0	0	0	-1	0	0	0	-1	0	0	0	0	0	1	0	0	0	1	0	0	0	0	4.2300
$V_{u(5,4,6)}$	0	0	0	0	-0.566	-0.223	0	0	0	0	-1	0	0	0	0.566	0.223	0	0	0	0	1	0	0	0	1.9341
$V_{u(6,4,3)}$	0	0	0	0	0.566	0.223	0	0	0	0	0	-1	0	0	-0.566	-0.223	0	0	0	0	0	1	0	0	7.9859
$V_{u(1,5,6)}$	0	0	0	0	-0.193	-0.499	0	0	0	0	0	0	-1	0	0.193	0.499	0	0	0	0	0	0	1	0	2.5317
$V_{u(6,5,4)}$	0	0	0	0	-0.193	-0.499	0	0	0	0	0	0	0	1	0.193	0.499	0	0	0	0	0	0	0	-1	5.5417
r	0	0	0	0	0.604	1.818	0	2	0	2	2	2	2	0	1.396	0.182	2	0	2	0	0	0	0	2	-68.7786

The resulting simplex table shows that in the processing of geodetic network using this method, the optimal solution was found, since all elements of the vector r are positive and the table of resulting corrections

(Tab. 8) points out to the fact that this method identified two outlier measurement at the points where the angles are loaded by multiples of the standard deviation of the measured angle. The graphical interpretation of the results obtained is shown through the confidence ellipses constructed at 95% and 99% probability (Fig. 10).

Tab. 8 Comparing the corrections in the geodetic network loaded by outlier measurements

Angle	l~ Theoretical value of angle	l Measured value of angle	v~ Actual measurement error	v LSM	v Simplex
L-S-P	[g]	[g]	[cc]	[cc]	[cc]
2-1-6	31.04550	31.04790	-24.00	-20.86	-20.33
6-1-5	86.04331	86.04290	4.08	0.94	0.36
3-2-6	57.35031	57.35090	-5.90	-9.24	0.00
6-2-1	82.30387	82.30587	-20.00	-16.66	-25.86
4-3-6	46.63242	46.63190	5.21	-0.71	0.00
6-3-2	43.61620	43.61630	-0.98	4.94	4.23
5-4-6	94.62331	94.62290	4.11	7.32	1.93
6-4-3	32.88348	32.88290	5.81	2.60	7.99
1-5-6	71.37301	71.37290	1.07	4.67	2.53
6-5-4	54.12859	54.12900	-4.08	-7.68	-5.54

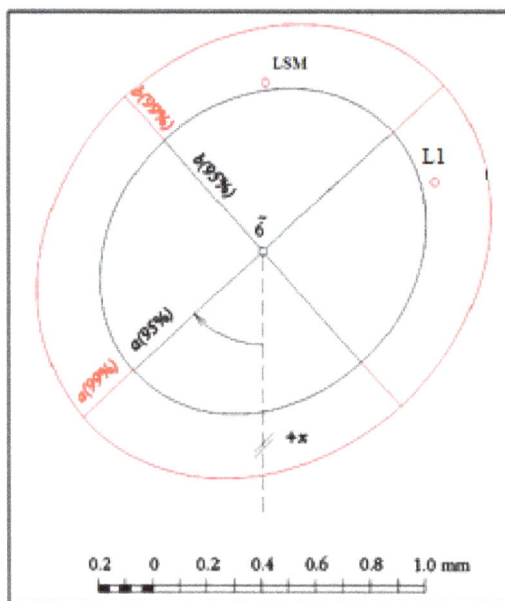

Fig. 10 Graphical interpretation of the estimation methods for adjusting the geodetic network loaded by outlier measurements

5   CONCLUSIONS

The estimation methods that were used and applied to the example of a regression line and a geodetic network showed the mutual tightness of the achievements. In the present contribution, simple but all the more illustrating examples demonstrating the positive attributes of alternative estimation methods were deliberately chosen. Among the many such methods published in foreign literature [8], the simplex method was presented in this paper which also allows to solve the problem of infiltration of outlier measurements to the set of processed data. The minimization problem of the L1-norm is normally solved using a linear programming tabular form; however, the paper presents a simpler and more efficient way to solve a linear programming problem using matrix relations.

REFERENCES

[1]   BAJTALA, M. – SOKOL, Š.: Odhad variančných komponentov z meraní v geodetickej sieti. In: *Acta Montanistica Slovaca*. ISSN 1335-1788, 2005, roč. 10, č. 2. s. 68-77.

[2]   BASSET, P.: *Some properties of the least absolute error estimator*. Ph.D. Dissertation, University of Michigan, Ann Arbor, 1973.

[3]   BÖHM, J. – RADOUCH, V. – HAMPACHER, M.: *Teorie chyb a vyrovnávací počet*. 2 vydanie. Praha: Geodetický a kartografický podnik, s. p., 1990. ISBN 80-7011-056-2.

[4]   CASPARY, W.F.: *Concepts of network and deformation analysis*. First edition. Kensingthon: School of surveying The University of New South Wales, 1987. 187p. ISBN 0-85839-044-2.

[5]   GAŠINCOVÁ, S., GAŠINEC, J., WEISS, G., LABANT, S.: Application of robust estimation methods for the analysis of outlier measurements. In: *GeoScienceEngineering*. Vol. 57, no. 3 (2011), p. 14-29. ISSN 1802-5420.

[6]   GAŠINCOVÁ, S.: *Spracovanie 2D sietí pomocou robustných metód* : doktorandská dizertačná práca / školiteľ Janka Sabová. - Košice, 2007. - 92 s.

[7]   IŽVOLTOVÁ, J. *Teória chýb a vyrovnávací počet I. Príklady ku cvičeniam*. Žilinská univerzita, 2004, p. 121,. ISBN 9788080702397.

[8]   JÄGER, R., MÜLLER, T., SALER, H.   SCHVÄBLE, R.: *Klassischle und robuste Ausgleichungsverfahren*, Herbert Wichmann Verlag, Heidelberg, 2005.

[9]   LABANT, S., WEISS, G., KUKUČKA, P.: Robust adjustment of a geodetic network measured by satellite technology in the Dargovských Hrdinov suburb. In: *Acta Montanistica Slovaca*. Roč. 16, č. 3 (2011), s. 229-237. - ISSN 1335-1788 Spôsob prístupu: http://actamont.tuke.sk/pdf/2011/n3/7labant.pdf...

[10]   SOKOL, Š., BAJTALA, M., LIPTÁK, M. Creation of a Surveying Base for an Ice Rink Reconstruction Project. *In: Acta Montanistica Slovaca.* ISSN 1335-1788, 2011, vol. 16, no. 4, p. 312-318.

[11]   STAŇKOVÁ, H., ČERNOTA, P., NOVOSAD, M. *Problematika rovinných transformací na území ovlivněném hornickou činností.* In: GeoScience Engineering. ISSN 1802-5420, 2012, vol. 58, no. 3, p. 1-12.

[12]   WEISS, G., LABANT, S, WEISS, E., MIXTAJ, L. : *Detection of erroneous values in the measurement of local geodetic networks. In: Acta Montanistica Slovaca.* Roč. 15, č. 1 (2010), s. 62-70. - ISSN 1335-1788 Spôsob prístupu: http://actamont.tuke.sk/pdf/2010/n1/13weiss.pdf...

[13]   WEISS, G., ŠÜTTI, J.: *Geodetické lokálne siete I.* 1st ed. Košice, Vydavateľstvo Štroffek, 1997. 130p. ISBN 80-967636-2-8 .

RESUMÉ

Záverom si dovoľujeme konštatovať, že použitie alternatívnych odhadovacích metód má   nesporne svoje opodstatnenie, pretože geodetické merania sa nezriedka realizujú v náročných a neštandardných podmienkach, ktoré môžu byť príčinou skutočnosti, že do súboru meraných dát preniknú odľahlé merania. Z opakovaných meraní na stanovisku a z obmedzených doplnkových fyzikálnych meraní teploty, tlaku a vlhkosti má geodet len obmedzenú možnosť posúdiť, či a do akej miery sa do súborov meraných geometrických veličín infiltroval vplyv predovšetkým tepelného poľa, prípadne iných rušivých fenoménov, o ktorom môže prijať závery zväčša až na základe štatistického spracovania po návrate z terénu. Pokiaľ sa jedná o jedno meranie, je identifikovateľné napr. štatistickými testami, v prípade, že súbor je kontaminovaný viacerými odľahlými meraniami, štatistické testy nemusia byť úspešné. V tomto prípade alternatívne odhadovacie metódy, kde nesporne patria aj metódy lineárneho programovania dokážu identifikovať túto množinu a potlačiť ich vplyv na výsledky vyrovnania.

6

IMPACTS OF MEASURING AND NUMERICAL ERRORS IN LSM ADJUSTMENT OF LOCAL GEODETIC NETWORK

Silvia GAŠINCOVÁ [1], Dušan KNEŽO [2], Ladislav MIXTAJ [3], Peter HARMAN [4]

[1] *MSc., PhD., Institute of Geodesy, Cartography and Geographic Information Systems, Faculty of Mining, Ecology, Process Control and Geotechnologies, Technical University of Košice*
Park Komenského 19, 043 84 Košice, Slovak Republic
e-mail: silvia.gasincova@tuke.sk

[2] *prof., RNDr., CSc., Department of Applied Mathematics, Faculty of Mechanical Engineering,Technical University of Košice*
Letná 9, 042 00 Košice, Slovak Republic
e-mail: dusan.knezo@tuke.sk

[3] *MSc., PhD., Institute of Geotourism, Faculty of Mining, Ecology, Process Control and Geotechnologies, Technical University of Košice*
Letná 9, 042 00 Košice, Slovak Republic
e-mail: ladislav.mixtaj@tuke.sk

[4] *MSc., Institute of Geodesy, Cartography and Geographic Information Systems, Faculty of Mining, Ecology, Process Control and Geotechnologies, Technical University of Košice*
Park Komenského 19, 043 84 Košice, Slovak Republic
e-mail: peter.harman@tuke.sk

Abstract

The Local Geodetic Network (LGN) adjustment is affected by various errors, whether arisen from measuring or calculating works. The measuring errors occurred in the consistent LGS measuring can be detected by control calculations in dL using redundant measurements. The numerical errors may be detected, or the accuracy of the formulas used can be verified by control calculations in a configuration matrix A and control relations, which should show zero values.

Abstract

Na vyrovnanie lokálnej geodetickej siete (LGS) vplývajú rôzne chyby, ktoré vznikli či už pri meracích alebo počtárskych prácach. Meračské chyby vzniknuté pri dôslednom zameraní LGS sa dajú odhaliť kontrolnými výpočtami v dL pomocou nadbytočných meraní. Výpočtové chyby sa môžu odhaliť, resp. overiť správnosť použitých vzorcov pomocou kontrolných výpočtov v konfiguračnej matici A a kontrolných vzťahov, ktoré by mali vykazovať nulové hodnoty.

Key words: LGN, LSM, measuring and numerical errors.

1 INTRODUCTION

While measuring geometrical quantities needed for various calculations to determine the parameters and characteristics of a LGN, it is impossible to achieve, or ensure always error-free quality and reliability of performances and results. Is it realistic to expect, even when safeguarding carefully the measuring and processing activities, that some of the elements being measured can be for various reasons measured erroneously and then by miscalculations also the results, i.e. the correctness of the entire adjustment, affected.

Coordinate estimates of new established datum points (DB) in a LGN are a function of the joint impact of several different (by content and structure) matrices, containing both the given quantities (DB coordinates) and also the values of measured quantities with their quality assessment (cofactors) as well as the effects of the geometric structure of the generated LGN to its different resulting parameters.

It is therefore necessary to highlight the effect of individual argument matrices in the structure and values of the developed matrix $\hat{C}_{UB}$ even in a general view and the dependence of its properties on the impact of the matrices C_{UB}^0 as well as L, L^0, or $dL = L - L^0$ and Q_L.

2 THE EFFECT OF ERRORS ON THE LGN DEVELOPMENT

Let us consider a LGN, whose point field is created by:
- verified (compatible) points of the state point field used for the LGN as datum points (DB) with fixed coordinates in the state system,
- points "new", formed under the LGN purpose, i.e. determined points (UB), complementing the DB to the overall needful point field for the LGN.

In thus created LGN its necessary characteristics and properties are determined by the LGN adjustment (a bound LSM adjustment is assumed), from where we get the coordinate estimates $\hat{C}_{UB}$ by:

$$\hat{C}_{UB} = \begin{bmatrix} \hat{X}_{UB1} \\ \hat{Y}_{UB1} \\ \\ \hat{X}_{UBp} \\ \hat{Y}_{UBp} \end{bmatrix} = C_{UB}^0 + (A^T Q_L^{-1} A)^{-1} A^T Q_L^{-1} \cdot dL = \tag{1}$$

$$= C_{UB}^0 + N^{-1} A^T Q_L^{-1} (L - L^0) =$$
$$= C_{UB}^0 + d\hat{C}_{UB}.$$

As resulting from (1), the coordinates $\hat{C}_{UB}$ are formed by the prescribed interactive effect of matrices:

$A \atop (n,2u)$ configuration matrix of a complete LGN, where p is a number of UB,

$L \atop (n,1)$ vector of measurands in the LGN (lengths, horizontal and vertical angles and other quantities),

$L^0 \atop (n,1)$ approximate values of quantities in the LGN from calculations,

$Q_L \atop (n,n)$ cofactor vector matrix L,

$C_{UB}^0 \atop (p,2)$ approximate UB coordinates determined by measurements of appropriate quantities and by

$$\tag{2}$$

necessary calculations, and therefore the coordinates $\hat{C}_{UB}$ are a function of all geometric elements (2) in terms of:

$$\hat{C}_{UB} = f(A, L, L^0, Q_L, C_{UB}^0), \tag{3}$$

whose elements require for obtaining correct results an errorless structural content of all the matrices involved in the LGN adjustment.

Elements in the matrices (3), involved in the adjusting process, (functions of measured, or computed values d, z, ω, C^0, ...) either affect by their correctness the generation of correct results or when the matrices (3) are encumbered with errors, also the matrices produce erroneous results of the adjustment.

3 STRUCTURE OF ARGUMENT MATRICES IN THE LGN ADJUSTMENT

- Matrix $\underset{(n,2u)}{A}$:

contains in rows for each geometric quantity of the net (L_i) coefficients "a". E.g. in the row for d_{ij}^0 these are the coefficients: "a_{iX}" "a_{iY}" and "$-a_{iX}$" "$-a_{iY}$", some of which may be incorrect due to erroneous, or insufficiently "approximate" coordinates C_{UBi}^0, C_{UBj}^0 of terminal points of the measured length d_{ij}^0 (Fig. 1, Fig. 2).

		UB_i		...	UB_j		...	UB_k		DB_1	...
		X_i^0	Y_i^0		X_j^0	Y_j^0		X_k^0	Y_k^0		
$\underset{(n,2u)}{A}$											
$\vdots$											
d_{ij}^0		a_{iX}	a_{iY}		$-a_{jX}$	$-a_{jY}$					
$\vdots$											
ω_{jik}^0		a_{iX}	$-a_{iY}$		$-a_{jX}$	$-a_{jY}$		$-a_{kX}$	a_{kY}		
$\vdots$											

Fig. 1 *Coefficients "a" in the matrix A*

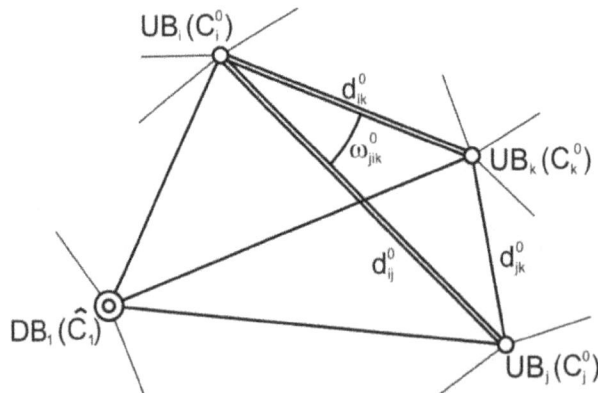

Fig. 2 *The geometry of lengths and angles in a LGN*

E.g. for the length d_{ij}^0 between the points UB_i, UB_j, for one terminal point UB_i the appropriate coefficients a_{iX}, a_{iY} in the matrix A are, as follows:

$$a_{iX} = \frac{\partial d_{ij}^0}{\partial X_i^0} = \frac{X_j^0 - X_i^0}{\sqrt{(X_j^0 - X_i^0)^2 + (Y_j^0 - Y_i^0)^2}} = \frac{X_j^0 - X_i^0}{d_{ij}^0} = \frac{f_{dX}(C_i^0, C_j^0)}{f_d(C_i^0, C_j^0)},$$

(4)

$$a_{iY} = \frac{\partial d_{ij}^0}{\partial Y_i^0} = \frac{Y_j^0 - Y_i^0}{\sqrt{(X_j^0 - X_i^0)^2 + (Y_j^0 - Y_i^0)^2}} = \frac{Y_j^0 - Y_i^0}{d_{ij}^0} = \frac{f_{dY}(C_i^0, C_j^0)}{f_d(C_i^0, C_j^0)}$$

and analogously for the other terminal point UB $_j$ of this length d_{ij}^0 :

$$a_{jX} = \frac{\partial d_{ij}^0}{\partial X_j^0} = -\frac{X_j^0 - X_i^0}{\sqrt{(X_j^0 - X_i^0)^2 + (Y_j^0 - Y_i^0)^2}} = -\frac{X_j^0 - X_i^0}{d_{ij}^0} = -\frac{f_{dX}(C_i^0, C_j^0)}{f_d(C_i^0, C_j^0)},$$

$$a_{jY} = \frac{\partial d_{ij}^0}{\partial Y_j^0} = -\frac{Y_j^0 - Y_i^0}{\sqrt{(X_j^0 - X_i^0)^2 + (Y_j^0 - Y_i^0)^2}} = -\frac{Y_j^0 - Y_i^0}{d_{ij}^0} = -\frac{f_{dY}(C_i^0, C_j^0)}{f_d(C_i^0, C_j^0)}. \tag{5}$$

Similar structures of coefficients a_{iX}, a_{iY}, a_{jX}, a_{jY} apply to all points UB and DB generating the measured lengths between them. Any errors in the values X_i^0, Y_i^0 and X_j^0, Y_j^0 thus generate incorrect, unrealistic values of the corresponding coefficients a_{iX}, ..., a_{jY} and thus also a defective structure of the matrix A.

The same is true for coefficients "a" of horizontal angles ω_{jik}^0 in the matrix A where the coefficients a_{jX}, a_{jY}, a_{iX}, a_{iY}, a_{kX}, a_{kY}, e.g. for the angle ω_{jik}^0 (Fig. 2) are, as follows:

$$a_{jX} = \left(\frac{\partial \omega_{jik}^0}{\partial X_j^0}\right)^0 = \frac{Y_j^0 - Y_i^0}{(d_{ij}^0)^2} = \frac{f_{dY}(C_i^0, C_j^0)}{f_d(C_i^0, C_j^0)},$$

$$a_{jY} = \left(\frac{\partial \omega_{jik}^0}{\partial Y_j^0}\right)^0 = -\frac{X_j^0 - X_i^0}{(d_{ij}^0)^2} = -\frac{f_{dX}(C_i^0, C_j^0)}{f_d(C_i^0, C_j^0)},$$

$$a_{iX} = \left(\frac{\partial \omega_{jik}^0}{\partial X_i^0}\right)^0 = \left(\frac{Y_k^0 - Y_i^0}{(d_{ik}^0)^2} - \frac{Y_j^0 - Y_i^0}{(d_{ij}^0)^2}\right) = \frac{f_{dY}, f_{dY}(C_i^0, C_j^0, C_k^0)}{f_{dij}, f_{dik}(C_i^0, C_j^0, C_k^0)},$$

$$a_{iY} = \left(\frac{\partial \omega_{jik}^0}{\partial Y_i^0}\right)^0 = \left(-\frac{X_k^0 - X_i^0}{(d_{ik}^0)^2} + \frac{X_j^0 - X_i^0}{(d_{ij}^0)^2}\right) = \frac{f_{dX}, f_{dX}(C_i^0, C_j^0, C_k^0)}{f_{dij}, f_{dik}(C_i^0, C_j^0, C_k^0)}, \tag{6}$$

$$a_{kX} = \left(\frac{\partial \omega_{jik}^0}{\partial X_k^0}\right)^0 = -\frac{Y_k^0 - Y_k^0}{(d_{ik}^0)^2} = \frac{f_{dY}(C_i^0, C_k^0)}{f_d(C_i^0, C_k^0)},$$

$$a_{kY} = \left(\frac{\partial \omega_{jik}^0}{\partial Y_k^0}\right)^0 = \frac{X_k^0 - X_i^0}{(d_{ik}^0)^2} = \frac{f_{dX}(C_i^0, C_k^0)}{f_d(C_i^0, C_k^0)}$$

and show by their structure, that due to the erroneous determination of the relevant coordinates $C^0 = [X^0 \ Y^0]$ of UB points, or erroneous determination of approximate values of horizontal lengths and angles, also the matrix A may enter into the adjusting process with a wrong structure.

- Matrix $\underset{(n,1)}{L}$:

contains the quantities needed in the LGN processing, obtained from the measurements: lengths d between pairs of LGN points reduced to the plane of projection, horizontal angles ω also reduced to the plane of map projection, zenith angles z and other necessary quantities, which in addition to the weighted with random errors may be contaminated also by the influence of latent systematic errors. Their effect is appropriately reflected in all measured quantities between the DB, UB points and between the points DB and UB.

However, it is to be expected also that e.g. in using inadequate measuring instruments, incorrect measurement procedures, due to various mistakes, some latently defective functions of devices and the like, also the measured values d, ω, z, ... may be significantly incorrect. The elements L_i encumbered with errors of the matrix L within the meaning of the dependence (3) then affect the results of the LGN adjustment (coordinate estimates $\hat{C}_{UB}$ and other parameters) to a different, distorted and devalued extent.

- Cofactor matrix Q_L :
(n,n)

The matrix Q_L of the vector of measured values L can be created by different procedures such as
$(n,1)$
using a priori variances σ_0^2 (of the manufacturer of the device) and aposteriori variances s_0^2 (obtained on the basis of performed measurements) by $q_i = (s_0^2)_i / \sigma_0^2$, the Minque , Bique methods and other procedures (Wolf 1968, Mikhail 1976, Höpcke 1980, Pelzer 1980, 1985, Ressmann 1980, Weiss et al., 2008). According to different methodological principles and experiences formed cofactors q_i however need not always significantly affect by their matrix $Q_L = diag(q_1,...,q_i,...,q_n)$ formed estimates $d\hat{C}_{UB}$ and therefore either the final results of the UB coordinates $\hat{C}_{UB}$. In justified cases (well measurement conditions and the homogeneity of observations, etc.), also unit cofactors $q_i = 1$, $i = 1, 2, ..., n$. are used for all L_i. For various measurands, i.e. measured lengths and angles, the cofactors are obtained, as follows: $q_{d(\omega)i} = \sigma_{d(\omega)i}^2 / \sigma_0^2$, where $\sigma_0^2 = (\Sigma\sigma_{dij}^2 + \Sigma\sigma_{\omega jik}^2)/n$. The creation of reliable cofactors, adequate to equipment features and current measurement conditions, represents in the LGN adjustment an important influencing factor to obtain real values of various LGN parameters.

- Matrix L^0 :
$(n,1)$
contains numerically determined (not by measuring) approximate values d^0, ω^0, z^0, C^0,..., of relevant variables, i.e. lengths, horizontal angles, zenith angles, approximate values of UB coordinates and other possible, or necessary quantities based on known procedures for their calculation.

Approximate values of L_i^0, $i = 1, 2, ..., n$ together with their measured values L_i create a matrix dL with elements $dL_i = L_i = L_i^0$, i.e. with "complements" to approximate values L^0 of variables in the net.
$(n,1)$
The elements dL_i on the basis of their significant or negligible value (typically: $dL < (1-3)$ mm, $dL < (1-5)^{cc}$) indicate either the numerical acceptance of the value dL_i or significant, unacceptable value dL_i, indicating a significant discrepancy between L and L^0.

- Matrices C_{UB}^0 and $\hat{C}_{UB}$:
coordinates $(C_{UB}^0 = [X^0 \ Y^0]_{UB})$ of UB points are for the LSM processing of the LGN determined (based on their orientation with simple structural links, as a system of rayons, intersections, etc. in relation to the DB) as coordinates without any adjustment and are declared as the approximate coordinates C^0 of UB points. These then within the LGN adjustment finally get their values $\hat{C}_{UB}$, by adding coordinate complements $d\hat{C}_{UB}$ to the respective values C_{UB}^0;

coordinates of datum points $(C_{DB} = [X \ Y]_{DB})$ in the LGN will be considered as non-encumbered with nonrandom errors. These points based on the results of their compatibility verification (Bill 1984, Weiss et al., 2004, Pukanská et al. 2007, Labant et al. ,2009, Weiss et al., 2010), with acceptable congruences of coordinate point and its physically measuring mark, may be regarded as datum points (DB), whose coordinates will not in the net adjustment adversely affect the resulting LGN parameters.

4 CONTROL RELATIONS IN THE ADJUSTMENT

To assess the structural and content correctness of the used matrices and quantities as well as the final results of the adjusting process their numerical accuracy is verified.

Typically, different valid relations are used, created from products of individual matrices, which provide in case of their correctness a zero output: $\underset{(1,1)}{0}$, i.e. zero value as a result for a given matrix product, or give acceptable values close to 0.

Checks of matrix solution accuracy are realized both in the structure of individual matrices (e.g. in A), namely in terms of accuracy of numerical values in rows, or in columns. E.g. row controls in A in which zero sums of coefficients are to be created, i.e. in each row j ($j = 1, 2, ..., n$) the sum

$$a_{i,1} + a_{i,2} + ... + a_{i,2n} = \underset{(1,1)}{0}.$$

Other forms of adjustment controls are different product blocks of suitable matrices $A, Q_L, dL, V, d\hat{C}, C^0$, created on the basis of their property of binding suitability. If the product blocks are correct, they are equal to 0 (in theory), or equal to small numerical values, which are negligible (in practice).

To the controls of the above method e.g. the following relations can be applied:

$$\underset{(1,n)}{dL} \cdot \underset{(n,n)}{Q_L^{-1}} \cdot \underset{(n,2u)}{A} \cdot \underset{(2u,1)}{d\hat{C}} = \underset{(1,1)}{0},$$

$$\underset{(1,n)}{dL} \cdot \underset{(n,n)}{Q_L^{-1}} \cdot \underset{(n,1)}{dL} = \underset{(1,1)}{0},$$

$$\underset{(1,n)}{V} \cdot \underset{(n,n)}{Q_L^{-1}} \cdot \underset{(n,1)}{dL} = \underset{(1,1)}{0},$$ \hfill (7)

$$\underset{(1,n)}{dL} \cdot \underset{(n,n)}{Q_L^{-1}} \cdot \underset{(n,1)}{V} = \underset{(1,1)}{0},$$

$$\underset{(2u,n)}{A} \cdot \underset{(n,n)}{Q_L^{-1}} \cdot \underset{(n,1)}{V} = \underset{(2u,1)}{0},$$

and other appropriate similar structures of numerical values (Wolf 1968, Reissmann 1980, Höpcke 1980, Böhm et al. ,1990, Weiss et al., 2009).

If the control relations show instead of the theoretical values "0" non-zero values, which is always actual, and these will be smaller than 0.001 (or other limits), the results of the adjustment may be accepted, otherwise they are rejected and it is necessary to explore their genesis in the adjusting process.

REFERENCES

[1] Bill, R. (1984): *Eine Strategie zur Ausgleichung und Analyse von Verdichtungsnetzen*. Veröff. D. Deutsch Geod. Komm. R.C, H. 295, München.

[2] Böhm, J., Radouch, V., Hampacher, M. (1990): *Teorie chyb a vyrovnávací počet*, GKP Praha.

[3] Heck, B. (1981): *Der Einfluss einzelner Beobachtungen auf das Ergebnis einer Ausgleichung und die Suche nach Ausreissern in der Beobachtungen*, Allgem. Verm. Nachr. 88, 1, 17-34.

[4] Höpcke, W. (1980): *Fehlerlehre und Ausgleichungsrechnung*, W. de Guyter, Berlin.

[5] Labant, S., Kalatovičová, L., Kukučka, P., Weiss, E. (2009): *Precision of GNSS instruments by static method comparing in real time*. Acta Montanistica Slovaca, 14, 1, 55-61.

[6] Mikhail, E. M. (1976): *Observations and Least Squares*. IEP-Dun-Donnelley Publisher, New York.

[7] Pelzer, H. (Hrsg) (1980, 1985): *Geodätische Netze in Landes-und Ingenieurvermessung I., II*. Wittwer, Stuttgart.

[8] Pukanská, K., Weiss, G. (2007): *Precision the points position with use technology GPS* In: Uhlí - Rudy - Geologický průzkum. 14, 9, 30-35.

[9] Reissmann, G. (1980): *Die Ausgleichungsrechnung VI g.f. Bauwesen*, Berlin.

[10]   Weiss, G., Gašinec, J., Engel, J., Labant, S., Rákay, ml., Š. (2008): *Effect incorrect points of the Local Geodetic Network at results of the adjustment.* Acta Montanistica Slovaca, 13, 4, 485-490.

[11]   Weiss, G, Jakub, V., Weiss, E. (2004): *Compatibility of geodetic points and their verification.* TU Košice, F BERG.

[12]   Weiss, G., Labant, S., Rákay ml, Š. (2009): *Numerical controls in an adjustment of the local positional geodetic network (LGN).* Acta Montanistica Slovaca, 14, 1, 38-42.

[13]   Weiss, G., Labant, S., Weiss, E., Mixtaj, L. (2010): *Detection of erroneous values in the measurement of local geodetic networks.* Acta Montanistica Slovaca, 15, 1, 62-70.

[14]   Weiss, G., Šütti, J. (1997): *Geodetic local networks I.* TU Košice, F BERG.

[15]   Wolf, H.: *Ausgleichungsrechnung nach der Methode der kleinsten Quadrate.* Dümmler, Bonn 1968.

RESUMÉ

Predkladaný článok sa zaoberá vplyvom rôznych chýb, ktoré vznikli či už pri meraciach alebo počtárskych prácach, na vyrovnanie lokálnej geodetickej siete (LGS). Je reálne očakávať aj pri starostlivom zabezpečení meračskej a spracovateľskej činnosti, že niektoré z meraných prvkov môžu byť z rôznych dôvodov chybne zamerané a potom chybnými výpočtami ovplyvnené aj výsledky t.j. korektnosť celého vyrovnania.

Uvažujeme LGS, ktorej bodové pole tvoria: 1. overené (kompatibilné) body bodového poľa použité pre LGS ako dátumové body DB s fixnými súradnicami, 2. body „nové", založené podľa účelu LGS, t.j. tzn. určované body UB, doplňujúce DB na celkové potrebné bodové pole pre LGS. Na základe vyrovnania LGS, z ktorého odhady súradníc $\hat{C}_{UB}$ sa vytvárajú predpísaným interakčným pôsobením matíc:

$$\hat{C}_{UB} = f(A,\ L,\ L^0,\ Q_L,\ C^0_{UB}),$$ ktorých prvky vyžadujú pre korektné výsledky bezchybnú štruktúru matíc pôsobiacich vo vyrovnaní LGS.

Matica dizajnu A obsahuje v riadkoch pre jednotlivé geometrické veličiny siete (L_i) koeficienty „a", ktoré sú parciálnymi deriváciami príslušných približných dĺžok d_{ij}^0 a uhlov ω_{jik}^0 podľa príslušných súradníc $X_i^0\ Y_i^0$. V dôsledku chybných určení súradníc $C^0 = [X^0\ Y^0]$ bodov UB, resp. chybných určení približných hodnôt vodorovných dĺžok a uhlov, aj matica A môže vstupovať do vyrovnávacieho procesu s chybnou štruktúrou.

Matica nameraných veličín L obsahuje veličiny potrebne v spracovaní LGS, získané z meraní: dĺžky d a vodorovné uhly ω redukované do roviny kartografického zobrazenia, zenitové uhly z i ďalšie. Ich pôsobenie sa primerane prejaví vo všetkých meraných veličinách medzi bodmi DB, bodmi UB a medzi bodmi DB a UB. Pri použití neadekvátnych meracích prístrojov, nesprávnych meracích postupov, v dôsledku rôznych omylov, je tiež očakávateľný negatívny vplyv na výsledky vyrovnania LGS.

Kovariančná matica Q_L vektora meraných hodnôt L sa môže vytvoriť rôznymi postupmi ako napr. s použitím apriórnych variancií σ_0^2 (od výrobcu prístroja) a aposteriórnych variancií s_0^2 (získaných na základe vykonaných meraní) podľa $q_i = s_{0i}^2 / \sigma_0^2$, a inými postupmi. Tvorba spoľahlivých kofaktorov, ktoré sú adekvátne vlastnostiam prístrojov a aktuálnym podmienkam merania, je vo vyrovnaní LGS dôležitým ovplyvňujúcim faktorom získania reálnych hodnôt rôznych parametrov LGS.

Matica L^0 obsahuje numericky určené približné hodnoty d^0, ω^0, z^0, C^0,... príslušných veličín. Približné hodnoty veličín L_i^0, $i = 1, 2, ..., n$ spolu s ich nameranými hodnotami L_i vytvárajú maticu doplnkov $dL_i = L_i = L_i^0$. Prvky dL_i na základe ich významnej alebo zanedbateľnej hodnoty signalizujú numerickú prijateľnú hodnotu alebo významnú, neprijateľnú hodnotu dL_i, ktorá oznamuje rozpor medzi L a L^0.

Matica C_{UB}^0 sú súradnice bez vyrovnania a deklarujú sa za približné súradnice C^0 bodov UB. Tieto v rámci vyrovnania LGS získajú potom definitívne svoje hodnoty $\hat{C}_{UB}$, pridaním súradnicových doplnkov $d\hat{C}_{UB}$ k príslušným hodnotám C_{UB}^0. Body na základe výsledkov z ich overenia na kompatibilitu s prijateľnou kongruenciou súradnicového bodu a jeho fyzicky meračskej značky, možno považovať za dátumové body (DB), ktorých súradnice nebudú vo vyrovnaní siete negatívne ovplyvňovať výsledné parametre LGS.

V štruktúre jednotlivých matíc (napr. v A) sa realizujú kontroly správnosti maticového riešenia a to z hľadiska správnosti numerických hodnôt v riadkoch, resp. v stĺpcoch. Napr. riadkové kontroly v A, v ktorých sa majú vytvárať nulové súčty koeficientov, t.j. v každom riadku j ($j = 1,\ 2,\ ...,\ n$) súčet $a_{i,1} + a_{i,2} + \ ... \ + a_{i,2n} = 0$. Iné formy kontrol vyrovnania predstavujú rôzne súčinové bloky z vhodných matíc A, Q_L, dL, V, $d\hat{C}$, C^0, vytvorené na základe ich vlastností a väzbovej vhodnosti. Ak sú súčinové bloky korektné, sú rovné 0. V prípade nenulových hodnôt sa zamietajú a je potrebné pátrať vo vyrovnávacom procese po ich vzniku.

7

MODELLING THE UNCERTAINTY OF SLOPE ESTIMATION FROM A LIDAR-DERIVED DEM: A CASE STUDY FROM A LARGE-SCALE AREA IN THE CZECH REPUBLIC

Ivan MUDRON[1], Michal PODHORANYI[2,3], Juraj CIRBUS[1], Branislav DEVEČKA[1], Ladislav BAKAY[4]

[1] *Institute of Geoinformatics, Faculty of Mining and Geology, VSB-TU OSTRAVA, 17.listopadu 15/2172, 70833, Ostrava, Czech Republic*
mud023@vsb.cz, juraj.cirbus@vsb.cz, branisalv.devecka@vsb.cz

[2] *IT4Inovation Centre of Excellence VSB-TU OSTRAVA, 17.listopadu 15/2172, 708 33, Ostrava, Czech Republic*
michal.podhoranyi@vsb.cz

[3] *Department of Physical Geography and Geology, Faculty of Science, University of Ostrva, Chittussiho 10, 710 00, Ostrava, Czech Republic*

[4] *Department of Garden and Landscape Design, Slovak university of Agriculture, Trieda A. Hlinku 2, 949 76, Nitra, Slovak Republic*
bakay@is.uniag.sk

Abstract

This paper summarizes the methods and results of error modelling and propagation analyses in the Olše and Stonávka confluence area. In terrain analyses, the outputs of the aforementioned analysis are always a function of input. Two approaches according to the input data were used to generate field elevation errors which subsequently entered the error propagation analysis. The main goal solved in this research was to show the importance of input data in slope estimation and to estimate the elevation error propagation as well as to identify DEM errors and their consequences. Dependencies were investigated as well to achieve a better prediction of slope errors. Four different digital elevation model (DEM) resolutions (0.5, 1, 5 and 10 meters) were examined with the Root Mean Square Error (RMSE) rating up to 0.317 meters (10 m DEM). They all originated from a LIDAR survey. In the analyses, a stochastic Monte Carlo simulation was performed with 250 iterations. The article focuses on the error propagation in a large-scale area using high quality input DEM and Monte Carlo methods. The DEM uncertainty (RMSE) was obtained by sampling and ground research (RTK GPS) and from subtraction of two DEMs. According to empirical error distribution a semivariogram was used to model spatially autocorrelated uncertainty in elevation. The second procedure modelled the uncertainty without autocorrelation using a random $N(0,RMSE)$ error generator. Statistical summaries were drawn to investigate the expected hypothesis. As expected, the error in slopes increases with the increasing vertical error in the input DEM. According to similar studies the use of different DEM input data, high quality LIDAR input data decreases the output uncertainty. Errors modelled without spatial autocorrelation do not result in a greater variance in the resulting slope error. In this case, although the slope error results (comparing random uncorrelated and empirical autocorrelated error fields) did not show any statistical significant difference, the input elevation error pattern was not normally distributed and therefore the random error generator realization is not a suitable interpretation of the true state of elevation errors. The normal distribution was rejected because of the high kurtosis and extreme values (outliners). On the other hand, it can show an important insight into the expected elevation and slope errors. Geology does not influence the slope error in the study area.

Abstrakt

Táto práca zhŕňa metódu a výsledky modelovania chýb a analýzu šírenia chýb vo výpočte sklonov z DMR získaných LIDAR-om v skúmanej lokalite okolia sútoku riek Olše a Stonávka. V terénnych analýzach výstupy uvedenej analýzy sú vždy funkciou vstupu. Na generovania pola výškových chýb boli použité dve rozdielne metódy podľa vstupných dát. Modelované chyby v nadmorských výškach následne vstupovali do analýzy šírenia chýb. Hlavným cieľom práce bolo tak ako aj poukázanie na význam kvality vstupných dát vo výpočte sklonov a odhad šírenej chyby z nadmorských výšok v sklonoch tak aj identifikácia chýb v DMR a ich dopad. Závislosti chýb boli vyhodnotené hlavne pre lepší odhad chyby v sklonoch. V simuláciách boli použité 4 vstupné DMR s rozlíšením 0.5, 1, 5 a 10 metrov s RMSE chybou do 0.317 metra (10 m DMR). Všetky DMR boli získané z mračna bodov získaných LIDAR metódou zberu dát. Šírenie chýb bolo modelované pomocou stochastickej simulácie Monte Carlo s 250 iteráciami. Článok sa zameriava na šírenie chýb z vysoko presných vstupných dát na malom území. RMSE chyba bola získaná v prvom prípade z dát získaných terénnym prieskumom (RTK GPS) a v druhom prípade z porovnania dvoch kvalitatívne rozdielnych DMR. V prvom prípade sa vypočítali chyby vo výškach pomocou náhodného generátora chýb bez autokorelácie chýb. V druhom prípade sa s pomocou semivariogramu namodelovalo autokorelované pole chýb vo výškach. Použitím vhodných štatistík boli odvodené výsledky simulácie a overené stanovené hypotézy. Tak ako sa očakávalo chyby v sklonoch sú vyššie s zvyšujúcou sa chybou v nadmorských výškach. Tiež závislosti chýb od vypočítaných sklonov boli preskúmané, kde sa potvrdila závislosť chýb na sklonoch. Na druhej strane geológia nemala žiaden vplyv na chybu v sklonoch. Chyby namodelované bez autokorelácie nevedú vo väčšine prípadov k štatisticky významnej odchýlke. Vzhľadom však k rozmiestneniu chýb v priestore (vysoká autokorelácia, zamietnutie normálneho rozdelenia pre vysokú špicatosť a extrémne hodnoty) nie je táto metóda vhodná. Napriek tomu dáva dobrú možnosť nahliadnutia do očakávanej chyby v sklonoch a nadmorských výškach.

Key words: Uncertainty, Error propagation, Monte Carlo simulation, LIDAR-derived DEM, Slope estimation.

1 INTRODUCTION

Although many studies in the field of digital elevation model uncertainty and its error propagation were carried out, still there are some unacceptable assumptions about the expected error. Firstly, the DEM error disappears with more precise data acquisition and an optimal interpolation algorithm. Secondly, the DEM error is thought to be as small as not affecting the outputs of the analyses using a DEM input. Last but not least, DEMs are assumed and used as error-free models of reality, even though the existence of elevation uncertainty and gross errors are widely recognized [38], [19]. In the last decades, geomorphometry based on fine topscale DEMs have become popular in environmental science [35]. The accuracy of a digital elevation model is particularly important with its intended use [35]. So the misjudgements increased the importance of solving DEMs uncertainty and the error propagation problem. The awareness that uncertainty propagates through spatial analyses and may produce poor results that lead to wrong decisions triggered a lot of research on spatial accuracy assessment and data quality management in GIS (e.g. [33], [10] , [36], [2]) [34]. The information on the uncertainties in results from Geographic Information Systems (GIS) is needed for effective decision-making. Current GISs, however, do not provide this information [10], [14], [23]. Furthermore, there is the demand for presenting a level of accuracy (precision) [23]. Thus the long term vision in the research in spatial data uncertainty, accordingly DEM as well, was to develop a general purpose "error button" for generating information systems (GIS) [2]. There are two main ideas how to implement this button. GIS could incorporate the button into the product metadata [30] or in a more sophisticated solution the button is seen as user-dependent, which offers various possibilities for refining the error model according to the user's level of expertise [32]. The first steps towards the vision became a reality with building a data uncertainty engine, which implements the general framework for characterising uncertain environmental variables with probability models [34]. According to the authors, many other research groups worked on the design of an 'error-aware GIS', but very few have reached the operational stage. After the call for the development of geographical information systems that can handle uncertain data lasting at least for twenty years, Heuvelink, developing the Data Uncertainty Engine (DUE) engine, filled the gap [34]. Just the first step towards the solution of the error propagation problem was made. The DUE must be further elaborated and improved. The sustained development of science and technology brought and will bring new methods of data collection and processing. The DUE as another potential software application, using different or the same approaches, has to adjust to the changes. The usage of massive high-resolution DEMs based on the airborne light detection and ranging (LIDAR) renewed some assumptions. Two important factors appear to explain the lack of scientific knowledge about the use of LIDAR DEMs in an uncertain-aware terrain analysis. Firstly, it was commonly believed that the high quality of LIDAR DEMs [13], [1], [20] will make the uncertainty-aware terrain analysis unnecessary. Secondly, uncertainty propagation studies typically made use of simulation methods, such as simulated annealing and sequential Gaussian simulations [31], that are unsuitable for massive data sets because of their poor scalability [38], [10]. The aim of this paper is to analyse the aforementioned problems.

2 DEM ERRORS

Spatial uncertainty is defined as the difference between the contents of a spatial database and the corresponding phenomena in the real world. Because all contents of spatial databases are representations of the real world, it is inevitable that differences will exist between them and the real phenomena that they purport to represent [27]. An error is defined as the difference between reality and a representation of reality. In practice, errors are not exactly known. At best, the distribution of values is known. The chances that the error is positive or negative are equal [12]. The paper follows the taxonomy in which an error is a measurable and well-defined (no ambiguities and vagueness in data) part of uncertainty [25]. This is a justifiable choice because the semantics of elevation do not suffer from conceptual ambiguities which are common in, for example, defining the error in area-class maps [38]. The detailed process, by which the errors in a DEM are created, depends on the type of DEM and how it was created. Whatever method is used, DEM estimates are affected by several error sources, which can be grouped generally under three main classes: accuracy, density and distribution of data, surface characteristics, and interpolation algorithms [11] [9]. Uncertainty in DEMs originates from two sources, errors in the lattice (gross, systematic, random) and accuracy loss due to the lattice representation of the terrain [37]. There is a difference between positional and attribute uncertainty. The attribute uncertainty represents the deviation from true state of height and the positional uncertainty the shift in the object's position. Understanding the uncertainty is essential to correct modelling. The most frequent error in standard DEM products is reported as the Root Mean Squared Error (RMSE). Various methods have been used for estimating the RMSE. Most recently it is supposed to be estimated by comparison of elevations between well located sites in survey of higher accuracy with the elevation recorded in DEM at a minimum of 20 test points. The test points may be contour lines, bench marks, or spot elevations [8]. RMSE is based on the following formula:

$$RMSE = \sqrt{\frac{\sum (z - h)^2}{n}} \tag{1}$$

where z is the elevation recorded in the DEM; h is the elevation measured with higher precision and n is the total number of tested locations (at least 20). The Gaussian error model (a mean is the estimate of true values and a standard deviation is a measure of the uncertainty) makes only the most general assumptions about the processes by which the error accumulates. [15]. To achieve an improved estimate of the error for any particular area, a set of measurements made with higher precision is required, at best having another DEM of the same area with higher precision. In this case, it is possible to compare all values [9]. The spot heights and DEM or both DEMs have to be constructed separately; the independence is strictly required. When additional information is available about the structure of errors in the data set, the Gaussian model should be replaced with a substituting more accurate pattern of error (non-stationary or stationary spatial dependent random error field). According to previous studies (e.g. [7] [10] [15] [17] [24] [32] [36]), DEM errors are spatially correlated; autocorrelation is a natural characteristic of the error data. Hunter distinguished three cases of spatial dependences. Case one is spatial independence (r = 0). The elevation of each point is considered to be spatially independent of its neighbours (r = 0). In other words, the knowledge of the error present at one point provides no information on the errors present at neighbouring points, even though the elevation may have similar values. The elevation realization h at a x, y location is achieved by disturbing each observed elevation z at the same location by an independent disturbance term N (0, RMSE), which is a normally distributed random variable with a mean 0 and standard deviation RMSE (Eq. 2):

$$h_{(x,y)} = z_{(x,y)} + N(0, RMSE) \tag{2}$$

Case two is spatial dependence (limit r =1). At the other extreme, spatial autocorrelation reaches its maximum. All errors are perfectly correlated, and there is only 1 degree of freedom in effect in the disturbance field being applied to the DEM. It is unlikely that any DEM production process would generate a systematic error in elevations. Case three is spatial dependence (0 < R < 1). The case of positive correlation less than 1 is clearly most realistic [15] and the disturbance N(0,RMSE) is spatially correlated to a certain range following the fitted error model. Exponential and Gaussian [38] spatial autocorrelation models were selected to represent the correlation of the DEM error in the DEM uncertainty propagation studies. First exponential and later Gaussian models were found to be realistic and suitable for topography [31]. The study investigates the type of the model, range and the spatially independent random error pattern [10].

2.1 Error propagation analysis

There are two main approaches to the error propagation of a continuous variable: the analytical and the numerical error propagation. The analytical error propagation method uses an explicit mathematical model to describe the mechanisms of error propagation for a particular multi-criteria decision rule [6]. In numerical methods, the calculations are not made with exact numbers. Numerically generated random data sets are used instead of exact numbers. Usually they are generated on a computer in case of too complicated data or a physical model for analytical approach. In this study, the simulation of error was made stochastically using a Monte Carlo simulation. This method is further subdivided into unconditioned and conditioned models [5]. Unconditional error simulation models are based on the number of realizations of random functions. At their most basic level, they comprise an algorithm to select independent and uncorrelated values drawn from a normal distribution which can be added to the original DEM. The problem with unconditioned simulations is that they still make the assumption that the pattern of error is uniform over the study area or a wider region. Conditional error models directly honour observations of error at the sample locations. Such observations might have been obtained by comparison between the DEM and a higher accuracy reference data set collected from the same area [5]. In else, the parameters of an error model vary depending on the specific location. Comparing the results of using different methods of error modelling, the best method, which gives widely implementable and defensible results, is that based on a conditional stochastic simulation [9]. The most common uncertainty propagation analysis approach makes use of a Monte Carlo stochastic simulation [22]. The utilisation of a Monte Carlo simulation, which is the most flexible method for investigating the propagation of uncertainty in terrain analysis, is time-consuming [17]. Despite this drawback the unconditional Monte Carlo simulation was used to propagate the error. Tab. 1 shows the computation time cost for one simulation [10] modelled by the software R.

Tab. 1 Computational time for modelling one error pattern

DEM resolution	Number of points	Elapsed Time
10 x 10	263 520	2 min 21 sec
5 x 5	1 051 997	40 min 57 sec
1 x 1	26 289 516	17 days 2 hrs
1 x 1	1275630	1 hr 1 min 38 sec
0.5 x 0.5	5051130	16 hrs 54 min 33 sec

Although the area is relatively small (11.26 km^2 respectively 1.25 km^2) and the relative difference in elevation less than 45 meters, the empirical error pattern was investigated to find out an anomaly or a trend within. None of it was found in the error pattern. The outline of the Monte Carlo simulation is shown in Fig. 1 (used SW ArcGIS, own programming in C++ to calculate statistics). In simulations the initial DEM was used (with a resolution of 0.5, 1, 5, 10 m). This DEM was considered as an error free representation of the true state of elevation. Next the "error free slope" slope estimate was calculated. Then DEM error patterns were generated according to the initial DEM and error model attributes. The initial DEM was perturbed with the generated random error field (with and without autocorrelation). The resulting DEM had the essential properties of both the error pattern and the initial raster. Thus 500 realizations of DEM (250 both with and without autocorrelation) were generated and subsequently slope estimates were derived from alternative DEMs. The set of error patterns in slopes was calculated as the difference between the error free slope and the particular alternative slope. Using appropriate statistics the results of the simulation were derived. In some cases the absolute error value had to be used instead of the error value [10].

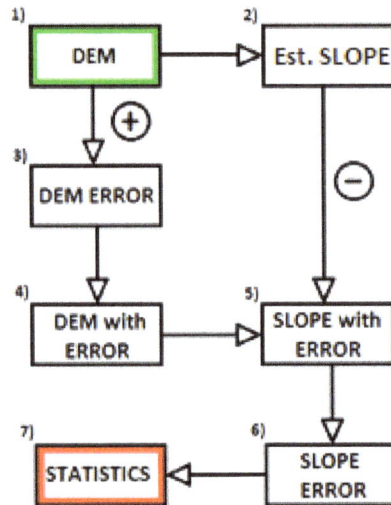

Fig.1 Outline of Monte Carlo simulation, here 1) denotes input DEM, 2) SLOPE calculated from 1 3) generated DEM ERROR, 4) Alternative DEM, 5) Alternative slope, 6) Error in slope, 7) Statistics.

2.2 Algorithm of slope computation

A variety of methods can be used to estimate the slope from DEM. Weighted least squares fit of a plane to a 3x3 neighbourhood centred on each point is the most amenable to a mathematical analysis of error propagation [15]. Most of the GIS SWs (including the most used ArcGIS) use this method to compute the slope from a DEM. In this paper, we decided to follow the aforementioned method's algorithm. The output slope derivate can be calculated in degrees (angular unit Eq. 8) or percentage (Eq. 7). The chosen units were degrees. The slope in degrees is calculated multiplying the slope in radians with 57.29578. The slope calculation (Fig. 2) is based on the change of height (rise) in the direction of x and y direction (run) - mathematically the first partial derivation of z in x and y axes. Thus the slope (Eq. 5) is determined by the rate of change (*Beta*) in both horizontal (*HD* Eq. 3) and vertical (*VD* Eq. 4) directions from the centre cell (*E*).

$$HD = \frac{\partial z}{\partial x} \tag{3}$$

$$VD = \frac{\partial z}{\partial y} \tag{4}$$

The approximation of the partial derivatives was made by a third-order finite difference method (Eq. 5 and 6) [18]. The method uses the 3x3 neighbourhood (Fig. 3) of the elevation values obtained in the raster around the centre cell. The distance between the elevation points denoted as *w* and represents also the cell (pixel) size of raster [10].

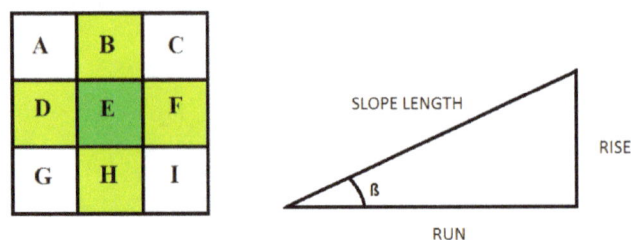

Fig. 2 Left the 3x3 neighbourhood window of the centre cell E and right the rise, run and beta description.

$$HD \approx \frac{(C+2F+I)-(A+2D+G)}{8*w} \qquad (5)$$

$$VD \approx \frac{(A+2B+C)-(G+2H+I)}{8*w} \qquad (6)$$

$$S = \sqrt{HD^2 + VD^2} \qquad (7)$$

$$\beta = \arctan\left(\sqrt{HD^2 + VD^2}\right) \qquad (8)$$

The influence of data precision on the derived slope is highly related to grid resolution. While using a high-resolution DEM (e.g. 1 m grid resolution), the influence of data precision becomes quite significant. DEM resolution determines the level of details of the surface being described. It naturally influences the accuracy of derived surface parameters. On the other side, usually the DEM error caused by data precision level is quite minimal, except in flat areas where the rounding errors could be significant [Zhou, Liu, 2004]. The precision significance was investigated as well, to prove or reject. We tried to minimize the rounding error because of flat areas [10].

3   STUDY AREA

The error propagation was carried out along a 5.9 km stretch of the Olše River and a 3.2 km stretch of the Stonávka River. Both river sections are located in the northeast region of the Czech Republic near its border with Poland [16].The area is located south of the town of Karviná in the north-eastern part of the Moravian-Silesian Region. The area is 5.544 km in length and 2.281 km in width spaced. After the area affected with gross error was eliminated, a total area of 11.262 km² remained. Because of gross errors and uncertainty in the data collection process caused by the atmosphere, three parts of the area (west) had to be clipped. Due the time-consuming computational method the 1.250 km² large study was used in case of a higher precision data input (Fig. 3). The elevation of the area varied between 211 and 256 (respectively 216 to 227 for small area) meters over the sea level. The slope varied from 0° to 85° (respectively 0 to 67 degrees). The average slope values (1.95° to 3.9° respectively 3° to 3.5°) and the data histograms revealed flat characteristics of the surface with few steep slopes [10].

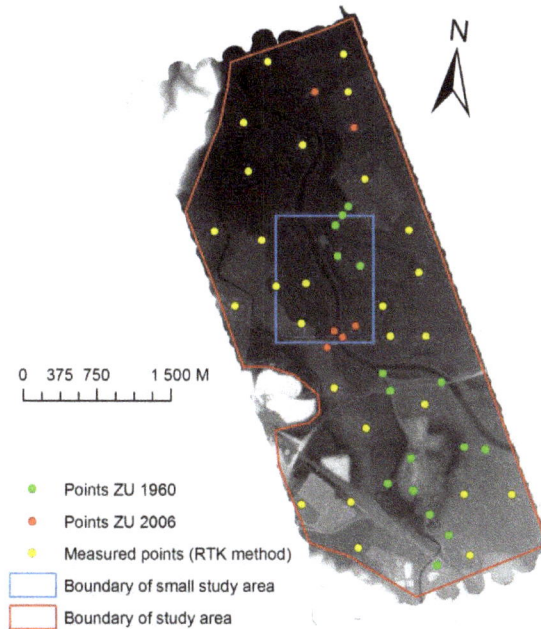

Fig. 3 Study area and measurement point locations for RMSE computation

4   DATABASE CREATION

The GIS database comes from various sources, each having its own level of uncertainty, depending on the specific technique used to acquire it [14]. The input data used to create the DEM in this study were obtained using the LIDAR method (Light Detection and Ranging). The Swedish company TopEyeAB, working with the MK-II laser system of its own design, carried out flights over the research area. The system consisted of a laser scanner with a 50 kHz frequency, the Inertial Navigation System (INS) and the Global Positioning System (GPS) systems. The optical portion of the scanner deviated the laser beam into circular traces. The system was equipped with the Rollei digital air camera with a 16-megapixel resolution (4080 x 4076 pixels). The scanning was carried out on the D-Hahn helicopter carrying the MKII-S/N 804 system at an altitude of 250 m [16]. The DEMs (0.5, 1, 5 and 10 m resolution) were computed independently of each other from a particular acquired LIDAR data point cloud. The density of all data points was 19 points per square meter. The density of terrain points was 9 points per square meter. The points were classified into three categories: terrain, vegetation under and over 3.5 meters high. The RMSE in input data were calculated two times for every DEM to make the comparison of possible inputs. First the error values were calculated subtracting the DEM from the DEM with higher precision (resolution). The 0.2m resolution DEM was used for the 0.5m resolution DEM. Then the RMSE (0.317 for 10 m, 0.156 for 5 m, 0.04 for 1 m and 0.035 for 0.5 meter resolution) was calculated from the error values of the whole area. This RMSE values were compared with the result of the second computation which was computed from 49 point measurements in the study area (Fig. 3). 22 of 49 points were created by CUZK (Land Survey Office of Czech Republic) without any given information of the data gathering method and accuracy. The second RMSE computation had a higher RMSE, which was effected by the location (sinking ground of mining area) of the 49 points. These are also not representative for the whole area and location. The 49 points were located often in error prone surfaces (roadsides, river bank sides). The 10m resolution RMSE difference takes 5.7 cm (0.374 for 49 points and 0.317 for LIDAR), which is 17 % of the total value of the LIDAR RMSE. In other cases, it was even worse (5 m – 14.1 cm, 1 and 0.5 m – 24.9 cm). It is necessary to mention that the LIDAR DEM of higher accuracy showed a certain uncertainty too. LIDAR RMSE results were taken to fit the spatially uncorrelated error pattern as a consequence of a better representation of the continuous empirical error pattern. The autocorrelated error pattern was made by investigating the empirical elevation error (Chapter 5.1).

4.1 Simulation of random fields

The input error field was made by the investigation of the empirical error pattern obtained with the aforementioned method (Chapter 2). The error propagation was modelled both with and without a spatially autocorrelated error field. The real state of nature was other than the expected theoretical state. First, there is an unjustified assumption that the mean error is zero [37]. The error mean statistics were close to zero, but all of them were rejected as statistical zeroes using a t-test hypothesis test in the Statgraphics software (Tab. 2).

Tab. 2 DEM error statistics (Number of Elevation Points (samplings), Error Mean [meters], Standard Deviation of Error [meters], and Maximum Absolute Error [meters])

DEM resolution	NUMBER OF POINTS	MEAN	STD. DEVIATION	MAX ABS ERROR
10 x 10	263 520	$-3.2 \ 10^{-2}$	0.692	11.942
5 x 5	1 051 997	$-1.2 \ 10^{-3}$	0.362	12.053
1 x 1	26 289 516	$-2.3 \ 10^{-3}$	0.085	9.567
0.5 x 0.5	83 963 724	$1.0 \ 10^{-5}$	0.008	1.597

The best fit of the elevation error pattern is to follow the empirical model [9]. If the difference between the elevation in the DEM and the actual surface (which equals the error surface) is done, the error surface should have a large positive autocorrelation [26] [28] [29] [30]. It is assumed that the RMSE over the study area is constant or spatially autocorrelated, which was confuted in previous researches (Fisher, Oksanen etc.). Although the total area is 11.262 km^2 small and according to the terrain surface and the aforementioned research results (RMSE should be constant), it was necessary to divide it into smaller subareas, where this statement was proved. Any significant difference in parameters (range, partial sill and nugget) was not found. The area was searched for trends. But none of them was found. The best fitted model was the Stable one. According to previous researches the Exponential and Gaussian models were chosen to fit the pattern as well. The Gaussian and Spherical models had almost the same results, but the Gaussian one better fitted the closest averaged values and that is why it was chosen (Tab. 3, Fig. 4, Fig. 5). The appropriate shape of the model was not so critical as the computed autocorrelation parameters.

Tab. 3 Gaussian error model parameters

DEM resolution	Lag Size [m]	Num. of Lags	Nugget [m]	Partial Sill [m]	Range [m]
10 x 10	10	12	0.254	0.163	52.925
5 x 5	5	12	0.068	0.042	31.100
1 x 1	1	12	$8.7\ 10^{-4}$	$3.3\ 10^{-3}$	8.178
0.5 x 0.5	0.5	12	$3.2\ 10^{-5}$	$2.7\ 10^{-5}$	3.897

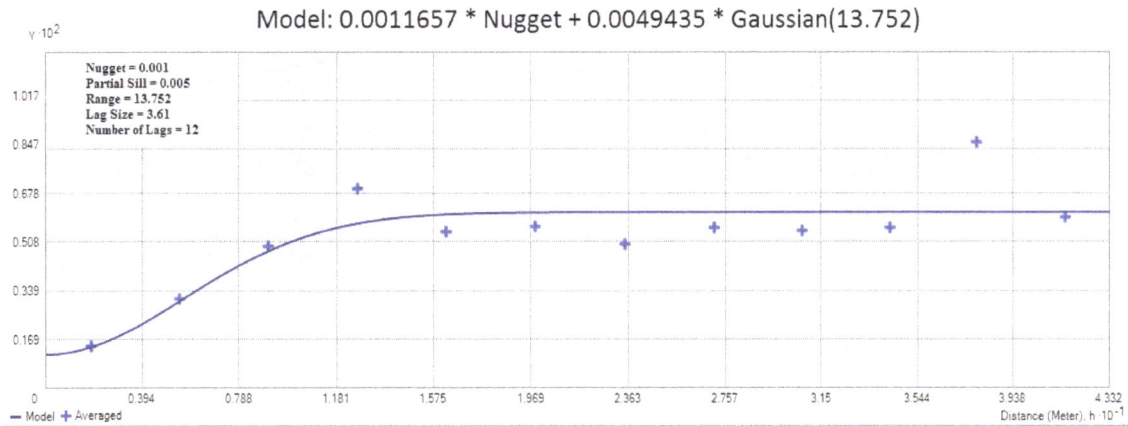

Fig. 4 Gaussian error model for 1x1m resolution DEM

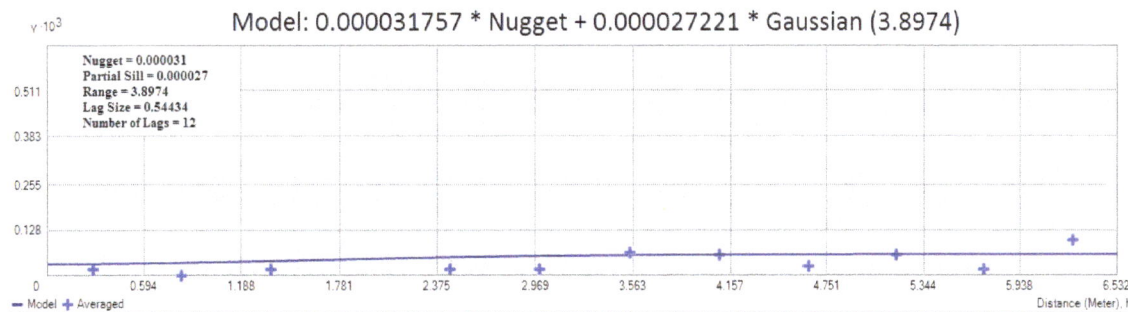

Fig. 5 Gaussian error model for 0.5x0.5m resolution DEM

The theoretical Gaussian models were used to model the fields; Fig. 8 depicts the difference between the spatially correlated and uncorrelated random fields (10 m DEM). The error fields were modelled 250 times for each DEM to perform the Monte Carlo simulation. The outputs of the aforementioned stochastic error propagation (Fig. 1) are mentioned in the following chapter results. The theoretical Gaussian error model of 0.5m resolution (Fig. 5) opens a question about the threshold; whether it is reasonable to use a spatially autocorrelated model or just white noise [10].

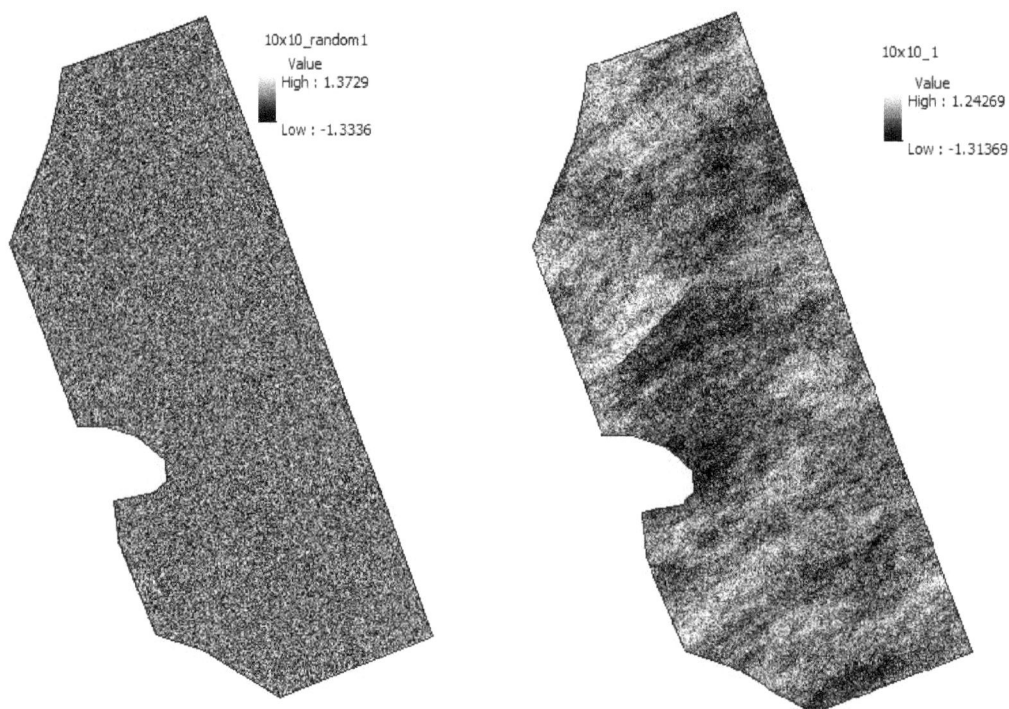

Fig. 6 Left uncorrelated white noise (10x10_rndom1) and right spatially correlated random error field (10x10_1) of 10 m DEM; randomness represented by granulation (left) and clustering of shades of grey (right) are obviously different instances and also inputs of error propagation analyses.

5 RESULTS

The error propagation results are summarized in Tab. 4. For example, in case of 10x10 m DEM the error input is expected to be 1.11° (respectively 0.66° without spatial autocorrelation) large slope error (the mean of the means in column 5). For 5x5 m it is 1.08° (0.64°), 1x1 m 1.24° (0.78°) and for 0.5x0.5m 2.18° (1.39°). The greatest difference was in case of the 0.5x0.5m DEM resolution. The results are represented in absolute values. The behaviour of the error when the value x and its opposite value $-x$ represent the same deviation from the real state of nature, made this representation possible. It is more natural to see the errors in positive values and it enables better interpretations. The most representative number in the evaluation of errors, then it is the mean. Fifty spot samples were randomly selected to prove the insignificant difference between the slope error result derived from the inputs with and without autocorrelation [10] (Statgraphics SW).

Every spot sample has 250 alternative values which were used to compute a mean and standard deviation. Two sample F test (standard deviation) and two sample t rest (mean) were used. The null hypothesis was set to: There is no difference in the standard deviations (respectively a means) and the alternative hypothesis to: There is statistically significant difference between the standard deviations (means). For example for the 1x1m resolution we discovered that 46 in 50 cases for the mean, respectively 43 in 50 for the standard deviation do not differ significantly. The errors without spatial autocorrelation do not result in a greater variance in the resulting slope error (Oksanen got the same results). Although some statistically significant deviations (small values of slope error means related to steeper surfaces and almost half of the values in case of 0.5x0.5m resolution) were found, it is possible to state that majority of the results computed from the elevation field without autocorrelation is slightly underestimated. Thus, it is possible that the use of less appropriate input data can lead to approximate the estimation of slope error, which is slightly underestimated [10].

The outliners have to be also incorporated in the error model which was not done due to the lack of time. The outliers were investigated only in the case of 10x10m resolution DEM (Fig. 9). They were connected with specific land cover types – steep slopes of roadsides, dump sides and river banks. The RMSE of the modelled elevation error pattern increased to 0.300 m by incorporating the outliners, which is close to the RMSE of the empirical elevation error pattern (0.317). The average mean decreased from -3.2 10^{-2} to -2.9 10^{-3}. The outliers were almost uniformly distributed with a mean value of -2.15 for the negative (respectively 2.06 for positive) outliers. Incorporating the outliers increased the average error slope from 1.11° to 1.24° [10].

The influence of elevation error was investigated comparing the LIDAR DEM with the photogrammetric DEM of the same 10x10m resolution and area. As expected the error in slopes increases with the vertical error in elevation. Using the LIDAR input for the 10m DEM the average slope error decreased to 78.36 % of the photogrammetric input [10].

Tab. 4 Error propagation results [m or °] (Particular DEM, Input DEM error standard deviation for all elevation values, DEM RMSE, Output Slope absolute error statistics according to cells – mean, min, maximum, standard deviation)

DEM resolution	Auto-correlation	Error in DEM		Output Slope absolute error [degrees]			
		s. d.	RMSE	Mean	Min	Max	standard deviation
10 x 10	Yes	0.692	0.317	0.61 – 2.34	0 – 0.37	1.85 – 8.44	0.43 – 1.40
10 x 10	No	0.692	0.317	0.38 – 1.30	0 – 0.28	1.17 – 4.93	0.28 – 0.86
5 x 5	Yes	0.362	0.156	0.38 – 2.27	0 – 0.37	1.43 – 8.92	0.29 – 1.48
5 x 5	No	0.362	0.156	0.22 – 1.29	0 – 0.32	0.63 – 4.55	0.18 – 0.85
1 x 1	Yes	0.085	0.04	0.94- 1.58	0 - 0.55	2.69 - 5.95	0.43 - 0.88
1 x 1	No	0.085	0.04	0.58 – 1.44	0 - 0.36	0.92 – 5.75	0.21 - 0.87
0.5 x 0.5	Yes	0.008	0.035	1.64 - 2.80	0 – 0.97	3.41 - 11.25	0.79 - 1.59
0.5 x 0.5	No	0.008	0.035	0.54 – 2.59	0 – 0.76	0.72 10.65	0.17 – 1.56

Every spot sample has 100 alternative values which were used to compute the mean and the standard deviation. Two sample F test (standard deviation) and two sample t rest (mean) were used. The null hypothesis set to: There is no difference in the standard deviations (respectively means) and alternative hypothesis to: There is statistically a significant difference between the standard deviations (means). For example for the 1x1m resolution we discovered that 49 in 50 cases for the mean, respectively 47 in 50 for the standard deviation do not differ significantly (Tab. 5 shows 5 examples). The errors without spatial autocorrelation do not result in a greater variance in the resulting slope error (Oksanen got the same results). Therefore, it should be challenged, if the error propagation without spatial autocorrelation represents sufficiently the true state of nature of the error representation. In else, we proved that the DEM error input without autocorrelation does not result (few exceptions) in a greater error estimate of slope. The 0.5x0.5m resolution DEM error input is critical; it leads to more inequalities. This phenomenon should be further investigated to understand the reason [10].

Tab. 5 Two sample F-test respectively t-test for 5 spots. The hypothesis (H_0) *(sigma1/sigma2 = 1.0)* concerning the ratio of the standard deviations of one spot sample of 100 observations for F-test, and the hypothesis concerning the difference between the means (mean1-mean2 = 0.0, sigma1 and 2 input needed too) for t-test. (both 95.0% confidence level, P-value 0.05 and less rejects H_0) [10].

Random Sample	Slope	Autoco rValue	White Noise Value	Hypothesis Test:		
				F(t) Statistics P-v.		Null Hypothesis
1 std. deviation	17.536°	0.322	0.314	(F) 1.052	0.803	Do not reject, ratio = 1
2 std. deviation	11.232°	0.396	0.400	(F) 1.051	0.803	Do not reject, ratio = 1
3 std. deviation	5.950°	0.407	0.416	(F) 0.957	0.828	Do not reject, ratio = 1
4 std. deviation	0.137°	0.442	0.392	(F) 1.271	0.234	Do not reject, ratio = 1
5 std. deviation	0.226°	0.502	0.322	(F) 2.435	1.10^{-5}	Do reject, ratio <> 1
1 mean	17.536°	0.798	0.795	(t) 0.067	0.947	Do not reject, difference = 0
2 mean	11.232°	0.787	0.778	(t) 0.200	0.841	Do not reject, difference = 0
3 mean	5.950°	0.769	0.817	(t) -0.825	0.410	Do not reject, difference = 0
4 mean	0.137°	1.117	1.155	(t) -0.643	0.521	Do not reject, difference = 0
5 mean	0.226°	1.008	0.894	(t) 1.911	0.057	Do not reject, difference = 0

Although the result of input error without autocorrelation did not show a greater aberration, it is not suitable for modelling the elevation error pattern. In fine topscale and microscale (Oksanen 2005) scales, the error patterns have large positive autocorrelation. Furthermore, in our case the outliers are responsible for the rejection of the Gaussian distribution. The outliners also have to be incorporated into the error model, which was not done due to the lack of time. The average variance and mean of the errors in slopes is not strictly increasing with steepness of the slope (e.g. Fig. 7). This causality should be further investigated; one of the reasons is the insufficient number of samples with a steeper slope. The prevailing spatial distribution of slopes in study is also partially captured in the mean slope error (Fig. 7, Fig. 8). The input based on the empirical elevation error (AC) describes better the error pattern and leads to a more realistic and accurate spatial distribution of slope errors according to the slope in the study area. The white noise (WN) input error field is closest to AC in a minimum slope error distribution. Linear planar surfaces (roads etc.) are inadequately propagated. Planar surface is the most error prone type. According to similar studies (Fisher, Goodchild etc.) using different DEM input data, the high quality LIDAR input data decreases the output uncertainty. In our case, the autocorrelated model fitted the error surface with exception of its outliers. Extreme values are higher in case of the theoretical model with autocorrelation; a random number generator produces smaller extreme values as well. Autocorrelation also expands the standard deviation of extreme values. On the one side, the extreme elevation error values were found to be clustered around the steepest slopes, on the other side, the steeper slopes has a smaller slope error result with the same elevation error input. The range of the fitted empirical error model (49.6 for 10x10, 31.1 for 5x5, 13.8 for 1x1 and 3.9 for 0.5x0.5) was decreasing with a higher resolution. We do assume that there should be a specific resolution limit value where the range is close to 0. Geostatistical modelling is very time consuming. We had to decrease the extent for the 0.5x0.5 and 1x1 meter resolution inputs. To compute one 1x1meter DEM resolution error pattern (21 983 304 values in 5964 rows and 3686 columns) took 12 days and 17 hours (using 30 GB RAM and 4 processors Intel(R) Core (TM2)2 Quad CPU Q9300, 2.5 GHz). This computation requires a super-computer.

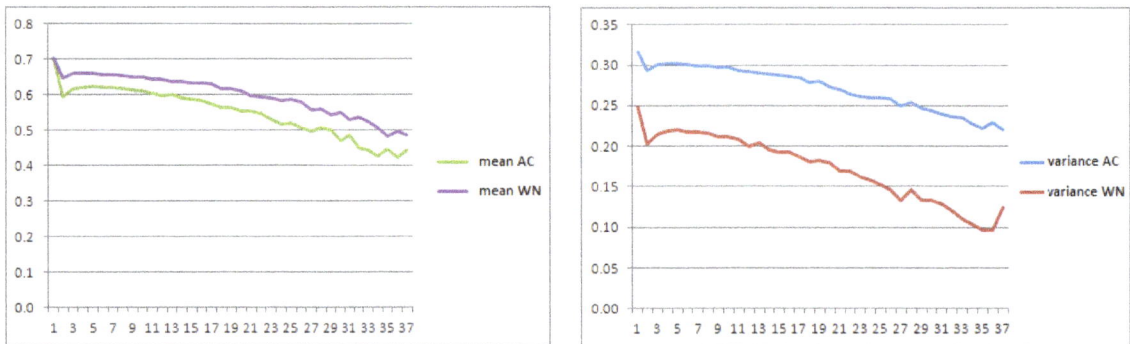

Fig. 7: Slope error dependent variable (vertical axis) vs. Slope independent variable (horizontal axis) (WN randomly generated white noise, AC autocorrelation input according to empirical error pattern) showing the decrease in slope error with increasing slope

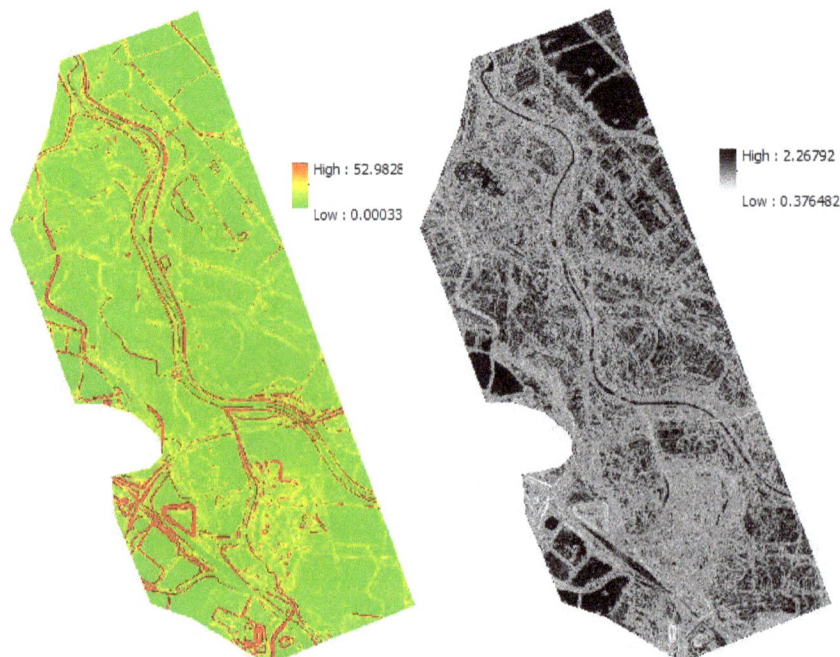

Fig. 8 Left the slope estimate (5x5 LIDAR DEM) and right the stochastic Monte Carlo result of an average slope error for cells in a 5x5m resolution. The flat plain areas are the most error prone surfaces and have black colour in the slope error image

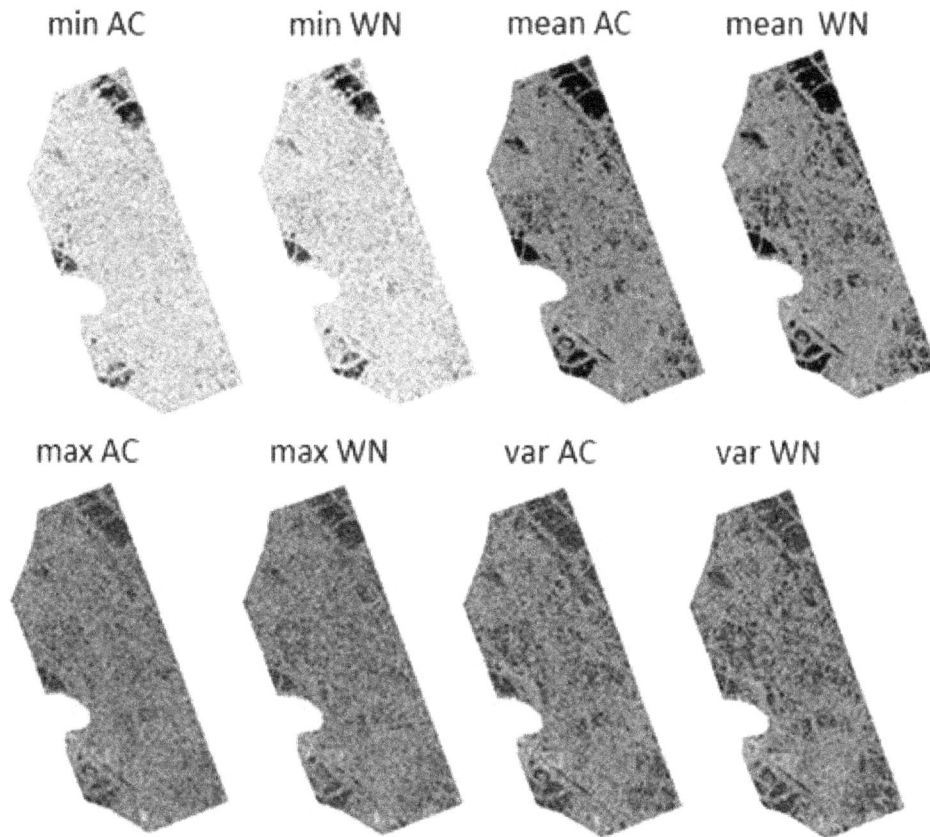

Fig. 9 The statistics important to reconstruct the slope error distribution - mean, a variance (var), minimum (min) and maximum (max) for 10x10m slope errors calculated from the autocorrelated input (AC) and the white noise input (WN); the darker the colour, the higher the slope error value and the more planar the surface (see fig. 6). The distribution of errors is normally distributed with the mean and variance (resp. standard deviation) value, outliers represent the maximum and minimum value. These inputs are necessary to the best description of the possible error in the result.

5 DISCUSSION

Although a lot of research has been made in the uncertainty and error propagation field over the last decades, still many questions left unanswered. In this study, we focused on clearing antagonistic results provided by Oksanen and Fisher. Oksanen declared that slope errors modelled without autocorrelation do not show worse results. In else, the slope derivate has not a maximum variation with a spatially uncorellated random error. On the other side, Fisher declared that the slope derivate computed from the uncorrelated random error is a worse result because of a poor input elevation error representation. We found out that Oksanen is right. Fisher is correct about the poor representation and that the research area should be always investigated before analysed. We were not able to completely ascertain the character of the pattern error. Definitely, the underlying error pattern was found. Some irregular outliners appeared which have to be incorporated. The next step should be the investigation of the outliners. The empirical error model and the modelled error model have to be subtracted and the product investigated (external data may help too – underlying geology, terrain roughness, land use, etc.). The resulting pattern is an addendum to the underlying error pattern. There can be more functions describing local shapes of error pattern. Sum of all functions (patterns) gives the resulting error pattern. We found that there should be a threshold value, which in case of high precision and resolution data do not require the usage of autocorrelation in error surface (in case of the high precision LIDAR data input and a relatively small area).

It is true that any given input data carry an error value significant enough to change the resulting slope – even the high precision micro-scale LIDAR DEM. The results obtained with DEM inputs of the same resolution and acquired with other methods (photogrammetric) could be used for a better comparison and calculation of the exact LIDAR improvement in slope error estimation. Other software tools should be used to prove the simulated reality with gstat. Because of the time demanding computational process, less consuming processes should be investigated for the error pattern simulation, e.g. a fuzzy approach. The software development and new supercomputers could be another solution. There is still a doubt, pros and cons, whether the unconditional Gaussian or sequential Gaussian simulation has to be used, how to model a non-stationary error field in larger areas and what it is dependent on?

It is necessary to remember the main reason for dealing with the uncertainty: decreasing the risk that the outcome will be incorrect and will lead to wrong decisions. This study was made as an error propagation background to inundation area delineation with a GLUE method in the area. The processing of airborne hyperspectral data introduces an uncertainty which is sufficient to change the product. To know the uncertainty in the result is important in crisis management and other fields. Sometimes even one degree in slope can change the situation and flooded area.

6   CONCLUSIONS

The main goals were fulfilled. Thus, the error assessment is an inevitable part of every result presentation. The deviations or the uncertainty of outputs, which we have to be taken into account, should be presented. Although there are high quality input data, they also introduce a certain uncertainty which can lead to a change in decisions and have further consequences. So the use of high quality data does not make the uncertainty analyses unnecessary.

Regarding to similar studies using different DEM input data, the high quality LIDAR input data decrease the output uncertainty. The comparison with photogrammetric data input in our study area proved and emphasised the statement that increased precision in input data decreases the uncertainty in result.

Although the result of input error without autocorrelation did not show a greater aberration, it does not interpret and reflect the properties of real error pattern. In fine topscale and microscale scales, the error pattern has large positive autocorrelation and its distribution is not the Gaussian one. In our case, the outliers (extreme values) are responsible for the rejection of the Gaussian distribution. These outliers were investigated and reasoned. The normal distribution was rejected because of the high kurtosis as well. Therefore, the realization of a random error generator is not suitable interpretation of the true state of elevation errors. On the other side, it can show an important insight into expected elevation and slope error. It is possible to improve the error result using dependencies (in our case between slope error and slope, elevation error outliers and specific land use types) and the fact that the error result (for random white noise input) is slightly underestimated. Geology does not influence the slope error in the study area.

The underlying error pattern has to incorporate the outliers too. If there are any of them, then the sources of them must be found. The simulated error pattern has to be as closest as possible to the empirical one. Error propagation is irrelevant without a proper reconstruction of the empirical input error pattern. The research area should be always investigated before analysed.

Acknowledgement

This paper was elaborated in the framework of the IT4Innovations Centre of Excellence project, reg. no. CZ.1.05/1.1.00/02.0070 supported by the Operational Programme 'Research and Development for Innovations' funded by the Structural Funds of the European Union and the state budget of the Czech Republic and the research project SGS - SPP SV51122M1/2101 (Vliv extrémních přírodních jevů a rizik na ekonomickou činnost člověka v krajině).

REFERENCES

[1]   C.P. Barber, A. Shortage, Lidar elevation data for surface hydrologic modeling: Resolution and representation issues, Cartography and Geographic Information Science, 2005, vol. 32 (4), pp. 401-410.

[2]   S. Openshaw, M. Charlton, S. Carver, Error propagation: A Monte Carlo simulation in Masser, Longman Scientific and Technical, London, 1991.

[3]   W. Shi, P.F. Fisher, M.F. Goodchild, Spatial Data Quality, Taylor & Francis, London, 2002.

[4]   S. Erdogan, Modelling the spatial distribution of DEM error with geographically weighted regression: An experimental study, Computers & Geoscience, 2010, vol. 36, pp. 34 – 43.

[5]   P.F. Fisher, N.J. Tate, Causes and consequences of error in digital elevation models, Progress in Physical Geography, 2006, vol. 30 (4), pp. 467-489.

[6]   J.R. Eastman, P.A.K. Kyem, J. Toledano, W. Jin, GIS and decision making, Explorations in Geographical Information Systems Technology, 1993, vol. 4.

[7]   P.F. Fisher, First experiments in viewshed uncertainty: the accuracy of the viewable area, Photogrammetric Engineering and Remote Sensing, 1991, vol. 57, pp. 1321-1327.

[8]   P.F. Fisher, First experiments in viewshed uncertainty: simulating fuzzy viewsheds, Photogrammetric Engineering and Remote Sensing, 1992, vol. 58 (3), pp. 345-352.

[9] P.F. Fisher, Improved modeling of elevation error with geostatistics, Geoinformatica, 1998, vol. 2 (3), pp. 215-233.

[10] Mudron, I., Podhoranyi M., Cirbus J.: Modelling the Uncertainty of Slope Estimation from LIDAR - derived DEM: A Case Study from Large-Scaled Area in Czech Republic in Sympozia GIS Ostrava 2012 Proceedings - Surface Models for Geoscience, Ostrava 23. - 25.1.2012, ISBN 978-80-248-2558-8

[11] J. Gong, L. Zhilin, Q. Zhu, H.G. Sui, Y. Zhou, Effect of various factors on the accuracy of DEMs: an intensive experimental investigation, Photogrammetric Engineering and Remote Sensing, 2000, vol. 66 (9), pp. 1113–1117.

[12] G. B. M. Heuvelink, Error-Aware GIS at work: real-world applications of the data uncertainty engine, Remote Sensing and Spatial Information Sciences, 2007, vol. 34.

[13] M.E. Hodgson, J. Jensen, G. Raber, J. Tullis, B.A. Davis, G. Thompson, K. Schuckman, An evaluation of lidar-derived elevation and terrain slope in leaf-off conditions, Photogrammetric Engineering and Remote Sensing, 2005, vol. 71 (7), pp. 817-823.

[14] D. Hwang, H.A. Karimi, D.W. Byun, Uncertainty analysis of environmental models within GIS environments, Computers & Geosciences, 1998, vol. 24 (2), pp. 119-130.

[15] G.J. Hunter, M.F. Goodchild, Modelling the uncertainty of slope and aspect estimates derived from spatial database, Geographical Analyses, 1997, vol. 29 (1), pp. 35-49.

[16] M. Podhoranyi, J. Unucka, P. Bobal, V. Rihova, Effects of Lidar DEM resolution in hydrodynamic modelling: model sensitivity for cross-sections, International Journal of Digital Earth 2012.

[17] J. Oksanen, T. Sarjakoski: Non-stationary modelling and simulation of LIDAR DEM uncertainty in Accuracy, in: 2010 Symposium, Leicester, UK, 2010

[18] A. K. Skidmore, A comparison of techniques for calculating gradient and aspect from a gridded digital elevation model, International Journal of Geographical Information Systems, 1989, vol. 3, pp. 309 – 318.

[19] K. Trolegårt, A. Östman, R. Lindgren, A comparative test of photogrammetrically sampled digital elevation models, Photogrammetria, 1986, vol. 41, pp. 1–16.

[20] J. Vaze, J. Teng, High resolution LIDAR DEM – How good is it?, Modelling and Simulation, 2007, pp. 692-698.

[21] Q. Zhou, X. Liu, Analysis of errors of derived slope and aspect related to DEM data properties, Computers & Geosciences, 2004, vol. 30, pp. 369–378.

[22] J. Beekhuizen, G.B.M. Heuvelink, I. Reusen, J. Biesemans, J. Uncertainty Propagation Analysis of the Airborne HyperspectralData Processing Chain, in: Hyperspectral Image and Signal Processing: Evolution in Remote Sensing, Whispers, 2009.

[23] P.A. Burrough, Principles of Geographic Information Systems for Land Resources Assessment, Clarendon Press, Oxford, 1993.

[24] J. Caers, Modeling Uncertainty in the Earth Science, John Wiley & Sons, Oxford, 2011.

[25] P.F. Fisher, Models of uncertainty in spatial data in Longley, John Wiley nad Sons, New York, 1999.

[26] M.F. Goodchild, Elements of Spatial Data Quality, Pergamon, Oxford, 1995.

[27] M.F. Goodchild, Imprecision and Spatial Uncertainty, Springer, 2007.

[28] G.B.M. Heuvelink, P.A. Burrough, A. Stein, Propagation of errors in spatial modelling with GIS, International Journal of Geographical Information Systems, 1989, vol. 3, pp. 303 – 322.

[29] M.F. Goodchild, Spatial autocorrelation, Geo Books, Norwich, 1986.

[30] M.F.Goodchild, A.M. Shortridge, P. Fohl, Encapsulating simulation models with geospatial data sets in Lowell, Ann Arbor Press, Chelsea, 2000.

[31] P. Goovaerts, Geostatistics for natural resources evaluation, Oxford University Press, New York, 1997.

[32] G.B.M. Heuvelink, Analysing uncertainty propagation in GIS: Why is it not so simple?, John Wiley and Sons, Chichester, 2003.

[33] G.B.M. Heuvelink, Error Propagation in Environmental Modelling with GIS, Taylor & Francis, London, 1998.

[34] G.B.M. Heuvelink, J.D. Brown, Uncertain Environmental Variables in GIS, Springer, 2007.

[35]   M. F. Hutchinson, J.C. Gallant, Digital elevation models and representation of terrain shape, Willey, New York, 2000.

[36]   J. Lee, P.K. Snyder, P.F. Fisher, Modeling the effect of data errors on features extraction from digital elevation models, Photogrammetric Engineering and Remote Sensing, 1992, vol. 58 (10), pp. 1461–1467.

[37]   Z. Li, On the measure of digital terrain model accuracy, Photogrammetric Record, 1998, vol. 12, pp. 873-877.

[38]   J. Oksanen, T. Sarjakoski, Error propagation of DEM-based surface derivates, Computers & Geoscience 2005, vol. 31, pp. 1015-1027.

RESUMÉ

Článok sa zaoberá šírením neistôt obsiahnutých vo výškových dátach. Na jednej strane článok predstavuje významný zdroj informácií o teórii šírenia chýb a zákonitostí neistôt v nadmorských výškach. Na druhej strane predstavuje významný zdroj informácií o skutočných odchýlkach v reálnych dátach na území Českej Republiky. Práve na základe výsledkov je možné urobiť si úsudok o možných chybách vo výškových dátach. Podobné informácie o presnosti dát sú veľmi dôležité a pritom sa bežne neuvádzajú ako vo svete tak aj v dátach publikovaných v českom a slovenskom regióne. V teoretickej časti je možné nájsť spôsob ako vypočítať neistotu a následne aj modelovať jej šírenie vo výpočte sklonov pomocou stochastickej metódy Monte Carlo. Tá sa dá jednoducho prispôsobiť aj vo výpočte ostatných charakteristík odvodených z DMR jednoduchou modifikáciou algoritmu. Výsledky tejto štúdie nemajú obdobu v regióne Česka a Slovenska (s výnimkou publikácií tohto autorského kolektívu na konferenciách SDH v Bonne a Sympóziu GIS Ostrava), aj keď podobné štúdie nie sú výnimočné vo svetovom meradle.

8

ESTIMATION OF AVALANCHE HAZARD IN THE SETTLEMENT OF MAGURKA USING ELBA+ MODEL

Martin BARTÍK[1], Matúš HRÍBIK[2], Miriam HANZELOVÁ [3], Jaroslav ŠKVARENINA [4]

[1]*Ing., Department of Natural Environment, Faculty of Forestry, Technical university in Zvolen*
T. G. Masaryka 2117/24, Zvolen, tel. (+421) 45 5206 219
e-mail bartikmartin@gmail.com

[2]*Ing.,PhD., Department of Natural Environment, Faculty of Forestry, Technical university in Zvolen*
T. G. Masaryka 2117/24, Zvolen, tel. (+421) 455206 551
e-mail vrchar@gmail.com

[3]*Ing., Department of Natural Environment, Faculty of Ecology and Environmental Sciences, Technical*
university in Zvolen, T. G. Masaryka 2117/24, Zvolen, tel. (+421) 45 5206 210
e-mail mirowka@gmail.com

[4]*Prof., Ing.,CSc., Department of Natural Environment, Faculty of Forestry, Technical university in*
Zvolen,T. G. Masaryka 2117/24, Zvolen, tel. (+421) 45 5206 209
e-mail skvarenina@tuzvo.sk

Abstract

In our study we focused on advanced software applications to allow simulation of an avalanche. We used the model ELBA+, by which we tried to assess the vulnerability of mountain environments around the old mining settlement called Magurka (1036 m a.s.l.), which lies below the main ridge of Low Tatras at the end of Ľupčianská valley. The avalanche in the Ďurková valley released on 14th March 1970 went down in history because it is still one of the largest avalanches recorded in Slovakia. Using archived data of the Avalanche Prevention Centre in Jasná, we tried its most faithful reconstruction. Then we tried to simulate an avalanche in the Viedenka valley using the same amount of snow in the release zone as it was supposed to be in 1970. Because a part of the Magurka settlement is situated at the mouth of this valley, which has more direct and shorter terrain than the valley of Ďurková, we examined the possibility of intervention by the avalanche.

Abstrakt

V našej štúdií sme sa zamerali na moderné softvérové aplikácie umožňujúce simuláciu lavíny. Použili sme model ELBA+, pomocou ktorého sme snažili zhodnotiť ohrozenosť horského prostredia v okolí starej banskej osady Magurka (1 036 m n. m.), ktorá leží v závere Ľupčianskej doline pod hlavným hrebeňom Nízkych Tatier. Do histórie sa zapísala hlavne lavína v doline Ďurková zo 14. marca 1970, ktorá dodnes patrí medzi najväčšie lavíny zaznamenané na Slovensku. S použitím archívnych údajov Strediska lavínovej prevencie v Jasnej sme sa pokúsili o jej čo najvernejšiu rekonštrukciu. Následne sme sa pokúsili o simuláciu lavíny v doline Viedenka pri použití rovnakej výšky snehu v odtrhovom pásme ako sa predpokladá pri lavíne v roku 1970. Keďže časť osady Magurka je situovaných pri ústí tejto doliny, ktorá má priamejší aj kratší priebeh ako dolina Ďurková, skúmali sme možnosť jej zásahu lavínou.

Key words:avalanches, model ELBA+, Magurka, Low Tatras Mts., upper forest limit

1   INTRODUCTION

Avalanches are a frequent phenomenon of our mountain landscapes during winter. They pose a threat to inhabitants of the mountains areas, skiers, climbers and tourists. In the past, people were looking for possibilities how to protect against avalanches. In order to avoid a formation of avalanches or minimize their destructive consequences, different measures are used. These might be in a technical, biological or organizational form [4]. In the process of making a decision about prevention and technical avalanche measures, software applications may help, mainly in the form of simulation programs. Their purpose is to faithfully describe the movement of an avalanche in an endangered area. The mentioned applications allow us to determine the parameters of potential avalanches (mainly the avalanche length) by using specific terrain and concrete conditions of snow cover. These help us to make decisions on evacuation, closure of the area or landscape planning. Furthermore, there are good accessories for dimensioning technical measures. It is necessary to say that the applications are only simulation models of upcoming situations. The degree of similarity between simulation and reality depends on the quality of input data. The models can be classified according to more aspects: one or more dimensional, fluent, powder or combination models. The best known applications are RAMMS, ELBA+, Samos AT, AVAL 1D, Alfa-Beta.

2   MATERIAL AND METHOD

2.1 Characteristics of area

Our research was carried out in Magurka, an old mining settlement. It is located at an altitude of 1 036 m.a.s.l., at the end of the Ľupčianská valley, below the main ridge of the Low Tatras Mts. (Slovakia) in the part of Chabenec - Latiborská hora Mt. (Fig.1). The settlement administration belongs to a town of Partizanska Lupča, but it is located 20 km away from it.

Fig. 1 Magurka settlement and its surroundings [4]

Surroundings of Magurka are from a geological aspect mainly made from rocks of Tatricum crystalline [10], on which cambi-podzolic soil and at higher altitudes modal and humus-iron podzolic soil [9] are found. The forest consists mostly of spruce (*Picea abies*) with addition of fir (*Abies alba*) and beech (*Fagus sylvatica*), at higher altitudes with addition of rowan (*Sorbus aucuparia*), alder (*Alnus viridis*) and dwarf pine (*Pinusmugo*) [8].

2.2 ELBA+ model

ELBA + is fully integrated into ArcGIS 10. When working with geographical data, it is not necessary to convert input or output data. The program development has been running for 15 years by the NiT-company located in Pressbaum, Austria. At present, we work with the 3th edition. The program creates a 2D simulation model and was used in countries such as Russia, Bulgaria and Switzerland. The utilisation for the purpose of education is free. For the data administration, it is necessary to create a database of programs. This is directly possible from ArcGIS, which is the most frequent GIS-application [7]. The input data for the simulation are represented by Catchment, Size of release area and Digital elevation model (DEM). The following layers were shown automatically during opening the database:

ELBA_CATCHMENT – definition of area

ELBA_CONTOURS – in these layers we can see the results of simulation, lines of selected avalanche characteristics (max. pressure, max. velocity, max. flow height, deposit, avalanche front ...)

ELBA_PROFILES – interesting lines as roads, power lines, tunnel…

ELBA_RELEASEZONES – release zone

ELBA_SIMCLIPS – layer with supplementing data for simulation, for example forest cover…

Catchment, Profiles, Release zones and Forest cover are loaded using the order editor. The next step is to import terrain in a form of DEM or TIN. The suggested pixel resolution of the DEM is 5-10 metres. Later, the following parameters are inserted for the simulation: Snow height in release zone, Height of snow entrainment in avalanche path, Roughness and Terrain model. These input parameters can be entered for the whole area, or possibly for a raster or a polygon. Now we can choose from three roughness models: Voellmy, Mohr-Coulomb and their combination. Here we can edit further parameters as Minimal flow height, Snow density and Critical stress.

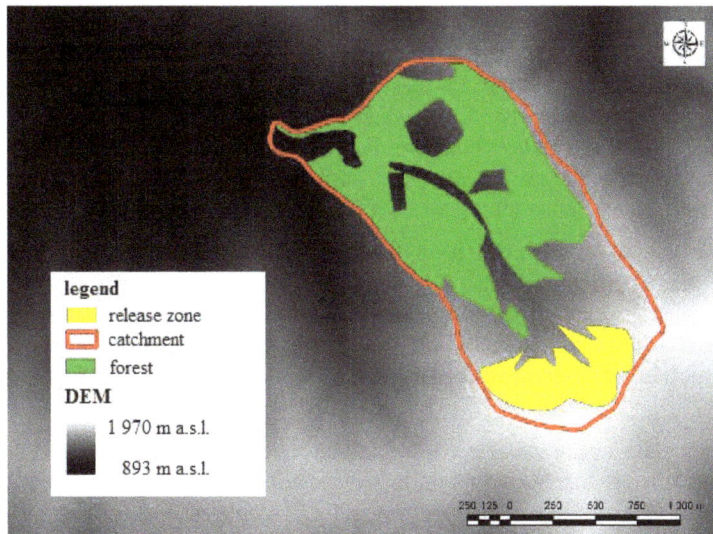

Fig. 2 Input data for avalanche simulation from year 1970

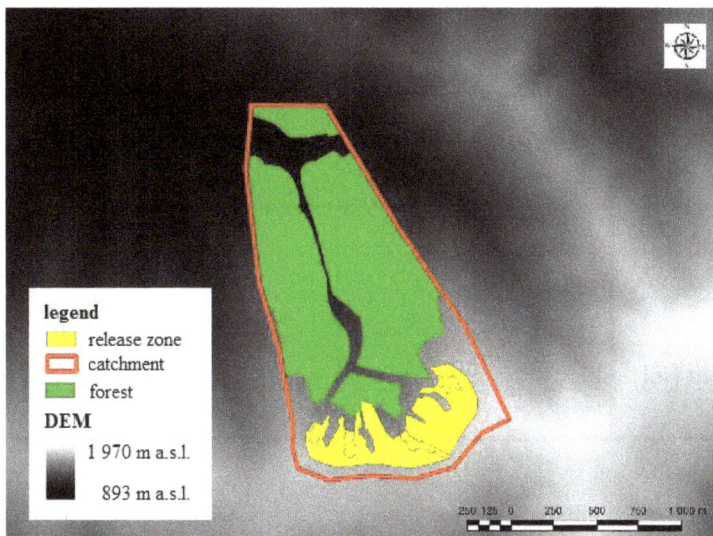

Fig. 3 Input data for avalanche simulation in Viedenka valley

In the first step of our work, we tried to create the most faithful reconstruction of the avalanche released on 14th March 1970. We used archived data from the Avalanche Prevention Centre in Jasná: the layer of avalanche, the data on deposit, snow density, weather, the snow height in the release zone and photos. The best model for this avalanche was the combined roughness model with an average snow height in the release zone of 1.7 meters, a roughness coefficient for the forest 1 and in an non-forest area 0.2, a minimal flow height of 1 metre and the height of snow entrainment in the avalanche path of 0.5 meter. Accordingly with these parameters, the potential avalanche in the Viedenka valley and the real avalanche released on February 2013 were simulated.

The data for simulation were obtained from an archive from the Avalanche Prevention Centre. The actual orthophotos, DEM, polygons of the release zone, the forest position in past were obtained from the Centre. The actual forest position was obtained from the orthophotos with a pixel size of 0.5 meters in a coordinate system S JTSK Krovak East North. DEM with a pixel size of 5 metres was generated from a contour line with the vertical interval of 5 metres. Figs. 2 and 3 show the input data of simulations.

3 RESULTS AND DISCUSSIONS

3.1 Reconstruction of avalanche from year 1970

We confront our results with another two sources: Milan and Kresák[5] and Biskupič et al. [2]. The first source consists of information from avalanche observers, employees of the Avalanche Centre who accurately recorded the avalanche path and the deposit. The avalanche released on 14th March 1970 was the third largest in Slovakia and it aroused much interest. Thus, at present, we have good information about the position of the release zone, the avalanche path and the deposit size. Biskupič et al.[2] used these data for the faithful reconstruction by using the RAMMS model. This two-dimensional numerical simulation tool is used to calculate the geophysical mass movements as snow avalanches, hillslope landslides and debris flows. The model was developed at the WSL Institute for Snow and Avalanche Research, SLF in Davos [1].

Our result of the avalanche reconstruction from the year 1970 is characterized by a high value of conformity, mainly in the avalanche length. Our simulated avalanche length is 8 metres shorter than the measured value of the avalanche length. In this avalanche with a length of 2 200 metres, the difference is 0.4 %. We remind that the pixel size in the raster result was 5 metres. We summarize that our simulation of the avalanche length corresponds exactly to reality. Biskupič et al. [2] who used the RAMMS simulation brings a similar conclusion.

When evaluating the avalanche cubature, we obtained a little underestimated result (- 10 %). This result was influenced by the parameters like Snow height in release zone and Snow entrainment in avalanche path. In our simulation, we used a snow height in release zone of 1.7 metres and an entrainment snow height of 0.5 metres. Especially these parameters show the most significant uncertainty. In an avalanche report, Milan and Kresák[5] estimated a minimal snow height in release zone of 1.8 metres, in glens 12 metres and an average snow height of 2.6 metres. The most accurate is the exact snow allocation in the release zone for each raster cell. This is not simple and it is possible only with exact weather data, such as wind speed, wind direction, snow fall amount during the time period before the avalanche fall.

Tab. 1 Comparison of model results from ELBA +, RAMMS with reality

Parameter	Measured value[5]	RAMMS result [2]	ELBA+ result	Difference between reality and ELBA+	Difference
Avalanche length [m]	2 200	2 221	2 192	8	-0.4%
Deposit length [m]	1 800	1 725	1 800	0	0 %
Deposit cubature [m³]	625 000	626 029	563 858	- 61 142	-9.8%
Avalanche front height [m]	20-25	4-5	4-5	-16-20	-80%
Total area [ha]	39.1	51.4	51.2	12.1	+30.9 %

The most significant undervaluation (80 %) we can see in the avalanche front height. In the first place, the possible deviation sources might be the non-ability of the model to simulate the entrainment of other materials into the avalanche flow. The forest covered lower and peripheral upper parts in the avalanche path (Fig. 2). In such large avalanche, the forest was totally destroyed in the avalanche path. Materials as broken trunks and boughs, which were in the avalanche flow, changed their inner friction. The avalanche decelerated and the force from backside uplifted its front. In the Elba+ simulation, the forest is only an external braked factor which changed from the value of 0.2 in the non-forest area to the value of 1 in the forest. Other materials (such as wood, rocks...) in the avalanche increase its destructiveness [3]. Milan[6] states that the avalanche destroyed 3.6 ha of the forest with 600 m³ of wood.

In our case, the total avalanche area is overestimated by about 31 %. Our simulated avalanche is more widespread in side, more diffluent. In reality, it was more located in the valley. The primary result was widespread, but after an adjustment of the minimal flow height from 0.1 m to 1 m, it reached better conformity with reality. The problem with too wide simulated avalanche was indicated by Volk [12] as well. Fig. 4 shows the differences between the deposit and the path in our simulation of the avalanche and the layout from the year 1970.

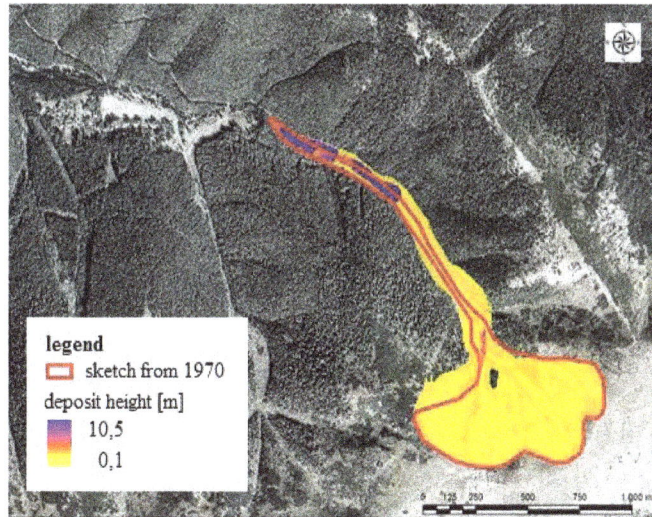

Fig. 4 Comparison of ELBA+ result with reality

3.4 Avalanche in Viedenka

We tried to simulate the avalanche in the Viedenka valley with the same parameters as in Durkova, with an average snow height in the release zone of 1.7 m. The simulated avalanche has a length of 2 230 m, and cubature of 610,000 m^3 and a total area of 58 ha. Because a part of the Magurka settlement is located at the mouth of this valley (Fig. 5), our result shows a possibility of hit.

Fig. 5 Mouth of Viedenka valley with part of settlement

Fig. 6 shows a potential deposit, because the valley has a direct shape and a shorter distance between Magurka and the main ridge of the Low Tatras Mts. From the view of historical sources, there are no observations of such large avalanches. The avalanche in the year 1984 took one victim, but its deposit was shorter and shallow [6]. Marks from the other avalanche can be found even today. It fell down from the left glen (downstream) and destroyed the forest of the opposite slope totally.

Our results show that if the avalanche hits the settlement, it could reach a flow height of 5-7 metres, pressure up to 70 kPa and velocity up to 15m.s^{-1}. These results can be much undervalued. The pressure and also the destroyed potential would be significantly higher, because the flow cannot exist without other materials originated from the destroyed forest by the avalanche.

At the end of February 2013, a medium to large avalanche fell down in this valley[11]. It probably came spontaneously from two glens in the release zone centre. We surveyed its area with a GPS-device. The deposit was 1-2 metres high and contained broken trees, wood and soil. The avalanche was 1 200 m long and destroyed the forest of 1,5ha (Fig.7). We estimated a one-meter snow height in the release zone. Immediately we tried to simulate this avalanche in Elba+. Figure 8 shows the difference between the simulation and our mapped path.

The deposit allocation simulated that the avalanche is very similar to its real state. Its length is 25 m longer, which differs by 2 % form the total length. In this simulation we used other parameters as in previous cases, because this avalanche was significantly smaller and fell down under other conditions (snow, weather …). We changed the friction coefficient μ in the avalanche path from the value of 0.155 to 0.3 and in the accumulation zone from the value of 0.35 to 0.4 for better conformity with a real avalanche. These parameters were determined by interpolation.

Fig. 6 Simulated avalanche in Viedenka valley

Fig. 7 Avalanche in Viedenka valley from February 2013

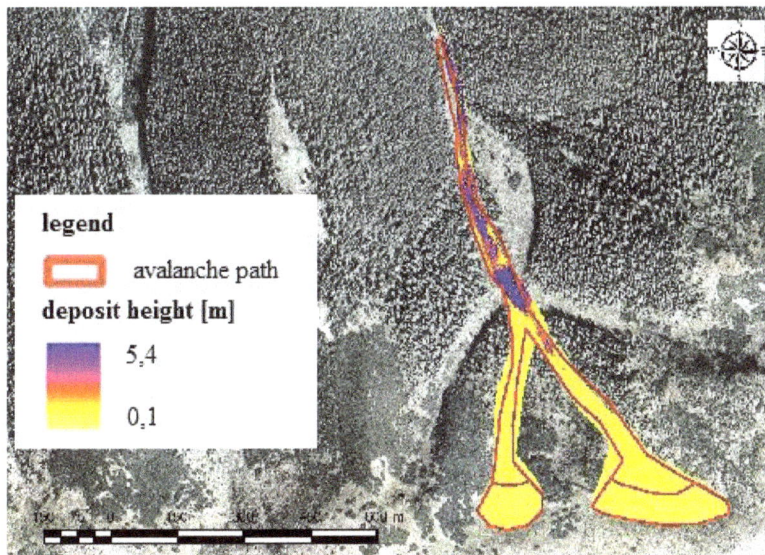

Fig. 8 Simulated avalanche from February 2013

4 CONCLUSIONS

The models used for the avalanche simulation give us a picture about the size and impact of potential avalanches. We can use them in avalanche prevention, projection and in dimensioning technical avalanche measures. The other usage area is landscape planning. In our work we used the model ELBA+ for the reconstruction of a historical avalanche and the evaluation of a possibility to hit Magurka, the old mining settlement located in the Ľupčianska valley below the main ridge of the Low Tatras Mts. Surroundings of this settlement were hit by avalanches many times. Our results are summarized in these points:

- We were able to successfully reconstruct the avalanche released on 14th March 1970 with differences in length of 0.4 % and cubature of 10 %.

- The front height of the simulated avalanche is significantly undervalued (80%) and the total area of the simulated avalanche is overestimated by about 31 %.

- In case that the avalanche falls from all release zones and with an average snow height of 1.7 m in Viedenka, it is possible that it hits a part of the settlement.

- This avalanche with its cubature and the area of 58 ha can overcome the avalanche in Ďurkova from the year 1970.

- In case of an impact on the settlement by an avalanche, the flow height could reach 5-7 metres and stops in the grass field below.

- For the evaluation of simulation results, there is a need of facticity to realize approximate calibration or research of a situation in environment marks and historical date.

REFERENCES

[1] BARTELT, P. et al. *RAMMS-User Manual v1.4 Avalanche.*2011. [cit.02-04-2012] Available at: http://ramms.slf.ch/ramms/downloads/ RAMMS_AVAL_Manual.pdf

[2] BISKUPIČ, M.et al. Rekonštrukcia historickej lavíny s využitím moderných nástrojov GIS. *Životné prostredie.* 2011, 45. 2. pp.83-88.

[3] BUKOVČAN, V.*Lavíny a lesy.* Bratislava: Slovenské vydavateľstvo pôdohospodárskej literatúry n. p. 1960. 196 pp.

[4] GOOGLE EARTH, 2012 [cit.02-04-2012].Available at: http://www.google.com/earth/index.html

[5] MILAN, L.&KRESÁK, K. *Správa o lavínovej udalosti zostavená dňa 10. 6. 1970.* Jasná: Archív SLP HZS,1970.15 pp.

[6] MILAN, L. 2006. *Lavíny v horstvách Slovenska.* Bratislava: Veda, 2006. 152 pp. ISBN 80-224-0894-8

[7]   NIT TECHNISCHES BUERO GmbH, *ELBA+*. 2013. [cit.02-04-2013]Available at: http://www.elba-plus.net/

[8]   NLC, *Lesnícky informačný systém*. 2012. [cit.02-04-2012] Available at: http://lvu.nlcsk.org/lgis/

[9]   ŠÁLY, R.& ŠURINA, B. Pôdy.*Atlas krajiny Slovenskej republiky*. Bratislava: MŽP SR, 2002.pp.106-107. ISBN 80-88833-27-2.

[10]   ŠGÚDŠ,*Geologická   mapa   SR.*   2008.   [cit.02-04-2012]   Available   at: http://mapserver.geology.sk/pgm_sk/mapviewer.jsf?width=1368&height=808

[11]   TIROL. LANDESREGIERUNG, *Ausbildungshandbuch der Tiroler Lawinenkommissionen*. Innnsbruck: Amt der TirolerLandesregierung, 2009. 290 pp.

[12]   VOLK, G.*ELBA+*. Pressbaum: NiT Technisches Buero GmbH, 2005. 94 pp.

RESUMÉ

V našej práci sme sa zamerali na moderné softvérové aplikácie umožňujúce simuláciu lavíny. Tieto programy využívame pri hodnotení ohrozenosti horského prostredia zásahom lavínou, ako aj pri dimenzovaní technických protilavínových opatrení. Na základe zvolených vstupných údajov, ku ktorým väčšinou patrí model terénu, tvar a rozloha pásma odtrhu ako aj údaje o snehovej pokrývke, nám tieto aplikácie podávajú obraz o možnom dosahu lavíny, ako aj charakteristikách jej priebehu, či už maximálny tlak, rýchlosť a výška toku. My sme vo svojej práci používali model ELBA+, pomocou ktorého sme snažili zhodnotiť ohrozenosť horského prostredia v okolí starej banskej osady Magurka (1 036 m n. m.), ktorá leží v závere Ľupčianskej doline pod hlavným hrebeňom Nízkych Tatier. Okolie osady bolo mnohokrát zasiahnuté lavínami, ktoré si vyžiadali obete na ľudských životoch, ako aj rozsiahle škody na lesných porastoch. Do histórie sa zapísala hlavne lavína v doline Ďurková zo 14. marca 1970, ktorá dodnes patrí medzi najväčšie lavíny zaznamenané na Slovensku. S použitím dochovaných archívnych údajov Strediska lavínovej prevencie HZS v Jasnej sme sa pokúsili o jej čo najvernejšiu rekonštrukciu. Následne sme sa snažili nasimulovať o simuláciu lavíny v doline Viedenka pri použití rovnakej výšky snehu v odtrhovom pásme ako sa predpokladá pri lavíne v roku 1970. Keďže časť osady Magurka je situovaných pri ústí tejto doliny, ktorá má priamejší aj kratší priebeh ako dolina Ďurková, skúmali sme možnosť jej zásahu lavínou. Lavínu z roku 1970 sa nám podarilo relatívne presne nasimulovať, čo sa týka hlavne dĺžky (rozdiel 0,4 %) a objemu (rozdiel 10 %). Po zhodnotení výšky nánosu a celkovej plochy bola skutočná lavína viac lokalizovaná v dolinke, ako naša výsledná namodelovaná lavína, ktorá sa vyznačuje výrazným postranným rozširovaním a tým aj nižšou výškou čela lavíny (rozdiel 80 %). Pri páde totožnej lavíny v doline Viedenka by pravdepodobne došlo k zásahu časti osady. Výška toku lavíny by mohla pri vniknutí do časti osady dosiahnuť 5 až 7 metrov a zastavila by sa pri náraze do protisvahu na lúke pod ňou.

9

APPLICATION OF GPR DURING INVESTIGATION CONCERNING CAUSES OF PAVEMENT FAILURE AND ROAD SUBGRADE QUALITY IN GRANITOID MASSIF NEAR SIMTANY

Luděk KOVÁŘ [1], Pavel POSPÍŠIL [2],

[1] *Ing., Ph.D., Institute of Geological Engineering, Faculty of Mining and Geology,*
VŠB – Technical University of Ostrava, tř. 17. listopadu 15/2172, 708 33 Ostrava-Poruba
Tel.: (+420) 59 732 3527
e-mail ludek.kovar@vsb.cz

[2] *Doc. RNDr., Ph.D., Institute of Geological Engineering, Faculty of Mining and Geology,*
VŠB – Technical University of Ostrava, tř. 17. listopadu 15/2172, 708 33 Ostrava-Poruba,
Tel.: (+420) 59 732 3527
e-mail pavel.pospisil@vsb.cz

Abstract

The survey of damaged engineering buildings is in many cases very demanding in terms of the selection of a right exploration method in relation to the results obtained for subsequent engineering works, time for survey implementation, and violations arising from survey activities. Heterogeneity of materials of a natural and anthropogenic origin is a fundamental axiom which can subsequently lead to either a distortion or a failure threatening statically the existence of a building structure. On the test object of a pavement, after some time of its use, severe deformations became evident whose causes and future evolution were not known. Within the design of survey techniques being able to quickly and efficiently uncover the causes of failures, the GPR (Ground Penetrating Radar) investigation was included which as an indirect, non-destructive survey method very quickly helped to clarify he causes of failures of the building structure.

Abstrakt

Průzkum porušených inženýrských staveb je v mnoha případech velmi náročný z hlediska volby správné průzkumné metody ve vztahu k získaným výsledkům pro navazující inženýrské práce, času na realizaci průzkumu a další porušení vznikající při průzkumné činnosti. Heterogenita materiálů přirozeného i antropogenního původu je základním axiomem, který může vést následně buď k deformacím, nebo až k porušení staticky ohrožujícímu existenci stavební konstrukce. Na zkoumaném objektu vozovky se po čase jeho užívání projevily závažné deformace, o kterých nebylo známo, co je způsobilo a jaký bude jejich vývoj v čase. V rámci návrhu průzkumných technik, schopných rychle a efektivně odhalit příčiny porušení byl zařazen průzkum georadarem-GPR (Ground Penetrating Radar), který jako nepřímá, nedestruktivní metoda průzkumu velmi rychle přispěl k objasnění příčin porušování stavební konstrukce.

Key words: GPR, engineering-geological investigation, failures, road structures, road subgrade

1 INTRODUCTION

The investigation works were required on site in order to find out the origin of pavement failures on the side of the road sloping south-west towards the pond located west of the village of Simtany. The pavement failures, especially in the area of road shoulders, have a character of a discrete local subsidence, rough potholes and cracks. One of hypotheses was the relationship between the pavement failures and potential slope deformations, or the sliding of subgrade.

With regard to the very limited possibility to perform more time-consuming survey works on a relatively narrow and busy road, which would lead to the reduction of traffic, the GPR (Ground Penetrating Radar) method was deployed as the main method to investigate the geological situation and determine possible inhomogeneity in the rock mass.

The GPR method is the worldwide most dynamically developing and applying method of research in the field of road and railway constructions. A number of works addresses GPR applications for road constructions. In particular, it is the applied research on the quality of pavements [1], [2], [4], [5], or materials used for the construction of roads [9]. Other publications are focused on the development and improvement of methodologies for GPR measurements on roads such as [8]. Only part of the works deals with a comprehensive evaluation of both the materials of road body and subgrade, on which the works [7], [10], [11], [12], [13], [14] are grounded. The Ministry of Transport of the Czech Republic also issued a special methodical technical prescription, describing the use of GPR for the survey of road constructions [22].

The area of interest is located in the Highlands Region, mainly in the cadastral area of Simtany (cadastral area No. 724653), at the boundary of the map sheets 23-214 and 23-223 of the base map at 1:25 000 scale. The studied area (Fig. 1, 2, 3) is located west of the village of Simtany and is part of the I/19 road connecting the village of Simtany with the town of Pohled north of the local pond.

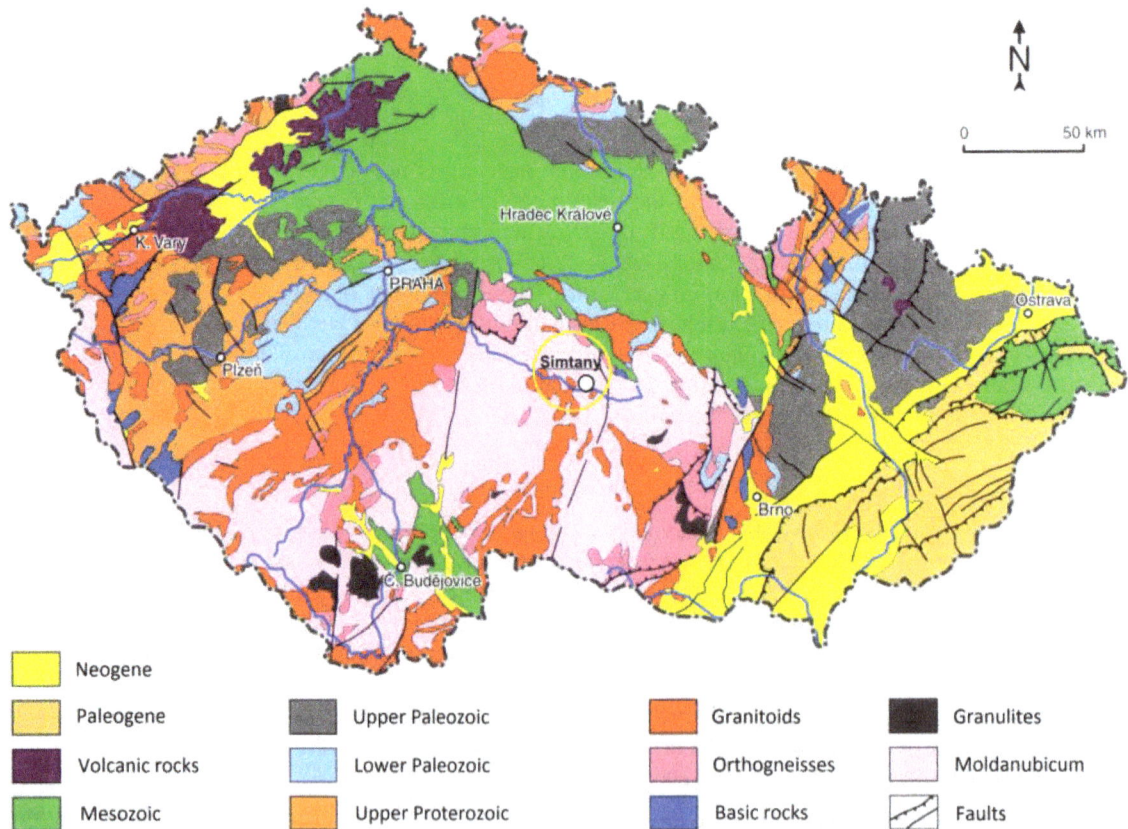

Fig. 1 The location of the area in question on the basis of a geological transparent map of the ČR compiled by ČGS (Czech Geological Survey) [18].

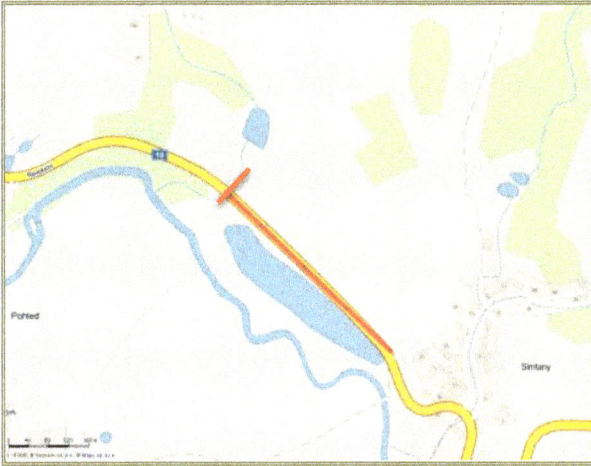

Fig. 2 The detailed map of the locality; the road section of interest, or measured profiles on which the measurements were made, in red [19].

Fig. 3 The aerial photograph of the locality; the road section of interest, or measured profiles on which the measurements were made, in red [16].

Right in the area of interest, no archival survey works are recorded by the map server of the Czech Geological Survey.

The earliest recorded works are registered at a distance of about 300 m and more of the area of interest.

For comparison of the results, probe profiles were used which are located in a considerable distance, but under similar conditions of the rock environment as the studied location. The probes of the following IDs (CGS-Geofond) were used: 398414, 684570 and 394824. Their location can be seen from Fig. 4.

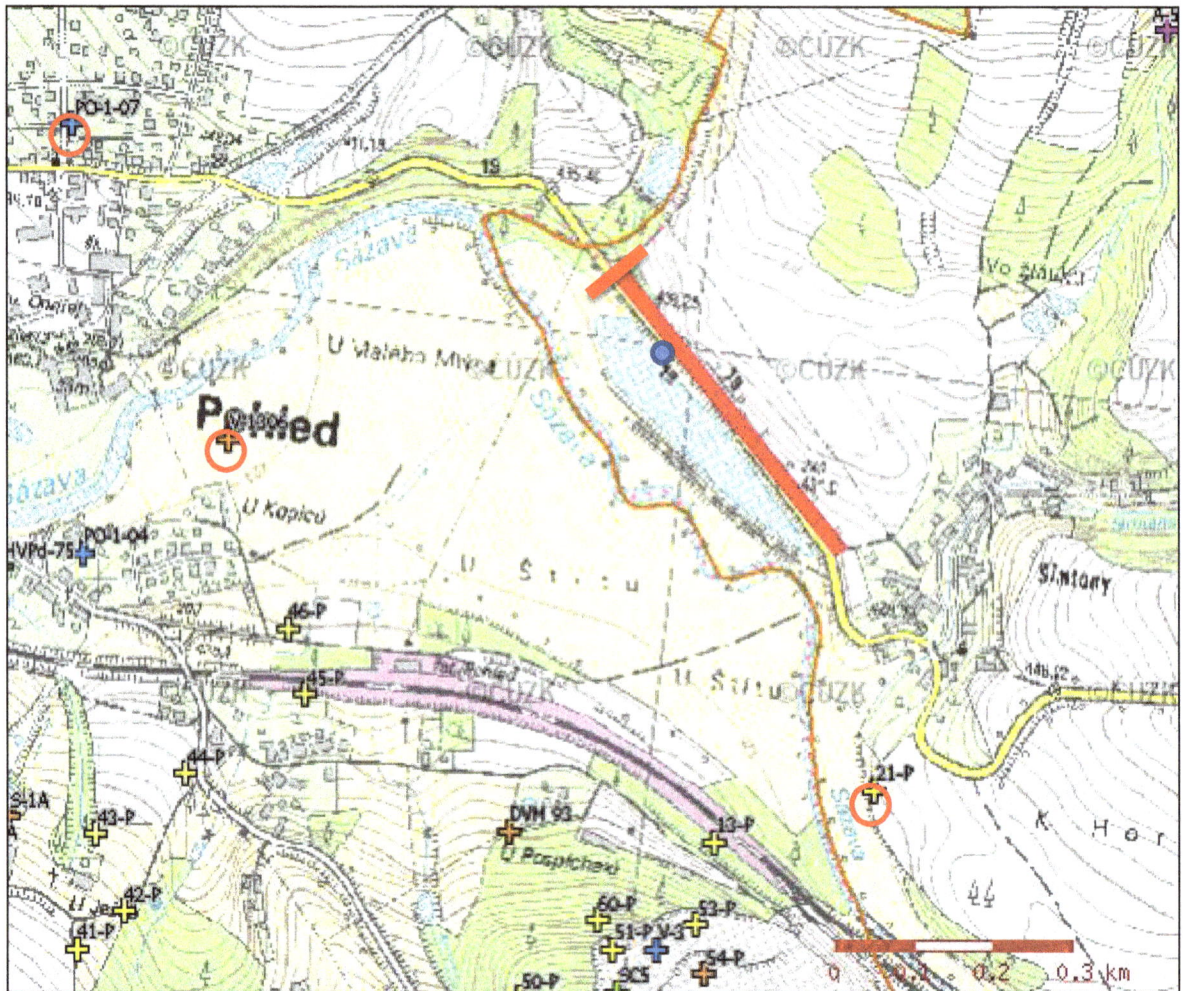

Fig. 4 The actual situation of the locality according to the mapping application of the CGS Geofond. The probes used for comparison are in red circles. The shown lines (in red) correspond to the measured georadar profiles, the location of the manual verification probe Rv-2 is indicated in blue [15].

Using a hand set (Eijkelkamp) with Edelmann drills, two shallow probes for the direct verification of results from georadar measurements were carried out. Drilling was made off the pavement with regard to the safety of the crew and the surrounding area as well.

2   NATURAL CONDITIONS

In the context of natural conditions, the attention was paid to the geomorphology of the area, the basic geological structure of the rock mass and the hydrological and hydrogeological conditions that are characteristic for the studied geological environment.

2.1   Geomorphological conditions

According to the geomorphological division of the Czech Republic [16], the area in question belongs to:

- Czech Highlands Province,
- Czech-Moravian System Subprovince,
- Bohemian-Moravian Highlands,
- Hornosázavská Upland,
- Jihlava-Sázava Furrow
- Pohledy Upland District.

The Pohledy Upland is characterized by rugged hills formed by plateaus and broadly rounded inter-valley ridges bordering the river valley cut of the Sázava River [3].

2.2 Geological conditions

From a regional geological point of view, the location of Moldanubicum belongs to the Bohemian Massif (Fig. 5) which is mainly composed of strongly metamorphosed rocks (parageneses) sporadically permeated by acidic intrusive rocks [6]. From a petrographic point of view, biotite granites of the Moldanubian Pluton Paleozoic age are represented across the entire documented area of the slope offcut. These are the rocks on the surface and in shallow near-surface parts disintegrated (residual soil - R6) to completely weathered (completely weathered - R5) to coarse-grained sands, in surface outcrops highly weathered rocks (highly weathered - R4), to a depth then acquiring a character of slightly weathered rocks (slightly weathered - R2-R3) – to describe the classes of rock, the classification was used that was stated in the ČSN 73 6133 standard [21] and in no longer valid ČSN 73 1001 [20], which, however, is in practice still used as an additional one.

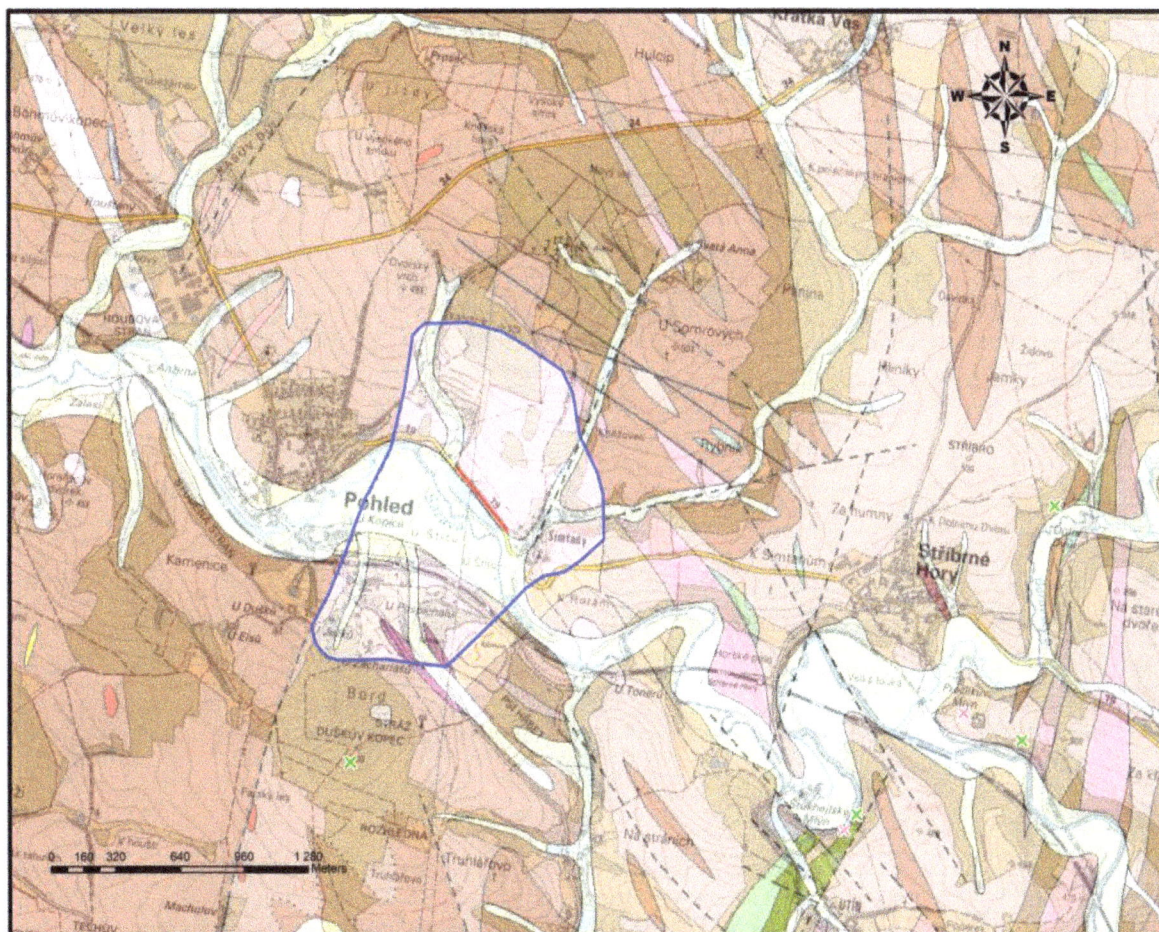

Fig. 5 Cut out from the geological map at a scale of 1:50 000 [18]. The plutonic body of biotite granite is bordered by the blue line, the road section of interest then by the red line.

The Quaternary cover is represented by a thin (about 1 m) complex of deluvio-eluvial sandy-nature soils over the road, then by fluvial or limnic sediments under the road. The road body itself and parts of the slopes with an incline to the pond are built by landfills of a variable composition and low thickness.

2.3 Hydrological and hydrogeological conditions

The Sázava River represents the main gathering channel in the area of interest. The area is therefore directly drained by the Sázava River, or by its small right-hand tributaries. The watershed of the first order is then formed by the Elbe River. Southwest of the area of interest there is a water reservoir - a pond.

According to the zoning of the Czech Republic to base layers, the locality is in the zone 6520 Crystalline turf in the Sázava River Basin [17].

Groundwaters in the area of interest are bound to the fissures and fault zones of the pre-Quaternary bedrock.

The backfill layer (structural layers of the road) is locally used as a collector of infiltrated vadose water. However, this type of saturation may be considered seasonal. It is a typical fluctuation of the underground water level in an unconfined aquifer with a considerable oscillation – from full saturation of the collecting environment to complete drying in precipitation-deficit periods.

The groundwater level was not reached by means of the carried out control, hand-drilled probes to a depth of 1.3 m below the ground.

3   ENGINEERING GEOLOGICAL CONDITIONS

The following geological profile was encountered in the area of interest:

- anthropogenic fills – structural layers and sporadically additional fills in slopes,
- deluvia,
- fluvial sediments (only in the alluvial part, i.e. not under the road itself)
- pre-Quaternary bedrock.

3.1 Anthropogenic fills

Due to the fact that the probes were not made directly in the pavement, we can only assume the composition of the anthropogenic layers. According to radargrams the structural layers achieve the thickness of 0.4 to 0.8 m. For the structural layers, local materials were probably used, which could distort the real thickness in the records. The radar-captured inhomogeneities in structural layers have no direct response in failures in the bituminous surface. We assume, therefore, that these are rather structures in which a migration of water through structural layers occurs. Locally, then piping or erosive processes can apply here, as is evident in particular in the eastern end of the section in question.

The slope from the pavement towards the pond also bears, in some places, signs of anthropogenic sediment, for which e.g. building rubble was used as well.

The ground surface is then outside the structural layer mostly covered by up to a 0.2 m thick layer of forest land.

3.2 Deluvia

Deluvia (often even runoffs of the same material) primarily cover the slope space above the road. In the vast majority, this is the transfer of residues from weathering of granites in the form of coarse sands and sandy loams.

Based on the ČSN 73 6133 standard [21], they mostly rank to the class S2 with possible transitions to the classes F2-F4.

3.3 Fluvial sediments

Fluvial sediments occur in the base of the slope where they are overlapped by pond sedimentation. The thickness of sediment deposits was detected by the radar on the profile P2 in the value of about 2.0 m, by the archival probe V-1306 situated almost in the middle of the floodplain of the Sázava River then in a value of 3.5 m.

3.4 Rocks of pre-Quaternary bedrock - granites

Granitic rocks, as already described above, are in a varying degree of weathering. On the surface and in shallow near-surface parts, they are almost completely disintegrated or completely weathered, locally then they are in some surface outcrops just highly weathered, to a depth then acquiring a character of slightly weathered rocks, however, broken by discontinuities.

Residues and sandy geests of these rocks are suitable, or at least conditionally suitable, for embankments and for the use in the core of the road subgrade. These are mainly non-frost-susceptible soils.

4   METHODOLOGY OF WORK

With regard to the very limited possibilities of implementation of time-consuming survey works on a relatively narrow and busy road, the GPR (Ground Penetrating Radar) method was deployed to investigate the geological conditions and to determine possible inhomogeneities. For proper orientation of radar waves transmitted to the subgrade, it was necessary to ensure full contact of the antenna with the ground, which was fulfilled in that locality.

The GPR method based on transmitting and receiving electromagnetic waves is very sensitive to environmental disturbances by the electromagnetic field induced by electric wires under high voltage or metal wires in general that affect the propagation of electromagnetic waves from the transmitting antenna to the receiver. This can be avoided by selecting shielded antennas during the measurement. In case of a need to show indirectly the environment through the depth in the order of meters, high-frequency antennas should be used with frequencies of hundreds of MHz and a short wavelength of propagated electromagnetic waves.

During the measurement at this location, the GPR set by the Swedish producer Malå was used. The set was formed by an antenna system of 250 and 500 MHz. The antenna system is designed as shielded against spurious electromagnetic interference from the environment. As a control unit, RAMAC Pro EX and as a display unit then View XV11 were used.

Measurements were performed across the entire section in question longitudinally in the roadside (profile P1) in 4 opposite direction travels (reverse profiles were used for control), always twice with both antennas on frequencies of 250 and 500 MHz. In the NW end of the section of interest, the profile perpendicular to the road I/19 ranging from the floodplain to the area of fields over the road, was subsequently measured. Much better results were achieved with the antenna system of 500 MHz. To evaluate the data in the form of radargrams, the software RadExplorer 1.41. was used.

Using the hand set (Eijkelkamp) with Edelmann drills, two shallow probes for verification of the results from georadar measurements were carried out. Drilling was made off the road with regard to the safety of the crew and the surrounding area.

5 MEASUREMENT RESULTS, MASSIF STABILITY ANALYSIS

5.1 GPR measurement results

On the basis of the measurements performed within the shoulders of the road, the likely level of a depth boundary of the surface of highly weathered granites (category 4 and less weathered) depicted by the red dashed line (Fig. 6 and 7) was established during interpretation.

In all the studied profiles, the probable interface of overburdens (both natural and anthropogenic) and highly weathered granites below the road in depths (from the surface of the pavement), ranging from about 1.3 to 1.6 m, was interpreted. In radargrams, this line is represented with a pink dotted line.

The depth of the interpreted interfaces is encumbered with an error quantified at about 10 % due to the variability of dielectric properties of the rock mass in the measured profiles.

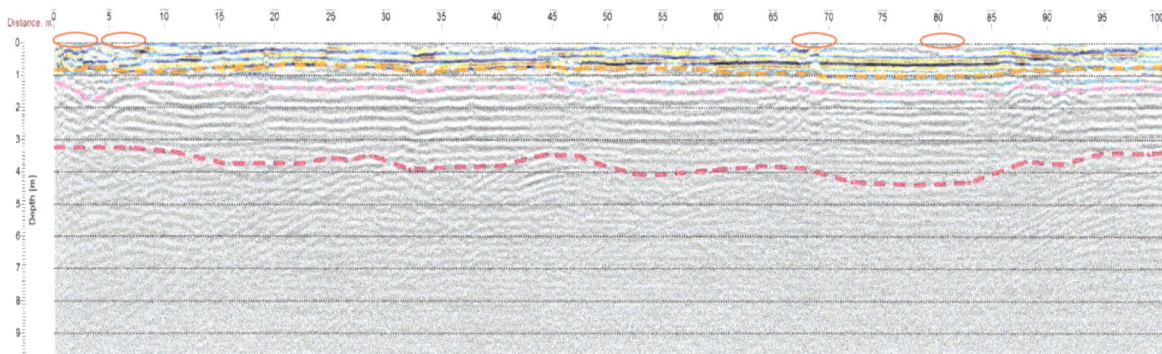

Fig. 6 The radargram of the selected portion of the longitudinal profile measured with GPR in the shoulder of the road with the interpretation of each interface (surface of granites - red, highly weathered granites - pink, eluvium - orange). The red ellipses highlight the examples of inhomogeneities in structural layers of the road.

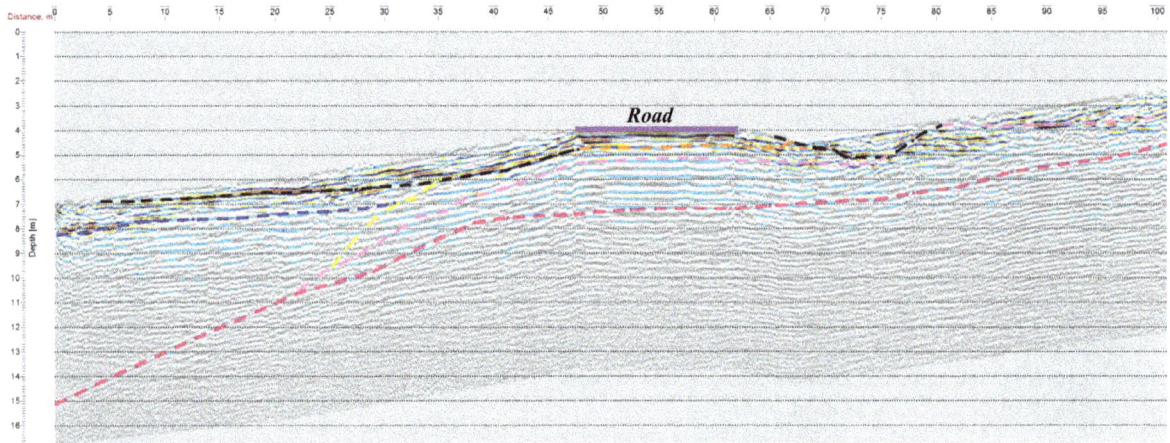

Fig. 7 The radargram of the selected part of the cross section area measured with GPR from the alluvial floodplain area near the pond against the slope NE with the interpretation of each interface (surface granites - red, highly weathered granites - pink, eluvium - orange, deluvium - yellow, alluvium - blue, anthropogenic layers - black).

5.2   Stability conditions and the principle of deformation occurrence in the local road

As follows from the results from the field reconnaissance and radar measurements, there is no reason to believe that roadside disorders have their origin in landslides in the true sense of the word (sliding). The hypothesis of landslides is not supported by either the geological structure of the area (granite massif) or signs of disturbances in the bitumnous surface of the pavement.

According to findings in situ, the road damage was caused in connection with the effects of flowing and leaking surface (rain) waters. In many places of the pavement and the adjacent slope, it is evident that during increased precipitations, an overflow of rainwater over the surface of the terrain from highly lying fields occurs. This is clearly obvious from the presence of minor erosion furrows and rain gullies in the slope and also from the presence of flushed (alluvial) sandy soils in the area of the road shoulder. Especially in the eastern part of the area of interest, also the places can be documented where infiltration of rain water in the shoulder of the road occurs, its migration through structural layers and re-discharge to the surface in the slope at the side of the shoulder towards the pond. These phenomena are then probably accompanied by piping as well. In radargrams, quite many inhomogeneities in structural layers of the pavement were interpreted, which can then be related to these phenomena.

6   CONCLUSIONS

The GPR method proved to be suitable for the given type of exploratory assignments and provided a reliable insight into the composition and condition of the rock environment under a pavement and base courses. This is a relatively cheap, quick and non-destructive method of survey which was as the only feasible without limitation of traffic on the road. The survey with the GPR method clearly defined the ground surface under the structural layers of the pavement. The compliance of GPR measurements with the reality was then established also using manually drilled test probes.

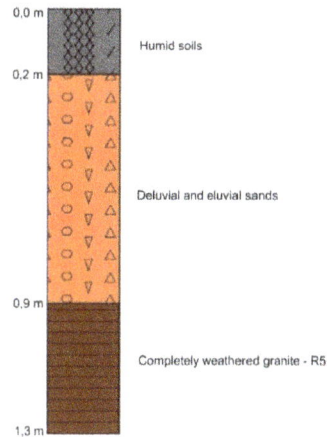

Fig. 8 A simplified profile of the manually drilled probe Rv-2

Fig. 9 The example of the yield of disintegrated (residual soil - R6) to completely weathered biotite granites (completely weathered - R5) to coarse-grained sands by hand drilling.

The survey and the evaluation of the survey results clearly confirmed the stability of the roadway subgrade and refuted the existence of potential slope failures (sliding) in the investigated section of the road. Objectively existing disorders are not caused by slope movements in the true sense of the word, i.e. landslides, but by the effects of surface water flowing on the road from above-lying parts of the slope. Part of the water flows over the surface of the pavement, and part of it then seeps in the "north" shoulder, migrates through structural layers and on the "southern" side of the road then rises from the structural layers. Thus, also piping phenomena occur here that weaken the structural layers by follow-up internal erosion. The pavement is also violated by the overflowing surface water that due to its erosive effects gradually takes away the material of the unpaved shoulder area and thus significantly reduces and weakens it locally. The shoulders and lanes of the road (in both directions), the closest to the shoulder, are then damaged (broken) first of all by freight transportation. Without the application of the GPR method, those results could not be obtained in such a short time and without destruction of the road.

REFERENCES

[1] ABDULLAH, R.S.A.R., SHAFRI, H.Z.B.M., and Bin ROSLEE, M. Data analysis of road pavement density measurements using Ground Penetrating Radar (GPR). *2008 International Conference on Computer and Communication Engineering*, Vols 1-3, 2008: p. 732-737.

[2] BENEDETTO, A. Prediction of structural damages of road pavement using GPR. *2007 4th International Workshop on Advanced Ground Penetrating Radar*, 2007: p. 270-274.

[3] DEMEK, J. et al. *Zeměpisný lexikon ČSR. Hory a nížiny*. Prague: Academia, 1987. 584 pp.

[4] FAUCHARD, C., et al. GPR performances for thickness calibration on road test sites. *Ndt & E International*, 2003. **36**(2): p. 67-75. ISSN 0963-8695.

[5] FAUCHARD, C., DEROBERT, X. and COTE, P. GPR performances on a road test site. *Gpr 2000: Proceedings of the Eighth International Conference on Ground Penetrating Radar*, 2000. **4084**: p. 421-426. ISSN 0277-786X.

[6]   CHLUPÁČ, I. et al. *Geologická minulost České republiky.* Prague: Academia, 2002. 436 pp. ISBN 80-200-0914-0.

[7]   JUNG, G.J. et al. Evaluation of road settlements on soft ground from GPR investigations. *Proceedings of the Tenth International Conference on Ground Penetrating Radar,* Vols 1 and 2, 2004: p. 651-654.

[8]   PORSANI, J.L., et al., Comparing detection and location performance of perpendicular and parallel broadside GPR antenna orientations. *Journal of Applied Geophysics*, 2010. **70**(1): p. 1-8. ISSN 0926-9851.

[9]   RAVASKA, O. and SAARENKETO, T. Dielectric-Properties of Road Aggregates and Their Effect on Gpr Surveys. *Frost in Geotechnical Engineering*, 1993: p. 17-22.

[10]   SAARENKETO, T., HIETALA, K. and SALMI, T. GPR Applications in Geotechnical Investigations of Peat for Road Survey Purposes. *Fourth International Conference on Ground Penetrating Radar*, June 8-13, 1992, Rovaniemi, 1992. **16**: p. 293-305.

[11]   SAARENKETO, T. and SCULLION, T. Road evaluation with ground penetrating radar. *Journal of Applied Geophysics*, 2000. **43**(2-4): p. 119-138. ISSN 0926-9851.

[12]   SAARENKETO, T., van DEUSEN, D. and MAIJALA, R. Minnesota GPR project 1998 - Testing Ground penetrating Radar technology on Minnesota roads and highways. *GPR 2000: Proceedings of the Eighth International Conference on Ground Penetrating Radar,* 2000. **4084**: p. 396-401. ISSN 0277-786X.

[13]   SAARENKETO, T. and VESA, H. The use of GPR technique in surveying gravel road wearing course. *GPR 2000: Proceedings of the Eighth International Conference on Ground Penetrating Radar*, 2000. **4084**: p. 182-187. ISSN 0277-786X.

[14]   SAARENKETO, T. Electrical properties of Road Materials and Subgrade Soils and The Use of Ground Penetrating Radar in the Traffic Infrastructure Surveys. Doctoral Thesis, Acta Universitatis Oluensis, A471, 2006. 127 pp. ISBN 951-42-8222-1.

[15]   www.geofond.cz, česká geologická služba – Geofond.

[16]   www.geoportal.gov.cz, portál veřejné správy České republiky.

[17]   www.heis.vuv.cz, hydroekologický informační systém VÚV T.G.M.

[18]   www.geology.cz, informační portál ČGS.

[19]   www.mapy.cz, mapový portal fy Seznam.

[20]   ČSN 73 1001 Základová půda pod plošnými základy.

[21]   ČSN 73 6133 Navrhování a provádění zemního tělesa pozemních komunikací.

[22]   TP 233 Georadarová metoda konstrukcí pozemních komunikací, Technické podmínky, Ministerstvo dopravy ČR.

RESUMÉ

Průzkum porušených inženýrských staveb je v mnoha případech velmi náročný z hlediska volby správné průzkumné metody ve vztahu k získaným výsledkům pro navazující inženýrské práce, času na realizaci průzkumu a další porušení vznikající při průzkumné činnosti. Heterogenita materiálů přirozeného i antropogenního původu je základním axiomem, který může vést následně buď k deformacím, nebo až k porušení staticky ohrožujícímu existenci stavební konstrukce. Na zkoumaném objektu (silnici) se po čase jeho užívání projevily závažné deformace, o kterých nebylo známo, co je způsobilo a jaký bude jejich vývoj v čase. V rámci návrhu průzkumných technik, schopných rychle a efektivně odhalit příčiny porušení byl zařazen průzkum georadarem-GPR (Ground Penetrating Radar), který jako nepřímá, nedestruktivní metoda průzkumu velmi rychle přispěl k objasnění příčin porušování stavební konstrukce. Při měření na této lokalitě byla využita sestava GPR švédského výrobce Malå. Sestava byla tvořena stíněným anténním systémem 250 a 500 MHz.

Z geologického hlediska tvoří podloží konstrukčních vrstev komunikace biotitické granity moldanubického plutonu paleozoického stáří v různém stupni zvětrání. Ve všech studovaných profilech bylo interpretováno pravděpodobné rozhraní pokryvných útvarů (jak přirozených, tak i antropogenních) a silně zvětralých granitů pod komunikací v hloubkách (od povrchu vozovky) v rozmezí cca 1,3 – 1,6 m. Dle radargramů dosahují vlastní konstrukční vrstvy komunikace proměnlivé mocnosti 0,4 - 0,8m. Hloubka interpretovaných rozhraní je zatížena chybou kvantifikovanou na cca 10%.

Jak vyplývá z výsledků terénní rekognoskace i radarových měření, není důvod se domnívat, že poruchy krajnice komunikace mají původ ve svahových pohybech v pravém slova smyslu (sesouvání). Radarem zachycené nehomogenity v konstrukčních vrstvách nemají přímou odezvu v poruchách v živičném krytu. Lze předpokládat, že jde spíše o struktury, ve kterých dochází k migraci vod konstrukčními vrstvami. Místně se zde pak mohou uplatňovat i sufozivní procesy.

Dle zjištění v terénu a po analýze radarových měření vzniklo tedy poškození vozovky v souvislosti s účinky proudících a prosakujících povrchových (srážkových) vod.

Nasazení georadaru jako hlavní průzkumné metody se ukázalo v daných podmínkách jako vysoce účelné a efektivní řešení (rychlost provedení bez nutnosti omezení či vyloučení dopravy, nízká nákladovost) s poměrně vysokou vypovídací schopností. Podstatně lepších výsledků z hlediska možností interpretace bylo dosaženo s anténním systém 500 MHz.

10

EVALUATION OF THE DATA QUALITY OF DIGITAL ELEVATION MODELS IN THE CONTEXT OF INSPIRE

Radoslav CHUDÝ[1], Martin IRING[2], Richard FECISKANIN[3]

[1]*Mgr., Department of Cartography, Geoinformatics and Remote Sensing, Faculty of Natural Sciences, Comenius University*
Mlynská dolina, 842 15, Bratislava, Slovenská republika, tel. (+421) 2 602 96 396
e-mail chudy@fns.uniba.sk

[2]*Mgr., Department of Cartography, Geoinformatics and Remote Sensing, Faculty of Natural Sciences, Comenius University*
Mlynská dolina, 842 15, Bratislava, Slovenská republika, tel. (+421) 2 602 96 396
e-mail iring@fns.uniba.sk

[3]*Mgr. Ph.D., Department of Cartography, Geoinformatics and Remote Sensing, Faculty of Natural Sciences, Comenius University*
Mlynská dolina, 842 15, Bratislava, Slovenská republika, tel. (+421) 2 602 96 396
e-mail feciskanin@fns.uniba.sk

Abstract

The contribution deals with the evaluation of the quality of geographic information in accordance with the ISO standards from the family of ISO 19100. The quality assessment was carried out on a sample of the data of the digital elevation model of the Slovak republic – DMR3. The selected data quality elements and sub-elements were evaluated using measures defined in the INSPIRE data specification for Elevation.

Abstrakt

Príspevok sa zaoberá hodnotením kvality geografických informácií v súlade s dostupným ISO štandardami z rodiny ISO 19100. Hodnotenie kvality bolo vykonané na príklade vzorky údajov digitálneho modelu reliéfu SR -DMR3. Boli hodnotené vybrané elementy, subelementy kvality pomocou mier definovaných v údajovej špecifikácii INSPIRE pre výškové modely.

Key words: INSPIRE, Geographic information – Data quality, digital elevation model, metadata

1 INTRODUCTION

The influence of georelief (relief of the Earth) and its geometrical properties on the spatial differentiation of processes in the geographical sphere is very significant. Digital elevation models and derived objects are ones of the basic sets of spatial data in the vast majority of spatial analyses. Therefore there are important not only the information on a spatial distribution of heights, but also the information about the quality of the information. A crucial role in determining the applicability of the model has the accuracy of the information that can be derived on the basis of the model.

Currently there is already a wide range of models available, which entirely cover the territory of the Slovak Republic. A permanent problem for these models is missing or only partial information about their quality. Since the quality of geographic data is one of the key parameters of data sets, which determines its usability and hence their price, it is necessary to solve the problem of missing quality metadata.

Every interested person solving the problem of missing metadata in our geographic area meets with the need of an INSPIRE directive application. The issue of the terrain models in detail covers the INSPIRE data specification Elevation, which is currently (5/2013) at a high level of elaboration. However, this specification is already implementable in its present state. It focuses on the data representation (a grid, a vector, a triangulated irregular network – TIN) for modelling different types of surfaces (a digital terrain model – DTM vs. a digital surface model – DSM). The specification also defines the quality requirements.

The main objective of this work was to verify the applicability of the data specification Elevation to a real digital elevation model – DMR3. We are primarily focused on addressing the quality of the data and metadata from the data specification. We published the results of the quality assessment by the metadata. As the author of this work known, this is the first work in our territory, in which the authors attempted to verify the possibility of applicability of the mentioned data specification and we will be happy if the results of our work will help to others.

2 INSPIRE DATA SPECIFICATION ELEVATION

The INSPIRE directive defines the topic of elevation as:

„Digital elevation models for land, ice and ocean surface. Includes terrestrial elevation, bathymetry and shoreline."

This theme includes:

- Digital Terrain Models (DTM) describing the three-dimensional shape of the Earth's surface (ground surface topography).

- Digital Surface Models (DSM) specifying the three-dimensional geometry of every feature on the ground, for example vegetation, buildings and bridges.

- Bathymetry data, e.g. a gridded sea floor model.

In terms of the spatial representation the data specification defines three models:

- Gridded data modelled as continuous coverages compliant with the standard ISO 19123 – Coverage geometry and functions which use a systematic tessellation based on a regular rectified quadrilateral grid to cover its domain.

- Vector objects comprise spot elevations (spot heights and depth spots), contour lines (land elevation contour lines and depth contours), break lines describing the morphology of the terrain as well as other objects which may help in calculating a Digital Elevation Model from vector data (void areas, isolated areas).

- TIN structures according to the GM_Tin class in ISO 19107 – Spatial schema. This is a collection of vector geometries (control points with known Elevation property values, break lines and stop lines).

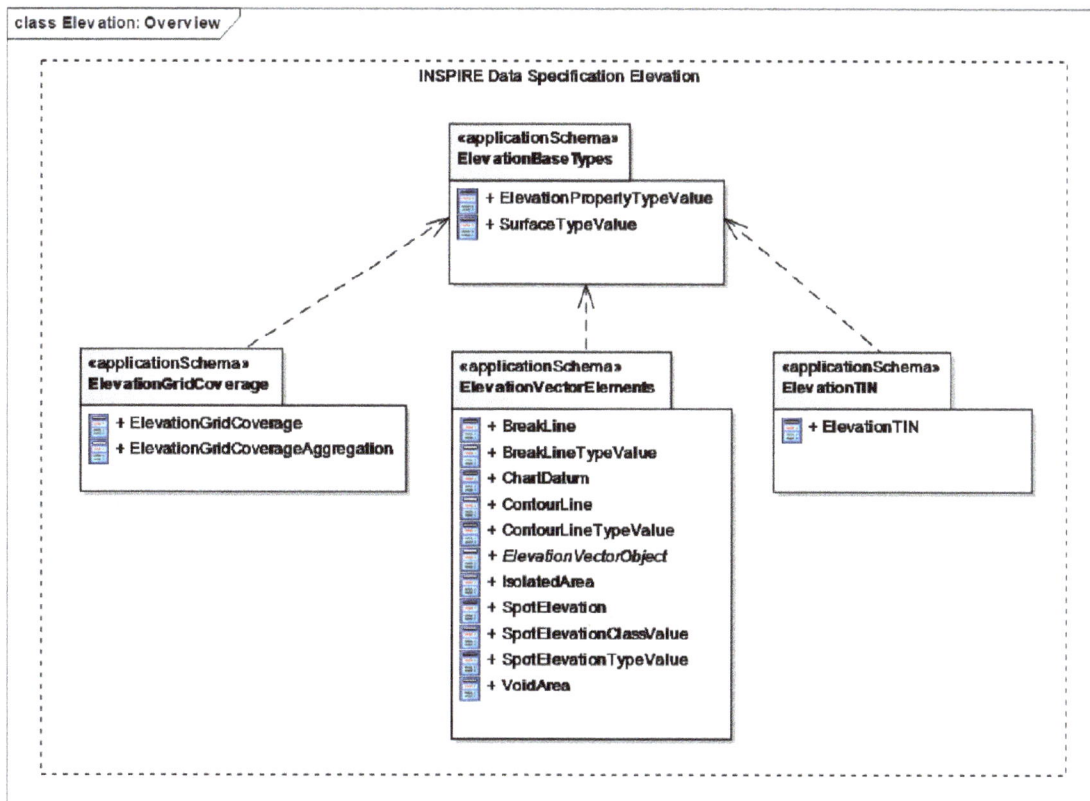

Fig. 1 Overview of Elevation application schemas [2]

3   DATA QUALITY OF GEOGRAPHIC INFORMATION

Quality is a summary of the characteristics of geographic data, which have an impact on their ability to meet established or implied requirements (STN EN ISO 19101). Quality management of geographic data should be carried out in conformity with the standards of the quality of geographic information. Standardization in the field of the quality of the geographic data has already taken place and in this day it is represented by international standards: STN EN ISO 19113, STN EN ISO ISO 19114, 19138 and particularly by ISO 19115. In the future, the standards STN EN ISO 19113, STN EN ISO 19114, ISO 19138 should be replaced by the ISO standard ISO19157 (4/13, it is still in the process of finalization and official publication), which will deal with the spatial data quality comprehensively.

The principle of the data quality evaluation is determined by a set of standard data quality components used to express the quality of geographic data. The components are divided into two basic groups. The first group contains a set of quality elements of geographic data and deals with the quantitative aspect of quality. The second group is made up of a set of elements of review of geographical data quality and deals with the qualitative aspect of quality.

The standard ISO 19157 defines the following data quality elements:

- Completeness
- Logical consistency
- Spatial accuracy
- Temporal quality
- Thematic accuracy
- Usability

If the standard set of elements does not cover all aspects of quantitative quality, it is possible to define own data quality elements. For the expression of quantitative data quality in more details its own standard sub-elements are defined for each element. If these do not reflect all aspects of the quality, it is possible to proceed to the definition of other custom sub-elements.

Completeness

- Commission
- Omission

Logical consistency

- Conceptual consistency
- Domain consistency
- Format consistency
- Topological consistency

Spatial accuracy

- Absolute accuracy
- Relative accuracy
- Gridded data position accuracy

Temporal quality

- Accuracy of time measurement
- Temporal consistency
- Temporal validity

Thematic accuracy

- Classification correctness
- Non-quantitative attribute correctness
- Quantitative attribute accuracy

Each sub-element of the data quality must be applied by using of assessments of seven quality descriptors:

- Data quality scope
- Data quality measure
- Data quality evaluation
- Data quality result
- Data quality type
- Data quality measure type
- Data quality date

The standard STN EN ISO 19113 defines three review elements of the non-quantitative data quality of geographic data

- Purpose
- Usage
- Lineage

If standard elements do not cover all non-quantitative requirements, it is possible to define other new data quality elements. The scope of data quality must be defined for each element.

3.1 INSPIRE data quality requirements for elevation models

The INSPIRE directive and its implementing rules require the evaluation of the quality of harmonized spatial data. The data specification involves the requirements defined in chapter 7. The chapter contains a definition of the elements of quality, the minimum quality requirements and recommendations for the quality of the data.

The following elements are defined for each application scheme of quality.

Tab. 1 Elements and sub-elements of Elevation data quality

Data quality element/sub-element	Evaluation scope	Application schema		
		Vector	Grid	TIN
Completeness /Commission	dataset /dataset series	*		*
Completeness /Omission	dataset /dataset series/ spatial object type	*	*	*
Logical consistency /Conceptual consistency	spatial object /spatial object type	*	*	*
Logical consistency /Domain consistency	spatial object /spatial object type	*		*
Logical consistency /Format consistency	dataset /dataset series	*	*	*
Logical consistency /Topological consistency	spatial object type / dataset / dataset series	*		*
Positional accuracy /Absolute or external accuracy	spatial object / spatial object type / dataset series / dataset	Horizontal component		
		*		*
		Vertical component		
		*	*	*
Positional accuracy /Gridded data position accuracy	spatial object / spatial object type / dataset series / dataset	Horizontal component		
			*	

Each data quality element has its own data quality measure. All measures are based on the standard ISO 19157. For Completeness/Commission the measure *Rate of excess items (measure num. 3 ISO/DIS 19157:2012)*

is proposed. It is a number of the excess items in the dataset in relation to the number of the items that should have been present. Element Completeness/Omission is evaluated by the measure *Rate of missing items (measure num.7 ISO/DIS 19157:2012)* – a number of the missing items in the dataset in relation to the number of the items that should have been present. Logical consistency/Conceptual consistency is evaluated by the measure *Non-compliance rate with respect to the rules of the conceptual schema (measure num. 12 ISO/DIS 19157:2012)* – a number of the items in the dataset that are not compliant with the rules of the conceptual schema in relation to the total number of these items supposed to be in the dataset. Logical consistency /Domain consistency must be evaluated by the measure *Value domain non-conformance rate (measure num. 18 ISO/DIS 19157:2012)* – a number of the items in the dataset that are not in conformance with their value domain in relation to the total number of the items. Logical consistency/Format consistency is evaluated by the measure *Physical structure conflict rate (measure num. 20 ISO/DIS 19157:2012)* - a number of the items in the dataset that are stored in conflict with the physical structure of the dataset divided by the total number of the items. For all this four measures, the evaluation scope is defined at the level of a data set or data set series.

For the element, Logical consistency/Topological consistency four data quality measures are defined. The first is *Rate of missing connections due to undershoots (measure num. 23 ISO/DIS 19157:2012)*. The measure defines the count of the items in the dataset, within a parameter tolerance, which are mismatched due to undershoots divided by the total number of elements in the data set. Missing connections exceeding the parameter tolerance are considered as errors (undershoots) if the real linear elevation features have to be connected. The tolerance parameter is the distance from the end of a dangling line in which it is possible to consider the line as to be continuous (Fig.2).

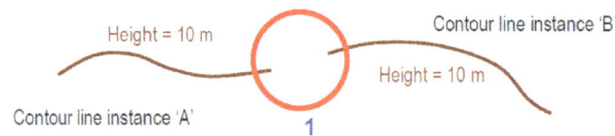

Fig. 1 Example of Rate of missing connections due to undershoots [2]

This parameter is specific for each data provider's dataset and must be reported as metadata using DQ_TopologicalConsistency – 102nd measure Description. The measure is applicable to the objects/feature classes from the application schema Vector – contour lines and break lines with the same height value.

The second measure for the evaluation of topological consistency is *Rate of missing connections due to overshoots*. It is the count of the items in the dataset, within the parameter tolerance, which are mismatched due to overshoots divided by the total number of elements in the dataset. The missing connections exceeding the parameter tolerance are considered as errors (overshoots) if the real linear elevation features have to be connected. The value of the tolerance parameter is a distance from the dangling end of the line in which the overshoots needs to be found. This parameter is specific for each data provider's dataset and must be reported as metadata using DQ_TopologicalConsistency - 102. measureDescription. The measure is applicable to the objects/feature classes from the application schema Vector – contour lines and break lines with the same height value.

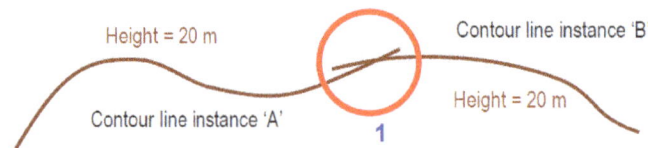

Fig. 2 Example of Rate of missing connections due to overshoots [2]

The third measure is *Rate of invalid self-intersect errors (measure num. 26 ISO/DIS 19157:2012)*. It is the count of all items in the data that illegally intersect with themselves divided by the total number of the elements in the dataset. The measure is applicable to the objects/feature classes from the application schema Vector – contour lines, break lines, void areas, and isolated areas.

Fig. 3 Example of Rate of invalid self-intersect errors [2]

The last measure of the topological consistency is *Rate of invalid self-overlap errors (measure num. 27 ISO/DIS 19157:2012)*. It is the count of all items in the data that illegally show a self-overlap divided by the total number of the elements in the data set.

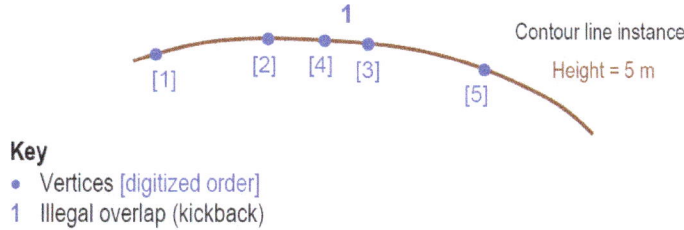

Fig. 4 Example of Rate of invalid self-overlap errors (taken from [2])

The measure is applicable to the objects/feature classes from the application schema Vector – contour lines, break lines, void areas, and isolated areas.

For the element Positional accuracy /Absolute or external accuracy two measures are defined to evaluate the data quality. The first measure is *Root mean square error of planimetry (RMSEP measure num. 47 ISO/DIS 19157:2012).* It is the radius of a circle around a given point, in which the true value lies with probability P.

Equation 1

$$\sigma = \sqrt{\frac{1}{n}\sum_{i=1}^{n}\left[(x_{mi} - x_t)^2 + (y_{mi} - y_t)^2\right]} \qquad (1)$$

where:
σ – Root mean square error of planimetry
x_t – true value of X coordinate
y_t – true value of Y coordinate
x_{mi} – measured value of X coordinate
y_{mi} – measured value of Y coordinate

The measure is applicable to the objects/feature classes from the application schema Vector as well as the whole dataset or dataset series.

The second measure is *Root mean square error RMSE* in a coordinate Z value. It is a standard deviation, where the true value is not estimated from the observations but known a priori. The measure is applicable to the objects from the application schema Grid – ElevationGridCoverage, feature classes from the Vector application schema– Spotelevetion, contour lines, break lines and the application schema Grid – ElevationGridCoverage at a level of a data set/data set series.

Equation 2

$$\sigma_z = \sqrt{\frac{1}{N}\sum_{i=1}^{N}(z_{mi} - z_t)^2} \qquad (2)$$

where:
σ_z – Root mean square error
z_t – true value of X coordinate
z_{mi} – measured value of X coordinate

The Positional accuracy/Gridded data position accuracy is also evaluated by the Root mean square error of planimetry at a level of an object, feature classes, a dataset, data set series from the Grid application schema.

The data specification doesn't define any minimum data quality requirements, only recommends the values for each of the proposed measures.

Tab.2 Recommended minimum data quality results for spatial data theme Elevation

Data quality element and sub-element	Measure name(s)	Target result (s)
Completeness/Commission	Rate of excess items	0%
Completeness/Omission	Rate of missing items	0%
Logical consistency/Conceptual consistency	Non-compliance rate with respect to rules of conceptual schema	0%
Logical consistency/Domain consistency	Value domain non-conformance rate	0%
Logical consistency/Format consistency	Physical structure conflict rate	0%
	Rate of missing connections due to undershoots	0%
	Rate of missing connections due to overshoots	0%
	Rate of invalid self-intersect errors	0%
Logical consistency/Topological consistency	Rate of invalid self-overlap errors	0%
Positional accuracy/Absolute or external accuracy	Root mean square error of planimetry (RMSEP)	Vector / TIN objects *Horizontal (m):* Max RMSEH = E / 10000 *Example.: For map scale 1: 10 000 is max RMSEH = 1*
	Root mean square error (RMSE)	Vector / TIN objects *Vertical (m):* Max RMSEv = Vint / 6 NOTE: Vint can be approximated by E / 1000. *Example.:For scale 1: 10 000 is max RMSEv = 1.67*
		Grid *Vertical (m):* Max RMSEv = GSD / 3 *Example.: Data with resolution 10 m is max RMSE = 3,34*
Positional accuracy/Gridded data position accuracy	Root mean square error of planimetry	Grid *Horizontal (m):* Max RMSEH = GSD / 6 *Example.: Data with resolution 10 m is max RMSE = 1,67*

3.2 Data quality metadata

The results of the data quality evaluation must be reported by metadata. The standard ISO 19115, which defines metadata elements, deals with the issues of metadata. For the evaluation of the data quality, ISO 19115 defines the package of metadata DQ_DataQuality. Each data quality element/sub-element has its own sub-package e.g. DQ_LogicalConsistency.

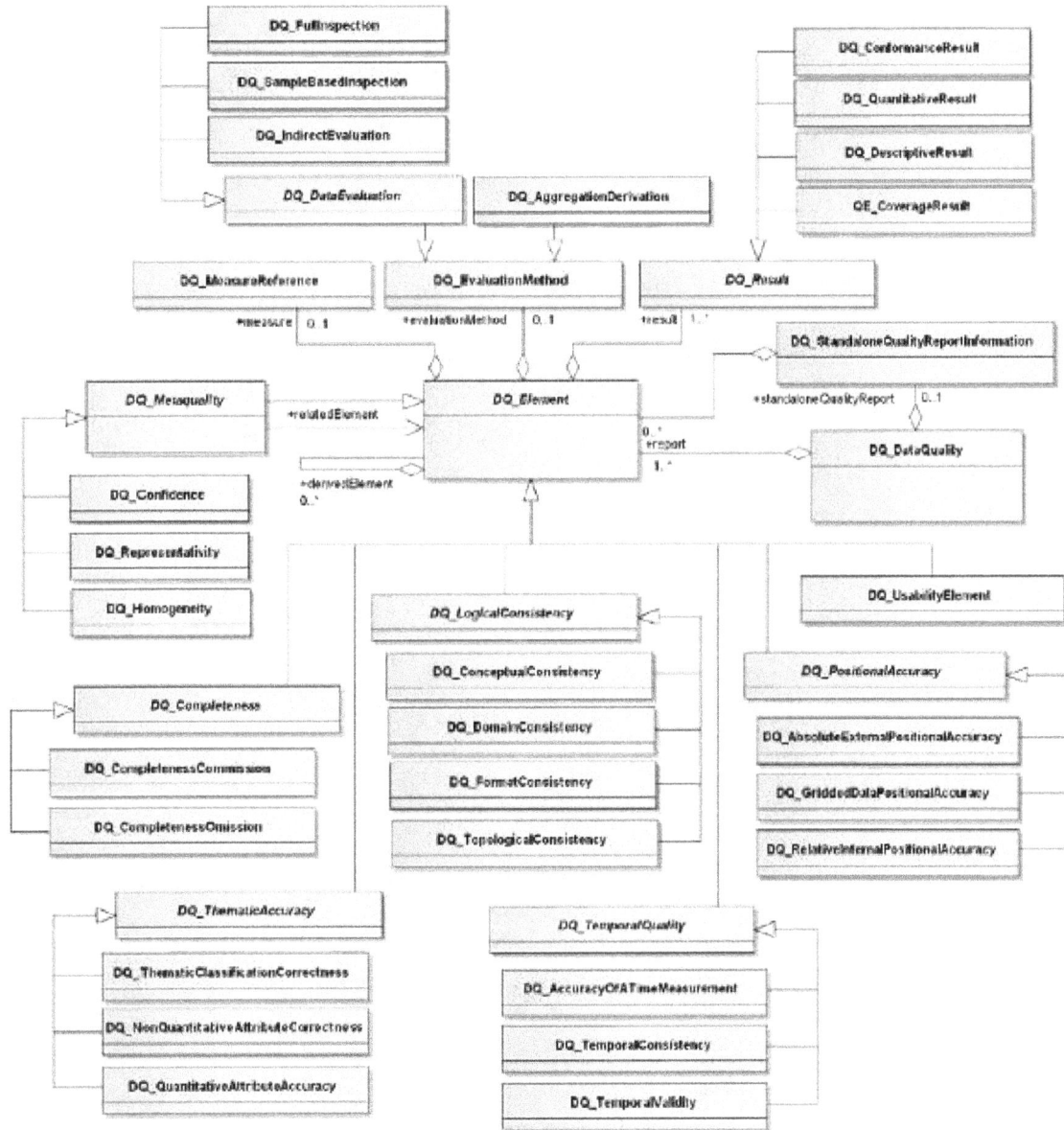

Fig. 6 UML model of data quality metadata [4]

The Foundation is an element DQ_Element, which carries all the elements for reporting the data quality. This element aggregates metadata packages for the data quality evaluation. DQ_MeasureReference is a collection of metadata elements which describes references on the used measure, DQ_EvaluationMethod is a collection of metadata used to describe the methods of the data quality evaluation, and DQ_Result is a collection of elements for reporting the results of the data quality evaluation. The Group DQ_Result includes several types of results. Compliance-DQ_ConformanceResult, the quantitative evaluation using DQ_QuantitativeResult, a text description of the result using DQ_DescriptiveResult and QE_CoverageResult to report the quality using a surface. For the quality evaluation it is necessary to specify the level at which the quality was evaluated in the metadata. This level is defined by the element DQ_DataQuality – DQ_Scope. It is recommended to use the data of values for the DQ_Scope-data set/data set series/feature class.

Under the rules of the INSPIRE data quality evaluation, it is necessary to use the quantitative evaluation using DQ_QuantitativeResult or the descriptive evaluation using DQ_DescriptiveResult. In the data specifications Elevation, the results of the evaluation of the quality are of the quantitative nature and must be reported by the metadata elements DQ_QuantitativeResult.

4   CHARACTERISTICS OF INPUT DATA

In the practical part of our work, we worked with the elevation data derived from the digital elevation model of the third generation – DMR3. For the testing area, we choose the territory on the border of Malé Karpaty (Little Carpathians Mountains) and Podunajská pahorkatina (Danube Wold) defined by the following geographical coordinates:

48°:34':06.978512'' N.

48°:05':54.007436'' N.

17°:42':48.66972'' E.

17°:00':29.420426'' E.

Those geographic coordinates correspond approximately to the extent of map sheet M-33-131-Db topographic maps in a scale of 1:25 000 (TM25).

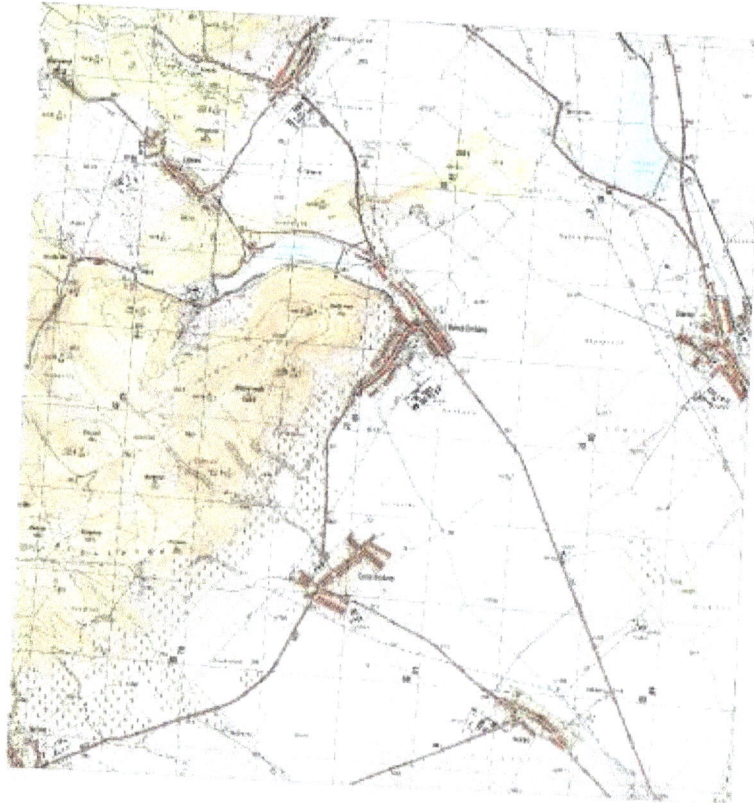

Fig. 7 Map sheet M-33-131-D-b TM25

DMR3 was created by the Topographic Institute Banská Bystrica in 2004 from altimetry map print base topographic maps at a scale of 1:10 000 (TM10) and 1:25 000 and some small parts from the basic maps at a scale of 1:10 000 (ZM10) [1]. Fig. 8 contains a sample altimetry map print base in the map sheet M-33-131-Db topographic maps at a scale of 1:25 000. Timeliness of DMR3 matches the state of TM10 and TM25, possibly ZM10. DMR3 has a form of a regular grid with a horizontal resolution of 10, 25, 50, or 100 m in the coordinate system S-JTSK. Models with a lower horizontal resolution were generated from the model with 10m resolution. Primarily, DMR3 is provided in the coordinate system S-JTSK and the height system Balt after adjustment, but there are versions in other coordinate systems (e.g. UTM34 and WGS84) as well. Since it is not publicly known which of these systems had been used in the process of building and which is the original version, while the remains are derived by transformation and resampling, we decided to use it as input DMR3 in S-JTSK. Inputs, the production process and also the data quality information about DMR3 were not published by their creator. The results of the independent evaluation for the whole territory of the Slovak Republic can be found in [7].

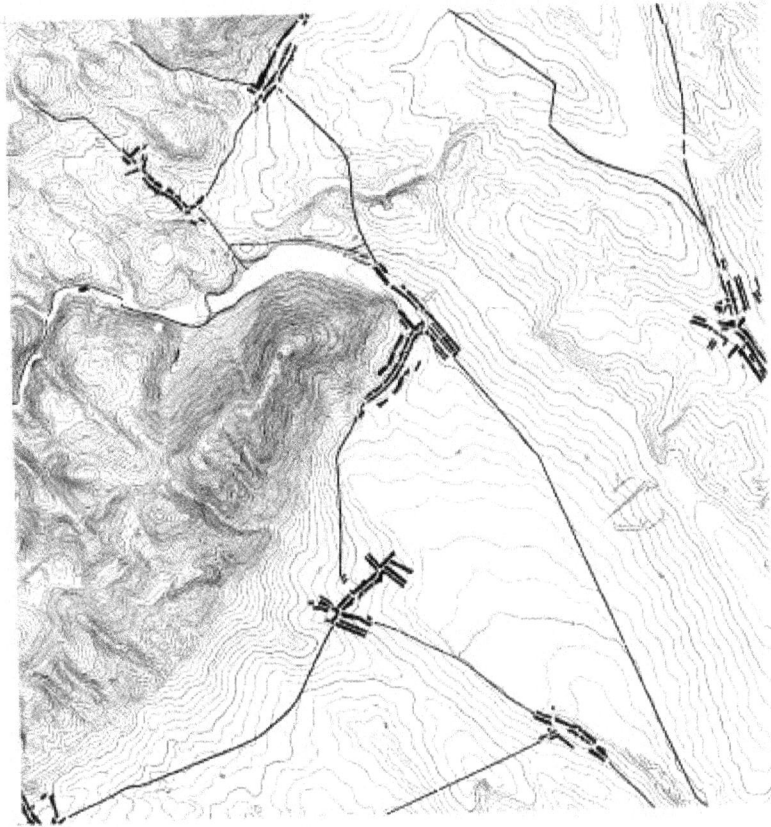

Fig. 5 Print pattern of hypsography of M-33-131-D-b TM25

DMR3 were used to prepare four layers according to the specification of the application scheme for the theme Elevation:

- Regular grid

- Triangulated irregular network - TIN

- Contour lines

- SpotElevation (Singular points of elevation)

4.1 Preparing data generation

The input DMR3 data have already been in the form of a regular grid, it was enough to transform them to a coordinate system ETRS89 and crop them by the vector representation of the map sheet M-33-131-Db. For the coordinate transformation from S-JTSK -> ETRS89 binding parameters from [11] were used. The error of the missing conversion of S-JTSK-> S-JTSK (JTSK03) could be neglected due to the nature and scale of the input DMR3 data. In order to minimize the distortion values of elevation, on the one hand, and efforts to preserve the smoothness of relief on the other hand, a resampling method of bilinear interpolation was used in the transformation process. The resulting regular grid resolution was 0°:00':00.41463". The model is shown in Fig. 9.

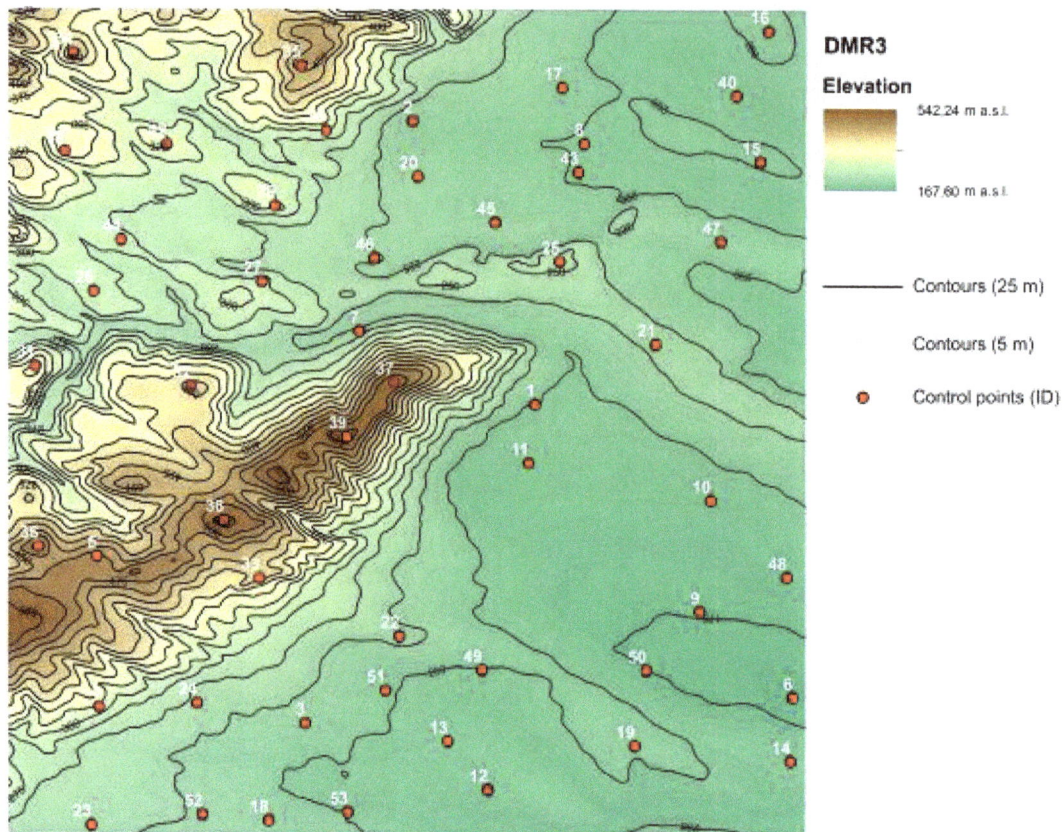

Fig. 9 Testing area with control points

In the following steps, the created regular grid was a basis for the derivation of three remaining layers.

The creation of an elevation model with the data representation in the form of a triangulated irregular network (TIN) was solved in two steps. At first it was necessary to create an input point field whose points will form the vertices of triangles. To utilize the ability of adapting the model (in this case, the distribution and local density of points) to the shape of the modelled surface, we chose a memoryless simplification method for reducing the model elements. This method works with a polyhedral model, reducing the elements by edge contraction. When contracting the edges, their two endpoints V0 and V1 are merged into a new vertex V. The algorithm minimizes partial changes in the model volume by the contraction as well as the total volume of the model. It proceeds from the edges, where the contraction creates the slightest change, which even in a small number of model elements maintains high geometric fidelity [6]. The number of nodes in the input grid was 754,292; we generated a set of 83,408 points which formed the vertices of triangles. We reduced the number of points to about 11 % of the original number.

The generated entry points were a base for the Delaunay triangulation. This method of triangulation is recommended by Directive ISO 19107:2003 for the geometric object GM_Tin. Triangulation constraints (Breakline, Stopline or maxLength) were not used. The resulting TIN created by Delaunay triangulation consists of 166,787 triangles.

The layer containing contours was created from a regular grid by the module for creating isolines in the GRASS GIS environment. A ten-meter contour spacing was chosen in a minimum of 160 m above sea level. Fig. 9 and Fig. 10 show the demonstration of the contour level representation, however it was necessary to modify the contour spacing to 25 meters for the main contour lines, and auxiliary contour lines spaced at 5 m intervals were added into flat areas because of clarity.

The layer of singular points was generated from a regular grid. There were calculated isolines of zero value of the first partial derivatives of the elevation in the direction of the axes X and Y. The singular points are determined by their intersections. We wanted to determine the type of singular points on the basis of positive resp. negative values of the second partial derivatives at those points. Whereas DMR3 does not allow to calculate partial derivatives of the second order in sufficient precision, this procedure has a high error rate, and we determined the type of singular points applying a professional manual approach. The resulting layer of the singular points is shown in Fig. 10.

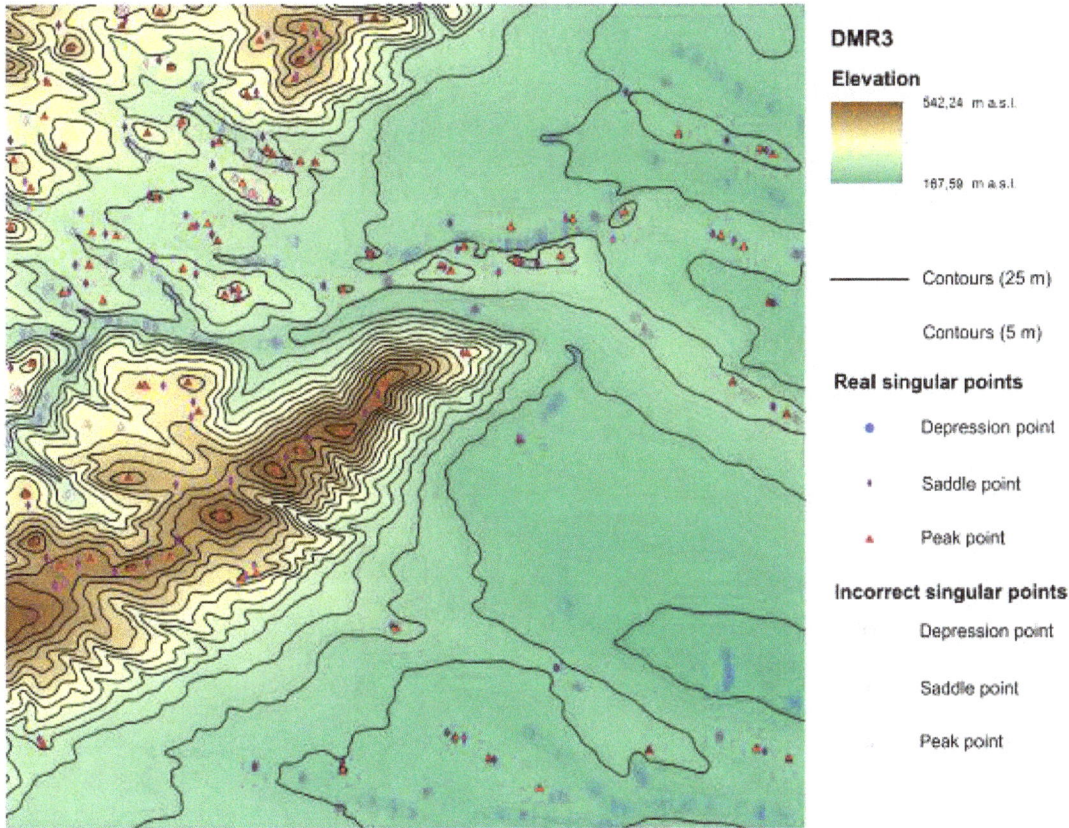

Fig. 10 Spatial distribution of elevation singular points

5 DATA QUALITY EVALUATION

The evaluation of data quality in the above layers was carried out in a defined set of quality components by applying all mandatory descriptors. In evaluating the quality parameters, we focused on the assessment of absolute positional accuracy, topological consistency and completeness of singular points. The quality in elements conceptual consistency, domain consistency and format consistency were not evaluated, because the raw data were not harmonized in accordance with the data model defined in the data specification. All layers were evaluated in all of the required components in the full territorial scope of the test area. In work [7] there was published the information on the procedure for the assessment and evaluation of results for the DMR3 in other thematic defined ranges (depending on the type of land cover and the degree of vertical relief segmentation).

Within the meaning of division of geographic data quality evaluation methods according to ISO 19114 a direct external-data-based method with a quasi-random variant or choosing control sites was used at all levels and in all evaluated elements. The method was applied mainly in an automated form.

The external data, for which the positional accuracy of the tested layers was evaluated, were a set of 52 geodetic survey points with the declared maximum mean error in position mxyz = 0:15 cm. The layout of control points in the test area is shown in Fig. 9.

5.1 Procedure of evaluation of absolute positional accuracy

For a regular grid in this element, we evaluated the quality of its vertical component. The actual evaluation was undertaken by obtaining values of altitude from DMR3 on the coordinates of control points and subtracting it from the control point attribute value of altitude.

By this procedure, we obtained a vertical error, for which a statistical method was subsequently used. For selecting the values of altitude from DMR3, a bilinear interpolation method was used. The interpolated surface of the value of the vertical error of DMR3 and the control points are given in Fig. 11:

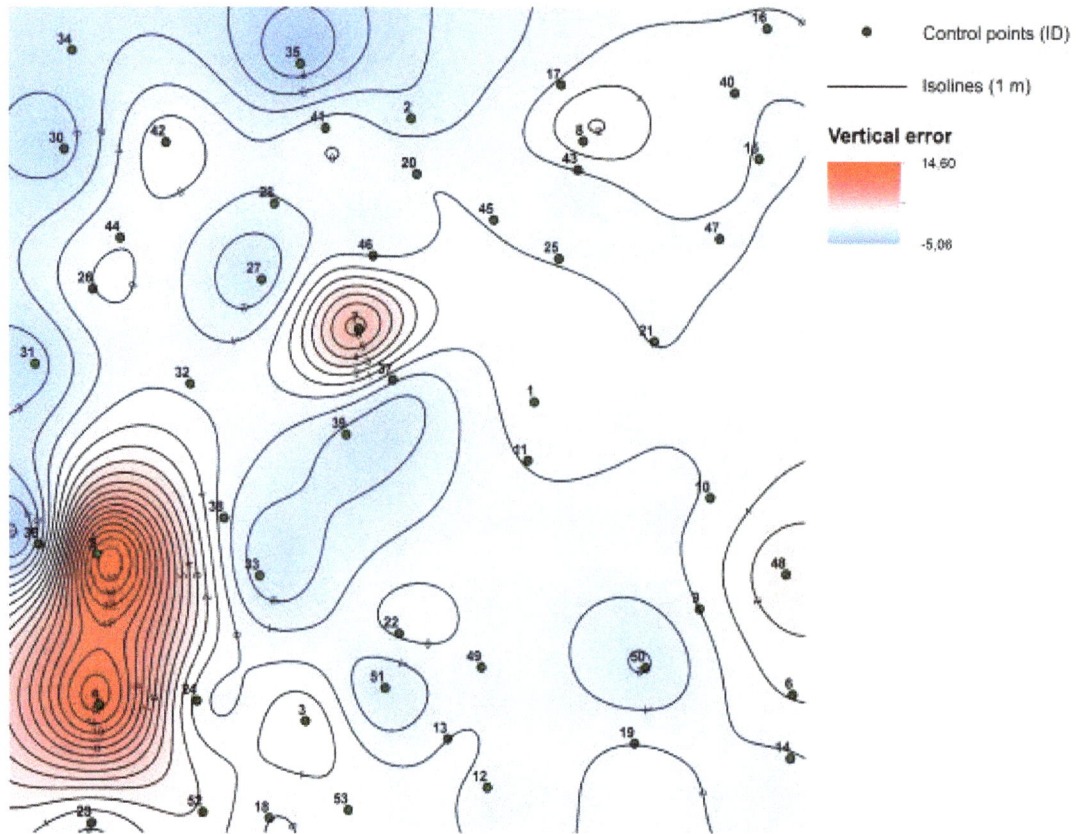

Fig. 11 Spatial distribution of values of the vertical errors for regular grid

For the TIN, we evaluated the quality in absolute positional accuracy of its vertical component. The actual evaluation was undertaken by obtaining the values of altitude from TIN on the coordinates of control points and subtracting it from the attribute value of altitude of control points.

By this procedure, we obtained the value of a vertical error, for which a statistical method was subsequently used. For selecting the values of altitude from TIN, a linear interpolation method was used at the vertices of the triangle which were spatially appertained to the checkpoint. The interpolated surface of the value of the vertical error of TIN and the control points are given in Fig. 12.

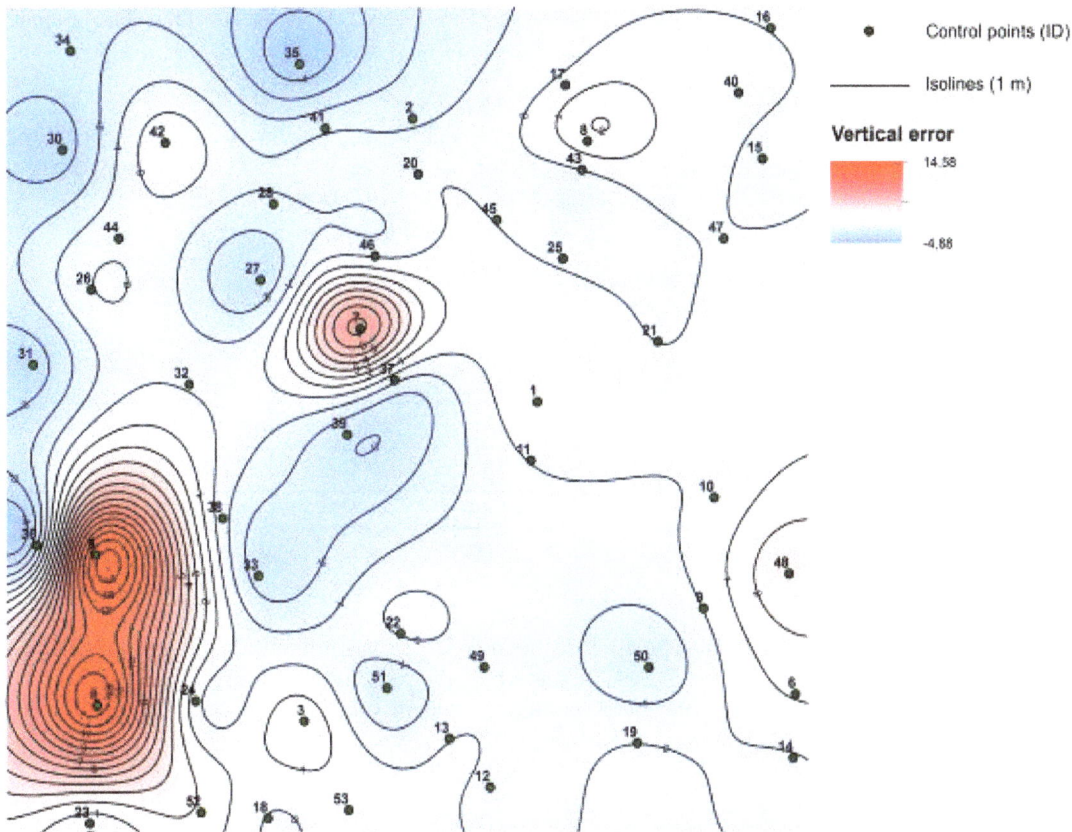

Fig. 12 Spatial distribution of values of vertical errors for TIN

For the isolines, we evaluated the quality in absolute positional accuracy of its horizontal component. The values of the horizontal errors for each control point were determined by a distance between the checkpoint and its appertained point on the isoline. We were not able to locate any appertained isoline for eleven control points and thus they were not used. They were the checkpoints in the top areas of the relief where DMR3 underestimated the height of relief.

5.2 Procedure of evaluation of singular points completeness

The layer of singular points was evaluated in two sub-components of completeness. Completeness - Omission and Completeness - Commission were evaluated on the basis of professional manual typing singular points. It was assessed whether the presence of singular points in DMR3 corresponds to reality. The basis for this was typing a contour with a small pitch. Most of the results of this evaluation showed DMR3 interpolation errors, which caused a large number of defective items in the valleys of depression and their related saddle points.

5.3 Procedure of evaluation of topological consistency

The evaluation of the quality in element topological consistency was performed for the isolines of layers and TIN, for which the topological consistency could be evaluated. The isolines were generated from the grid using GRASS GIS tools that help prevent creating the topologically incorrect data. Therefore, we further evaluated the topological consistency by other means, and consider it as correct. On the other hand, TIN was created out of the GRASS GIS environment, and after its import, a module was therefore launched for checking and correcting the topology, which revealed no topological error.

6 RESULTS OF DATA QUALITY EVALUATION

On the basis of the above procedures, we obtained the following results.

Absolute positional accuracy (vertical)

The most likely size of a vertical error of DMR3 in the test area for a regular grid has a value of 0.22 m and 0.25 m for TIN. The standard deviation of the vertical error of DMR3 in the test area is 3.04 m for a regular grid and 3.19 m for TIN. Ninety per cent of the surface of the tested area of DMR3 has the value of the vertical error in the range of -5.74 m to + 6.19 m for the regular gird and -6 m to +6.51 for TIN. The graphical representation of spatial distribution of the vertical error is shown in Fig.11 and Fig.12.

Absolute positional accuracy (horizontal)

The most likely size of a horizontal error of DMR3 contours in the test area has a value of 32.85 m. The standard deviation of the horizontal error of DMR3 in the test area has a value of 71.68 m. Ninety per cent of the surface of the test area has the value of the horizontal error in the range of 107.98 to 173.69 meters.

Tab. 2 Results of absolute positional accuracy evaluation

Data quality elements	Mean error [m]	RMSE [m]	90 % confident interval [m]
absolute positional accuracy – regular grid	0.22	3.04	<-5.74;6.19>
absolute positional accuracy - TIN	0.25	3.19	<-6.00;6.51>
absolute positional accuracy - Isolines	32.85	71.86	<107.98;173.69>

Completeness – singular points

The layer contains all existing singular points. It contains 472 points, from which 303 (64.2 %) doesn't exist in terrain. The graphical representation of the distribution of the existing and abundant singular points is shown in Fig. 10. The classification of types of singular points is shown in Tab. 4.

Tab. 3 Results of singular points completeness evaluation

Type of singular point	Valid		Invalid		Sum
	[abs.]	[%]	[abs.]	[%]	
Depression points	0	0.00	116	100.00	116
Saddle points	81	34.76	152	65.24	233
Peak points	88	71.54	35	28.46	123
All points	169	35.81	303	64.19	472

7   CONCLUSIONS

The evaluation of the quality of geographic information is an important element in building the infrastructure for spatial data. It contributes to the more efficient access of users to spatial data with the required quality. The INSPIRE directive and its implementing rules require the data providers to evaluate the data quality of individual data sets referred in Annexes I to III of the directive. Currently, this issue is not paid so much attention, as providers of spatial data are more focused on the process of harmonization of the data with the data models defined in implementing rules of the INSPIRE directive. Our aim was to point out the way how to approach the evaluation of quality in the context of INSPIRE. As an example, we chose a sample of data from the digital height model of the Slovak Republic – DMR 3. In view of the fact that we didn't harmonize the input data with the INSPIRE Elevation data model; we selected those data quality elements and measures, which didn't require any harmonized data model as well as we consider them as a crucial and very important in the process of the data quality evaluation of a digital height model. This was the absolute positional accuracy where we used mean error rates, a standard deviation and an error limit of 90% confidence interval. For a regular grid and TIN model, we evaluated the vertical accuracy and horizontal accuracy for contours. The resulting values reflect the way, in which DMR 3 was originated. We evaluated the quality of singular points of the relief by element completeness, where we identified the missing objects and objects in addition. In our case, we didn't want to highlight the resulting values, but rather the procedures of the spatial data quality evaluation. We summarized the results of the data quality evaluation by filling required data quality metadata elements; and the created metadata was published by the CSW service, which is available at http://gis.fns.uniba.sk/geonetwork.

Paper was supported by the project APVV-0326-11.

REFERENCES

[1] GKÚ 2010. Cenník produktov. Bratislava: GKÚ, 2010, [cit. 4/2013][online] Dostupné na internete:

[2] http://www.gku.sk/docs/cennik2010+dod_1-9.pdf

[3] INSPIRE Thematic Working Group ELEVATION, 2011. Data Specification on Elevation – Draft Guidelines [online]. 2013, version 3.0rc3

[4] ISO/TS 19138:2006, Geographic information -- Data quality measures.

[5] ISO/TS 19157:2012 Geographic information – Data quality(DRAFT)

[6] Smernica Európskeho parlamentu a rady 2007/2/ES zo 14.marca 2007, ktorou sa zriaďuje Infraštruktúra pre priestorové informácie v Európskom spoločenstve (Inspire)

[7] LINDSTROM, P.-TURK, G.: *Fast and Memory Efficient Polygonal Simplification. In: Proceedings of the conference on Visualization '98*. Los Alamitos, USA, Computer Society Press 1998, p. 279-286

[8] MIČIETOVÁ, E., IRING, M.: *Hodnotenie kvality digitálnych výškových modelov. Geodetický a kartografický obzor.* - ISSN 0016-7096. - Roč. 57/99, č. 3 (2011), s. 45-57 [1,8 AH]

[9] STN EN ISO 19113:2005, Geografická informácia – Princípy kvality

[10] STN EN ISO 19114:2005, Geografická informácia – Postupy hodnotenia kvality

[11] STN EN ISO 19115:2005, Geografická informácia – Metaúdaje

[12] Vyhláška č. 300/2010 Úradu geodézie, kartografie a katastra z 14. Júla 2009, ktorou sa vykonáva zákon Národnej rady Slovenskej republiky č. 215/1995Z.z. o geodézii a kartografii v znení neskorších predpisov [cit.4/2013][online] Dostupné na internete: http://www.pce.sk/kgk/Files/Vyhl.c.300_2009.pdf

RESUMÉ

Hodnotenie kvality geografických informácií patrí medzi dôležité elementy pri budovaní infraštruktúry priestorových údajov. Prispieva k zefektívneniu prístupu užívateľov k priestorovým údajov v požadovanej kvalite. Smernica INSPIRE a jej implementačné pravidlá kladú na poskytovateľov údajov požiadavky na hodnotenie kvality pre jednotlivé dátové sady podľa príloh smernice I až III. Tejto problematike sa v súčasnosti nevenuje až taká pozornosť, nakoľko poskytovatelia priestorových údajov sú viac zameraný na proces harmonizácie údajov do údajových modelov definovaných v implementačných pravidlách smernice INSPIRE. Našim cieľom bolo poukázať na spôsob akým pristupovať k hodnoteniu kvality v rámci INSPIRE. Ako príklad sme zvolili vzorku z údajov z digitálneho výškového modelu Slovenskej republiky - DMR 3 . Vzhľadom na fakt, že sme vstupné údaje nepodrobili harmonizácii do údajového modelu z údajovej špecifikácie Výška, vybrali sme tie elementy a miery kvality, na ktoré nemal nevykonaný proces harmonizácie vplyv a z hľadiska použiteľnosti DMR 3 v praxi sme ich pokladali aj za najdôležitejšie. Ide o absolútne polohové presnosti, kde sme využili miery stredná chyba, smerodajná odchýlka chyby a hranica 90 % intervalu spoľahlivosti. Pre pravidelnú mriežku a TIN model sme hodnotili vertikálnu presnosť a pre vrstevnice horizontálnu presnosť. Výsledné hodnoty reflektujú spôsobom akým DMR 3 vznikal. Kvalitu singulárnych bodov reliéfu sme zhodnotlili pomocou element úplnosť, kde sme identifikovali chýbajúce objekty ako aj objekty navyše. V našom prípade sme však nechceli poukázať na výsledné hodnoty, ale skôr na postup, ako pristupovať k hodnoteniu kvality a akým spôsobom ju vykázať. V našom prípade sme vykázali kvalitu údajov pomocou vyplnenia položiek v metaúdajoch zaoberajúcich sa kvalitou a výsledky sme sprístupnili pomocou katalógovej služby na http://gis.fns.uniba.sk/geonetwork.

11

ASSESSING RELATIONS BETWEEN WATER SUPPLY AND DEMAND IN THE ODRA AND MORAVA RIVER BASINS

Jan THOMAS [1], *Miroslav KYNCL* [2], *Silvie LANGAROVÁ* [3]

[1] *Ing., Ph.D., Institute of Environmental Engineering , Faculty of Mining and Geology, VŠB-Technical University of Ostrava, tel. (+420) 59 732 9391*
e-mail: jan.thomas@vsb.cz

[2] *prof. Dr. Ing, Institute of Environmental Engineering, Faculty of Mining and Geology, VŠB-Technical University of Ostrava, tel. (+420) 59 732 3556*
e-mail: miroslav.kyncl@vsb.cz

[2] *Ing., SmVaK Ostrava a.s., 28. října 169, Ostrava, tel. (+420) 596 697 355*
e-mail: silvie.langarova@vsb.cz

Abstract

Periods of drought represent a serious problem in the management of water resources. Currently used climatic models assume the onset of major climatic changes and periods of drought. Irrespective of whether the forecasts will be fulfilled or not, it is essential to prepare measures to ensure the supply of drinking water in dry periods. This paper deals with the preparation of water balances for the areas of the Odra and Morava River basins and the prediction of relationships between water supply and water demand in the given area.

Abstrakt

Suchá období představují závažný problém v nakládání s vodními zdroji. Současné klimatické modely předpokládají nástup velkých klimatických změn a s tím nástup suchých období. Bez ohledu, zda se tyto prognózy naplní nebo ne, je třeba připravovat opatření k zajištění zásobování pitnou vodou v suchých obdobích. Tento článek se zabývá zpracováním vodohospodářských bilancí vodárenských zdrojů oblastí povodí Odry a povodí Moravy a predikcí vazby těchto vodárenských zdrojů a samotných nároků na vodu v daném prostoru

Key words: Climate change, drought, water management, drink water, water supply

1 INTRODUCTION

The European Environment Agency has been emphasising changes in climate for a long time. The whole society has been considerably affected by climate changes. There is much interaction between physical and chemical phenomena in the atmosphere between the biosphere and other components which create a complex system of climate. This puts the complex process of water mass circulation into motion. A chemical response in the troposphere results in reactions influencing, in turn, changes in climate. Development of water management should be assessed in a rather wide context which is given by social economic parameters, changes in living environment and, last but not least, demands relating to water systems and individual sources of water supply. The Czech Republic depends entirely on rainfall – this is the consequence of the Czech Republic's position and hydrobiological balance. Lack of rainfall in the Czech Republic may deteriorate supply of water to people from water systems. [1,2]

In the Czech Republic, about 90 % of population is supplied with water from public water mains. These are, in particular, extensive water supply systems. The Czech Republic undertook to develop and improve the water system in line with the European Parliament and Council Directive No. 2000/60/EC. Attention is paid, among others, to securing sufficient and sustainable sources of water for water systems which should be able to supply water even in case of negative phenomena which shall accompany changes in climate. [3]

The goal of long-termed forecasts and planning is to find suitable measures being able to arrange that people and other consumers will be supplied with water in case of drought. Area water networks and group water networks which supply drinking water to people should be operated within an organisational and legislative framework and technical conditions so that the water could be supplied smoothly, even in a restricted scope. A wide range of measures should be taken to protect people against drought and it is necessary to have an overall strategy of protection as well. The proposed measures should be a general solution of the issue. In case of long-lasting drought, there might be a big difference between water supply and demand. In that case, it is essential to restrict consumption of water in order to keep basic quantities without restricting operation of important facilities. Changes in environment which affect quantity of water can, no doubt, influence quality, and vice versa. Knowledge of such measures and forecast should be the basis when making decisions on water demand and satisfaction of needs in terms of quality and quantity. [1]

We focused on water management relations in water supply systems which exist in the Odra and Morava River basins. Attention was also paid to the assessment of the data from water sources in terms of water supplies for people and industries. The goal was to identify areas where water deficit might occur in the future use of water sources.

2   SUSTAINABILITY IN MANAGEMENT OF WATER SOURCES AVAILABILITY

Sources of water rank among important elements which need particular responsible management in any place and in any point of time.

They are important for the development of all economic and other activities and influence balance between eco-systems (hydrological regimes). Because sources of water were regarded as sufficient ones for the satisfaction of needs of inhabitants and industry, an unbalance between the demand and availability of water in water distribution networks has been becoming more and more evident. In many regions, this unbalance reaches the critical level of sustainability.

The reason for this situation is, in most cases, a frequent alteration of periods of drought and those of flood. The unbalance of water in water sources is connected considerably with the quality of water. All those phenomena influence significantly territories where sources of water are supplied mostly from rainfall. Unbalance in water quantities and, in turn, consumption of water, play a major role in decision-making and management within water companies which take measures in water collection and waste water processing. The processing and treatment of water is a key prerequisite for maintaining the water balance for long-termed sustainable use of water. [4]

Those parameters influence water management and became the basis for creation of recommendations, the purpose of which is to ensure efficient and sustainable water management within EU. Sustainability and creation of conditions for efficient water management can result from the below described relation between potential water capacities, consumers and internal relations. Relations and cooperation of individual areas in water supply for heavily populated territories are shown in Fig. 1. [5] Assessing individual emergencies in water supplies should not be based on specific elements of the system such as a water supply system, catchment or water-treatment system. This should be a comprehensive assessment of the entire system which comprises water in catchment and urban water. Looking for joint goals should help evaluate efficient solutions in water supplies from sources of water. This should be the principle of cooperation for all stakeholders.

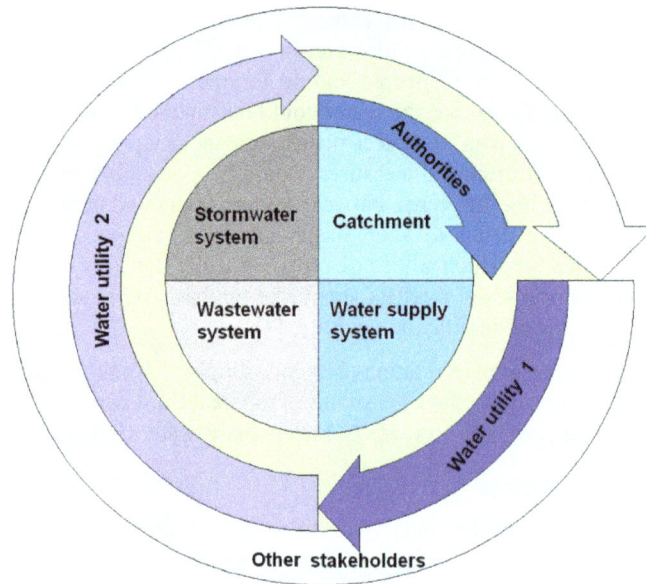

Fig. 1 Relations in a water supply system in heavily populated territories [5]

3   WATER BALANCE

Demands for sources of water have been becoming more and more extensive. One of key roles of the water management is the assessing of possible coverage of water needs from respective sources of water. Water balance characteristics can be defined as relations between potential consumption of water and demand for water. The relation is based on the fact that the demand for water is the sum of all demands for water in that place or territory in a specific point of time and for a specific quality and quantity of water. The quantity of water in a specific place and time can be defined, as Kos and Říha suggest, using the following formula [6]:

$$Zm = Zp + Zn + Zpr - Zt - Od \qquad (1)$$

where in [m³/s]:

Zm - total strength of water source,

Zp - strength of water in a place,

Zn - improvement thanks to tanks,

Zpr - improvement thanks to transfer,

Zt - losses,

Od - water takeoff.

The balance of water quantity in a specific place and time can be described as follows:

$$Bs = \frac{Zm}{(P + Om)} \qquad (2)$$

where in [m³/s]:

Bs - strength balance of a water source in a specific place and time,

Zm - total strength of water source,

P - requirements for use of water in water courses,

Om - water takeoff (water supply systems).

The analysis of relations above is an important part of the analysis of water quantity which is needed for water management. At the same time, the analysis provides conditions for a forecast of an available quantity of sources of water which are needed for the coverage of water demands.

4 PROGNOSES AND FORECASTS IN WATER MANAGEMENT

In most cases, the supplies of water which can be used for forecasts resulted from long-termed observations and analyses of the current situation. Several water management tasks have to rely on prediction, quantification and assessment of possible scenarios which could influence the decision-making process in complex water management systems. For purposes of the forecast, it is essential to evaluate and apply an educated qualitative estimate with quantitative data for current development of water needs and water consumption. As it is difficult to include all variables in individual models, it is necessary to make an early preparation and time-optimised planning of water demand for the sources of water. For this, all available means should be used. A theoretical basis for all estimates and plans is the prognostic work which considers trends in water economy and water supplies for that territory. An important part is the statistics which makes it possible to forecast the future situation with a certain degree of probability for a certain phenomenon. [4,7]

The mathematical-statistical approach comprises several alternative methods which can be used for forecasts. They can be divided into three groups:

- extrapolation (the normative method)

- synthesis (the morphological method)

- intuition (the theoretical method)

The most frequent combination is the method which combines normative and theoretical solutions. Detailed analyses and long-termed concepts use a number of basic statistical and prognostic tasks (such as projection, linear programming, regression, estimates...). Totals of all quantitative and qualitative parameters are taken into account as a total of a certain point of time. Fig. 2 summarises basic assumptions needed for the assessment of water management in a specific place and time, making the process more simplified. In order to make decisions, it is important to summarise trends in water demands from the point of view of all consumers. Available information indicates that in EU member states the intensity of use of the sources had been increasing gradually until the values became stable. There are also clear changes in individual EU member states. [4,7]. It is evident from historic trends in use of water in the EU member states, that the demand for water has been becoming stable and that the consumption is becoming stable as well.

This change is shown in Fig. 2 as a red frame field. This change is based on the development of a country, development of technologies and price of consumed water. It proves that the development of a country causes first the water consumption to increase. The reason are such pieces of equipment as dish washes (households, food processing industry), washers (households and industries), bathrooms or commercial car washes. Once the maximum value of water consumption is achieved, several scenarios are possible. The first scenario is a long-termed stagnation (trend 1). Changes in technology may result in savings and in different needs of and demand for water. The water consumption shows then a decreasing trend (trend 2). The development of technologies can be accompanied with socio-economic phenomena (changes in water tariffs, income of households). In that case, the decrease in total demand for water is the highest of all (trend 3). [4]

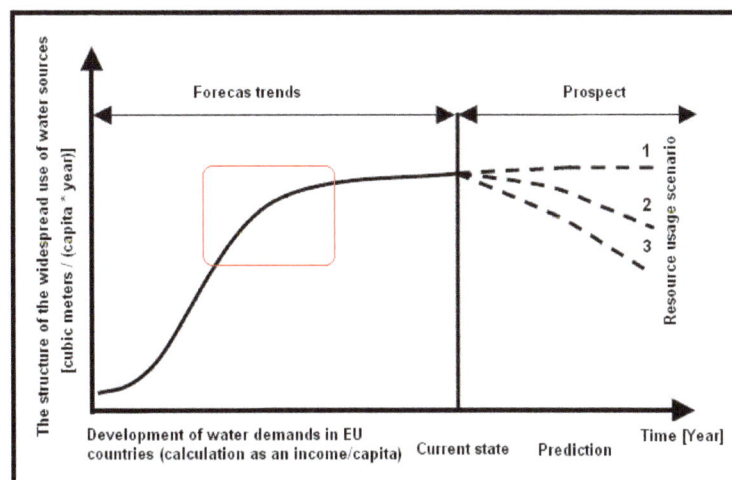

Fig. 2. Trend in use of water sources in EU [4,7]

5   RELATIONS BETWEEN WATER SUPPLY AND DEMAND IN CERTAIN TERRITORIES WITHIN THE ODRA AND MORAVA RIVER BASINS

The sources of water and requested quantity of water was analysed in order to summarise average sources of water in the water basins of the Odra and Morava Rivers, evaluating the average consumption of water from the point of view of water which is handed in the water network. The results are based on the observations made from 1990 to 2010. The values include the information about current conditions and maximum values which were taken from previous models and forecasts made in 1993. The following water supply systems and major takeoff of water were taken into account for the Morava River basin: the Vír Water Area Network, Hodonín Water Supply System, Boskovice-Blansko Water Supply System, Štítary, Výškov, Mikulov Water Supply System, Olomouc Water Supply System, Luhačovice Water Supply System, Prostějov Water Supply System, Přerov Water Supply System, Šumperk Water Supply System, Zlín Water Supply System and Kroměříž Water Supply System. The most important source of water which covers the water demand is the Vír Reservoir. Regarding the water basin of the Odra River, the following water systems were included into the analysis: the water system of the Ostrava Area Water Network and Bruntál water supply system. The most important sources of water are the Kružberk Dam, Šance Dam and Morávka Dam.

During the period from 1989 until 2010, there were two big significant changes in the approach to water supplies and water needs. Separate water management companies were established and the price of water increased considerably which, in turn, influenced the total consumption of water. There are studies which mention that the sources of available water (the water for water supplies) have been becoming more and more limited. Considering the geography of the Czech Republic, attention should be paid to those territories where few sources of water are available for water supplies.

The available information about consumption of water within the Odra and Morava River basins as well as the assessment of water supplies in those territories were used as a basis for the creation of two models of water needs and intensive use of water sources. The potential future scenarios were modelled using Statgraphics Plus which is able to evaluate data sets from the point of view of various alternatives.

Two models were chosen for the final assessment – they represented the critical condition and the condition which is compliant with current trends. The model evaluates the behaviour of a time sequence after the last known value with a reliability limit of 67 %. The reliability is regarded as the statistically minimum value of correlation of the model between the time sequences of average sources of water and sequences of water demand in individual areas. The confidentiality limit of 67 % makes it possible to optimise the model and make further comparisons.

After assessing the time sequences, the best models were taken from the predictions. Those models were the models resulted from the Random walk (R-w.) and S-Curve (S-c) relations. The R-w model takes into account the mean values of data and standard deviations in differences between the values. This model shows the maximum values of a trend drop in water sources and water needs in a time horizon of 40 years.

The S-c model shows curves which are placed in partial sections of monitored data, covering future periods. The S-c model is the best one for the input data from the point of view of cyclical changes in values and gradual balancing. The final data are again valid for a 40-year forecast. Tabs 1 and 2 show the forecast values for each model separately.

The figures below show the assessment of the water supply systems for each model. "The current state" in the chart is the border for the data of the water supply systems – it shows capacities of available water quantities for water supplies. For this development, the data for a period of 22 years was used. The values from now onwards represent the forecast values for the next 40 years.

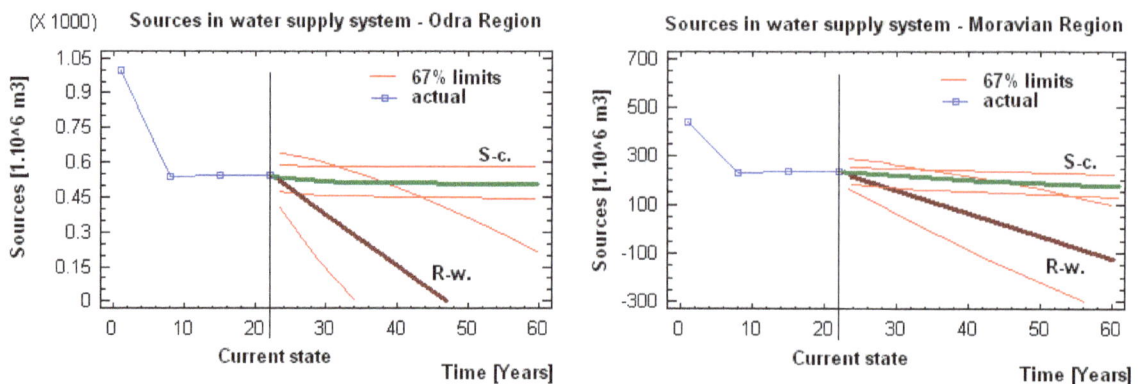

Fig. 3. Sources in water supply systems in the Odra and Morava River basins

Two separate trends are clear from the chart. The critical values were achieved in the R-c model where the values for the 20-year forecast indicate a permanent drop in sources in the water supply system. This fact correlates to the trend of drop of available water supplies which is mentioned in the "Climate change and global water resources". [8,9,10,11] Considering local conditions, the S-c trends seem to be more probable - they indicate gradual harmonic balancing of the values with a long-lasting descending trend. The harmonic correlation is rather typical for most water management tasks.

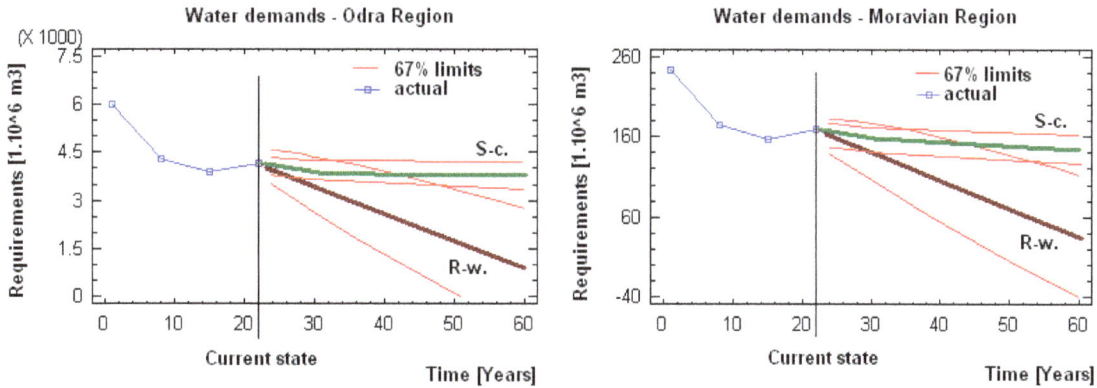

Fig. 4. Demands for water supply systems in the Odra and Morava River basins

Let us consider two scenarios of total demand for and consumption of water. The both trends show rather descending tendencies. In spite of this, it is possible to identify sections in the sources within the water supply system (this is, in particular, the task of R-c model) where critical water management is likely. This phenomenon is evident even if the demand for water is going down in both the Odra and Morava River basins. Tab.1 and Tab. 2 show the output values of the R-w and S-c models.

Tab. 1 Available sources of water in water supply systems in the Odra and Morava River basins

Period [year]	Sources in water supply system – Odra Region					
	Model: S-c Trend=exp(8.6 + 0.7/t) t [year]			Model: R-w		
	Forecast $[10^6 m^3]$	Lower and Upper Limits 67.0%		Forecast $[10^6 m^3]$	Lower and Upper Limits 67.0%	
29	526	501	552	390	177	603
36	524	499	549	238	< S.limit	539
43	522	498	548	86	< S.limit	455

Period [year]	Sources in water supply system – Moravian Region					
	Model: S-c Trend=exp(5.4 + 0.8/t) t [year]			Model: R-w.		
	Forecast $[10^6 m^3]$	Lower and Upper Limits 67.0%		Forecast $[10^6 m^3]$	Lower and Upper Limits 67.0%	
29	229	216	243	170	75	266
36	228	215	242	104	< S.limit	238
43	227	214	241	37	< S.limit	202

Tab. 2 Demand for sources of water in water supply systems in the Odra and Morava River basins

Period [year]	Water demands – Odra Region			
	Model: S-c Trend=exp(8.3 + 0.4/t) t [year]		Model: R-w	
	Forecast	Lower and Upper Limits	Forecast	Lower and Upper Limits

	$[10^6 m^3]$		67.0%	$[10^6 m^3]$		67.0%
29	403	376	432	356	275	437
36	402	375	431	295 Limit	181	410
43	401	374	430	235 Limit	94	375

Period [year]	Water demands – Moravian Region					
	Model: S-c. Trend=exp(5.1 + 0.4/t) t [year]			Model: R-w.		
	Forecast $[10^6 m^3]$	Lower and Upper Limits 67.0%		Forecast $[10^6 m^3]$	Lower and Upper Limits 67.0%	
29	164	152	176	145	112	178
36	163	152	176	120 Limit	74	167
43	163	151	175	96 Limit	39	152

6 CONCLUSIONS

The forecast above is based on average values for the locations within the water supply system. The values were prepared using the linear programming and development prediction by means of the S-curve and Random Walk methods. In this comparison, it is possible to address the water management in water supply systems from two different points of view. From the critical assessment point of view, it is possible to consider critical situations in water supply systems. If the descending trend in sources for water supply systems is taken into account, it is likely that critical scenarios will occur within 20 to 30 years in the both Odra and Morava river basins. Regarding the harmonic functions of the S-curves, it should be pointed out that if the current demand for water is maintained it does not necessarily mean that the decreasing quantity of available water from sources for the water supply systems will result in critical scenarios in water management.

Trends in EU show that consumers have started changing their behaviour in consumption of water sources. According to studies, the trend of the changes is similar and is often extrapolated by means of S-curves. The trends have been observed in EU countries for a long time. Considering the changes in the Czech Republic in the 1990s (in particular, from 1990 to 1993), the use of sources diversified and the consumption of water changed. The trend in water consumption and water demand correlates now within EU with the trend of the harmonic development of the S-curves. In case of the S-c forecast, the results of future periods converge now towards the current demand in terms of technology, consumption and available sources of water. This proves that the values within the Odra River basin indicate noncritical water management, if compared with Mediterranean regions. Regarding sub-regions within the Morava River basin, there are territories with a negative water balance for sources of water supply systems for the next 40 years and the deficit of water sources should become evident, with current need during the forecast period of 2012 - 2015. Those sub-regions include the Vír Area Water System, Olomouc Water Supply System and Kroměříž Water Supply System. This trend is stable in the forecast period and converges towards minimum capacities of resources for the water supply systems in a specific place and time.

REFERENCES

[1] POVODÍ ODRY, s.p.: *Zpráva o hodnocení množství povrchových vod v oblasti Povodí Odry*, odbor vodohospodářských koncepcí a informací, Ostrava, 2011

[2] EEA/WHO European Environment Agency: *Water resources and human health in Europe*, Environmental Issues Series, 1999

[3] PUNČOCHÁŘ, P.: Teze rozvoje oboru vodovodů a kanalizací pro „Koncepci vodohospodářské politiky ministerstva zemědělství pro období 2011-2015", SOVAK, Ročník 20, číslo 1, 2001

[4] FLORKE, M., ALCAMO, J.: *European Outlook on Water Use*, Center for Environmental Systems Research University of Kassel, Final Report, 2004, 3241/B2003. EEA. 5159

[5] SMEETS, P., ALMEIDA, M. d. C., STREHL, C., UGARELLI, R.: *Water cycle safety plans to prepare cities for climate change*, IWA-8139, World Water Congress 2012 Busan, Korea

[6] KOS, Z., ŘÍHA, J.: *Vodní hospodářství*, ČVUT, Praha, 2000, ISBN 80-01-02261-7, 142p

[7]   EEA European Environment Agency: *Towards efficient use of water resources in Europe*, EEA Report, Luxembourg, 2012, ISBN 978-92-9213-275-0

[8]   ARNELL, N., W.: *Climate change and global water resources*, Global environmental change, Issue 9, 1999, p.31-49, Elsevier, S 0 9 5 9 - 3 7 8 0 (9 9) 0 0 0 1 7 - 5

[9]   EEA European Environment Agency: *Urban adaptation to climate change in Europe*, EEA Report, Luxembourg, 2012, ISBN 978-92-9213-308-5

[10]   EEA European Environment Agency: *Water resources problems in Southern Europe* – An overview report, Topic Report 15, Inland Waters, Copenhagen, 1997

[11]   EEA European Environment Agency: *Sustainable Water Use in Europe* – Sectoral Use of Water, Topic Report 1, Inland Waters, Copenhagen, 1999

This paper has been prepared thanks to the support of the Czech Ministry of Agriculture within the project named "Investigating into measures aimed at supplies of drinking water in times of climatic changes" – ref. no. QI112A132.

RESUMÉ

Výše uvedené predikce vychází z průměrných hodnot pro jednotlivé lokality vodárenských soustav. Tyto hodnoty byly zpracovány na základě lineárního programování a predikce vývoje pomocí S-křivek a metody Random Walk. Uvedené srovnání umožňuje nahlížet na problematiku hospodaření s vodou ve vodárenských soustavách ze dvou pohledů. Z pohledu kritického vyhodnocení je možné uvažovat s krizovými situacemi ve vodárenských soustavách. V případě klesajících trendů zdrojů vod může docházet s výhledem 20 až 30 let ke krizovým scénářům nakládání s vodou jak v regionech povodí Odry, tak povodí Moravy. Z pohledu harmonických funkcí S-křivek je možné uvést, že v případě zachování současných nároků na vodu nemusí jednoznačně docházet při snižujících se množství disponibilní vody z vodárenských zdrojů ke krizovým scénářům vodního hospodářství.

Trendy v Evropské unii ukazují na fakt, že dochází ke změnám chování spotřebitelů ve smyslu využívání zdrojů vody. Tyto změny mají dle studií podobný trend, který je často extrapolován S-křivkami. Tyto trendy byly v zemích EU pozorovány po delší časové období. Vzhledem ke změnám v ČR v devadesátých letech (tj. především v první polovině v období 1990 – 1993), došlo k diverzifikaci využití zdrojů a k změnám spotřeby vody. V současné době je trend potřeby vody a nároků na vodu v porovnání s EU v korelaci, s trendem odpovídajícím harmonickému průběhu S-křivek. V případě S-c. predikcí je proto patrná oscilace výsledků budoucích období, konvergujících k hodnotám odpovídajícím současným nárokům na technologie, spotřeby a disponibilní zdroje. Zde se ukazuje, že hodnoty ve sledovaných oblastech povodí Odry jsou v současném okamžiku oproti středomořským regionům v režimu nekritického hospodaření s vodou. V rámci sub-regionů povodí Moravy je ale řada oblastí, která vykazuje záporné negativní hodnoty vodohospodářských bilancí vodárenských zdrojů s výhledem 40 let, s deficitem vodárenských zdrojů při současných potřebách na úrovni predikovaného období 2012 až 2015. Mezi takové oblasti patří vodohospodářská soustava Vírského oblastního vodovodu, Olomoucko, anebo Kroměřížsko. Tento trend je v predikovaném období stálý, konvergující k minimálním hodnotám kapacit vodárenských zdrojů v daném místě a čase.

12

RESEARCH ON PETROPHYSICAL PROPERTIES OF CHOSEN SAMPLES FROM THE POINT OF VIEW OF POSSIBLE CO$_2$ SEQUESTRATION

Martin KLEMPA [1], *Michal PORZER* [2], Petr BUJOK[3], Ján PAVLUŠ[4]

[1] *Ing., Institute of Geological Engineering, Faculty of Mining and Geology, VSB – Technical University of Ostrava*
17. listopadu 15, Ostrava Poruba, tel. (+420) 59 732 5496
e-mail martin.klempa@vsb.cz

[2] *Ing., Institute of Geological Engineering, Faculty of Mining and Geology, VSB – Technical University of Ostrava*
17. listopadu 15, Ostrava Poruba, tel. (+420) 59 732 5487
e-mail michal.porzer@vsb.cz

[3] *prof. Ing. CSc., Institute of Geological Engineering, Faculty of Mining and Geology, VSB – Technical University of Ostrava*
17. listopadu 15, Ostrava Poruba, tel. (+420) 59 732 3529
e-mail petr.bujok@vsb.cz

[4] *Ing., Institute of Geological Engineering, Faculty of Mining and Geology, VSB – Technical University of Ostrava*
17. listopadu 15, Ostrava Poruba, tel. (+420) 59 732 5487
e-mail jan.pavlus@vsb.cz

Abstract

Man-made CO$_2$ emissions (the so called anthropogenic CO$_2$ emissions) and their increasing trend can be, by some scientists, considered a serious menace for the sustainable development of mankind, and their reduction a prerequisite for the environment protection. Carbon dioxide is one of the most important gases that cause a greenhouse effect which warms up the earth surface as a consequence of a different heat flow between the earth and the atmosphere. Our laboratory measurements determined the porosity, permeability and grain density for clastic sedimentary rock samples which were drilled from an underground gas storage facility. Additionally, our results showed a reduction in porosity and permeability after a confining pressure was applied. We assume that this effect is caused by internal structure changes due to the repeatedly increased and decreased net pressure applied to the samples.

Abstrakt

Emise CO$_2$ vznikající lidskou činností – tzv. antropogenní emise CO$_2$ a jejich vzestupný trend, mohou být některými odborníky považovány za vážné nebezpečí pro udržitelný vývoj lidstva a jejich omezování za nezbytnou podmínku ochrany životního prostředí. Oxid uhličitý je významný z plynů způsobujících skleníkový efekt, který se projevuje oteplováním zemského povrchu v důsledku změn toků tepelného záření mezi zemí a atmosférou. Laboratorní měření poskytla hodnoty porozity a koeficientu propustnosti horninových vzorků, které byly odvrtány z podzemního zásobníku plynu. Naše měření vykázalo snížení kolektorských parametrů horninových vzorků, které bylo způsobeno změnou vnitřní struktury horniny díky opakovanému zvýšení a snížení tlaku na rostlou část vzorku.

Key words: carbon capture and storage (CCS); enhanced oil recovery (EOR); porosity; permeability; laboratory experiment

1 INTRODUCTION

Several projects that deal with theoretical and pilot research on CO_2 storage in geological formations are currently underway. These projects are addressed by national programmes in USA, Canada, Australia and Japan. One of the projects under the supervision of the European Union was the RECOPOL project which evaluated the CO_2 storage in coal seams of the Lower-Silesian Basin. The main objective of these projects is to find out whether the CO_2 storage in geological formations is economically feasible and environmentally safe. In the Czech Republic, the most perspective formations for storing this gas are connected with oil and gas reservoirs.

Potential storage spaces are the depleted and actively produced oil- and gas fields, in which it is possible to enhance oil recovery by 10 to 15 % by using the CO_2 injection into the reservoir. Oil fields are a favourable variant, because before they were produced, the hydrocarbons were stored inside them during geological time, and similarly the carbon dioxide can be stored there now. Another advantage they have is a well explored geological environment, and, therefore, an abundance of information on the selection of suitable storage locality, its utilization and long-term monitoring. The capacity of the CO_2 storage space in an oilfield depends on the pore space freed after oil production and the pore space that is filled with water under the oil bearing horizon. Depleted oil- and gas- fields represent suitable porous rock structures either for CO_2 sequestration or for underground storage of imported natural gas.

2 THEORETICAL ASPECTS OF CO_2 STORAGE

It is assumed that horizons used for the carbon dioxide sequestration will lie in depths below 800 m. At temperatures and pressures corresponding to the depths, lower than the above mentioned, carbon dioxide changes its phase behaviour, its density resembles liquids and its state is called a supercritical state. This transition into the supercritical state takes place under p,T conditions of 7.38 MPa and 31.1°C, respectively.

Carbon dioxide injected in a supercritical state occupies much less space than in a gaseous state. In the depth interval of 600-800 m, the CO_2 density increases with depth. From the depth of 1000 m, it reaches its maximum value and it doesn't change with depth any more. Under standard conditions (temperature of 25°C and pressure of 0.1 MPa), the density of CO_2 is 1.977 kg/m^3. That means that 1 tonne of CO_2 occupies a volume of 526 m^3. On the temperature and pressure conditions in a depth of 1000 m (35°C, 10 MPa), one tonne of CO_2 occupies a space of 1.5 m^3 (CO_2 density is 650 kg/m^3) [4].

In order to inject CO_2 effectively, its density should be in an interval of 600 to 800 kg/m^3 (on p,T conditions of 30°C and 8 MPa, resp.).

Mathematical modelling of a geochemical sequestration process is needed for developing a notion of how the injected CO_2 will behave in a reservoir. One of such models based on our laboratory experiments was created for a saline aquifer of the Upper Silesian Coal Basin conditions [2]. The Geochemist's Workbench 7 (GWB) simulator was used for modelling. The modelling process had two stages. The first stage aimed at observing the changes in the rock environment at the beginning of CO_2 injection. The second stage evaluated the changes caused by the CO_2 influence on permeable rocks after the injection.

A timespan of 20 thousand years was analysed in the model. During the first three years after the injection ended, a continuous increase in porosity takes place in the rock environment. Afterwards, this value stabilizes at a maximum level without further changes [2].

3 SIMULATION OF RESERVOIR ROCK CAPACITY ABILITY

For ascertaining the CO_2 storage capability, the knowledge of *porosity* and *permeability* values of a natural reservoir is essential. These parameters were measured by means of the apparatuses COREVAL 700 and Benchtop Relative Permeameter 350 (BRP 350) in the Laboratory of Wells and Hydrocarbon Deposits Stimulation at the Institute of Clean Technologies for Extraction and Utilization of Energy Resources, under the Faculty of Mining and Geology at the VŠB – Technical University of Ostrava.

3.1 Porosity and permeability of natural reservoirs

Pores can be defined as spaces of different shapes, size and origin in soil or between rock grains that are not filled with solid phase. We differentiate these porosities:

- absolute porosity;
- open porosity;
- effective porosity.

The porosity as well as the permeability is evaluated at our facilities by means of the automatic porosimeter and permeameter COREVAL 700. The device's method of work is based on API recommendation [5] which uses the *Boyle's Law Single Cell Method* for measurements of a free space. The method uses a reference cell filled with gas of a reference volume and pressure which is afterwards released into the pore volume of a given sample. The sample is placed into a core-holder and is fastened in an elastic sleeve which induces a confining (lithostatic) pressure. The whole experiment is isothermal. The determined value is the open porosity which is the ratio of the bulk sample volume to the volume of interconnected pores including dead-end pores.

The core samples porosity measurements show a variance of 0.1 cm^3 for a 50 cm^3 volume sample, the porosity margin is ± 0,2 % of the real value.

The measured parameters are:

- pore volume V_p (cm^3),
- sample porosity φ (%),
- bulk volume V_b (cm^3),
- grain volume V_g (cm^3),
- grain density Gd (g/cm^3),
- gas permeability K_g (mD),
- slip factor b (psi),
- initial resistance β (ft^{-1}),
- turbulence factor α (μm).

It is convenient to determine the above mentioned parameters at expected reservoir pressures. Next, it is necessary to determine a hysteresis (with an appropriate step – at minimum 6 values) up to the pressure that is 15 % higher than the expected reservoir pressure. Last but not least, it is desirable to determine an extreme hysteresis (at an appropriate step – at minimum 6 values) for approximately three times the value of the expected reservoir pressure.

__Permeability__ is a property of a porous medium and is a measure of its ability to transmit fluids. The reciprocal of permeability represents the viscous resistivity that the porous medium offers to fluid flow when low flow rates prevail.

A transient pressure technique for gases: Pressure-Falloff, Axial Gas Flow measurements [5]. Transient measurements employ fixed-volume reservoirs for gas. These may be located upstream of the sample, from which the gas flows into the sample being measured. The pressure falloff apparatus (Fig.1) employs an upstream gas manifold that is attached to a sample holder capable of applying hydrostatic stresses to a cylindrical plug of diameter D and length L. An upstream gas reservoir of calibrated volume can be connected to the calibrated manifold volume by means of a valve. Multiple reservoir volumes are used to accommodate a wide range of permeability values. The downstream end of the sample is vented to the atmospheric pressure. An accurate pressure transducer is connected to the manifold immediately upstream of the sample holder. The reservoir, manifold, and the sample are filled with gas. After a few seconds for thermal equilibrium, the outlet valve opens to initiate the pressure transient. The pressures and times are recorded. This technique has a useful permeability range of 0.1 to 5000 milliDarcys (mD).

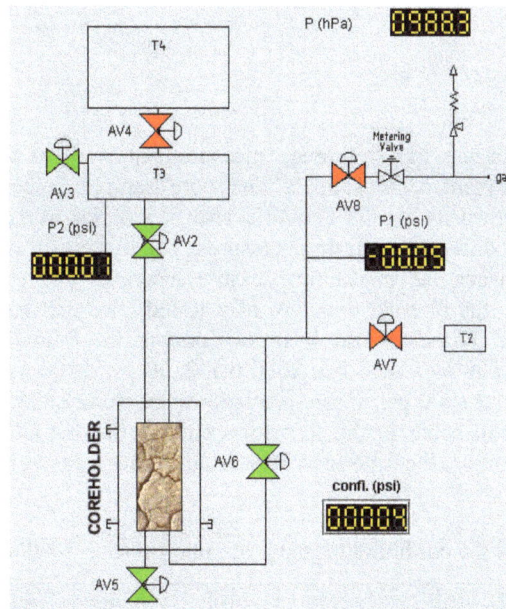

Fig. 1 The scheme of COREVAL 700 apparatus
(T2-4: Reference chambers; P,P1,P2,confi.: manometers;
AV1-8: valves)

Fig. 2 Tested core

3.2 Laboratory measurements of chosen parameters of core samples on Automatic porosimeter-permeameter

To understand the behaviour of rock massif considered for the CO_2 storage, real core samples were selected. Their petrographic composition corresponds to the potential storage formation. They are fine-grained sandstones with addition of clay. They were drilled out from one larger core perpendicularly to its axis. The parameters of the samples were (fig. 2):

- diameter – 38,4 mm,
- length – 68,2 mm,
- initial weight – 156,56 g.

The confining pressure of the first reference measurement was set to 1000 psi (6.895 MPa). The measurement outcomes are presented in Tab. 1. The core sample before inserted into the Automatic porosimeter-permeameter is shown in Fig. 2. The objective of a series of subsequent measurements was to determine hysteresis curves for different confining pressures. The hysteresis curves tell us how differently the measured parameters change when the confining pressure increases and subsequently decreases. Another objective was to ascertain how the internal structure of a tested core will look like after repeated rises and drops of the confining pressure. To portray the hysteresis curves, the following pressure steps were chosen: 1000 psi, 1200 psi, 1400 psi, 1600 psi, 1800 psi, 2000 psi, 2100 psi, 2200 psi, 2100 psi, 2000 psi, 1800 psi, 1600 psi, 1400 psi, 1200 psi, and 1000 psi. It is a pressure range from 6,895 MPa to 15,168 MPa. These are pressures that can be expected to occur in the formation suitable for the CO_2 storage. Because of assumed shape memory of a measured sample, the time span between different measurements was at least 24 hours.

Tab. 1 Measurement outcomes at the confining pressure of 1000 psi (6,895 MPa)

confining pressure	pore volume	porosity	bulk volume	grain volume	bulk density	grain density	confining pressure	air (N_2) permeability
Pc	Vp	φ	Vb	Vg	Bd	Gd	Pc	K [air]
(psi)	(cm^3)	(%)	(cm^3)	(cm^3)	(g/cm^3)	(g/cm^3)	(psi)	(mD)
1000	19.9167	25.2958	78.7226	58.8445	1.9789	2.6620	1000	41.1791

Hysteresis curves determination for chosen measurement parameters

The hysteresis curves were determined based on the measurement of chosen parameters in a pressure range from 6.895 MPa to 15.168 MPa (between 1000 psi and 2200 psi; **1 MPa = 145.0377 psi**) for the above mentioned pressure steps. After the first measurement, the weight of the core sample was 156.45 g. The weight loss after the first set of measurements was 0.11 g. The measurement outcomes are documented in the following charts for different measured parameters. During the measurement, the porosity and permeability of samples generally decrease up to the point of the highest used confining pressure and subsequently increase in general during the release of the confining pressure. In each case, the values of porosity and permeability are higher when determined at the starting pressure (1000 psi) than the values determined at the same pressure after the confining pressure was adjusted to the maximum value (2200 psi) and released to the value of 1000 psi.

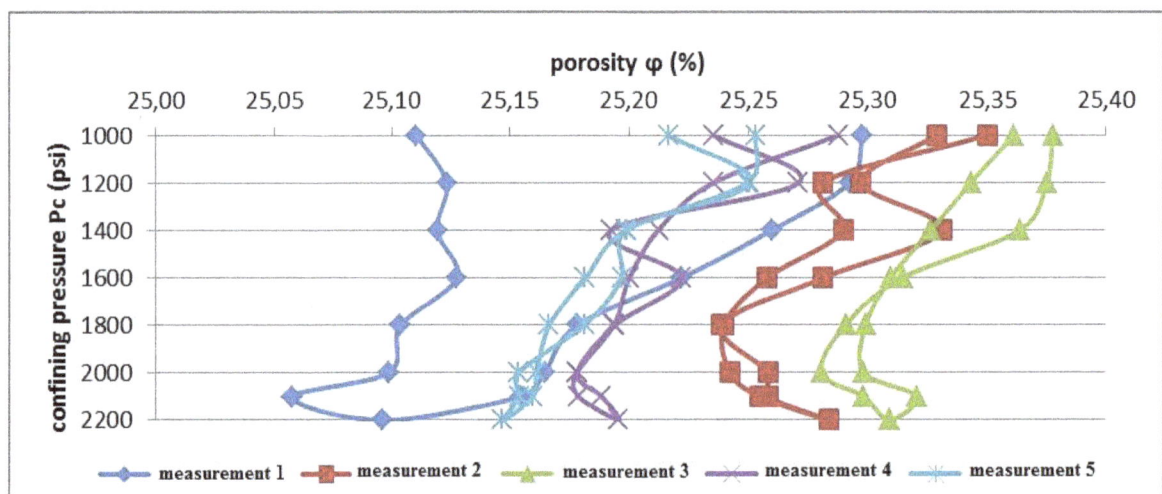

Fig. 3 Hysteresis curves of porosity for different pressure steps

Ad Fig. 3:

The comparison of the first measurement with other four measurements tells us that the internal structure of the sample had undergone significant changes. In absolute numbers, the changes are negligible in the order of one hundredth of a percent (the average value of porosity at 1000 psi is 25.2880 % before the pressure started to rise and 25.2756 % after it dropped; at a maximum pressure of 2200 psi, the porosity was 25.2062 %). Nevertheless, the shape of the curves confirms that after several measurements, changes in the internal structure took place, which indicates at least some damage of the internal structure. It can be assumed that during the last two measurements, a partial internal sample consolidation took place - proportionally between effective and closed porosity.

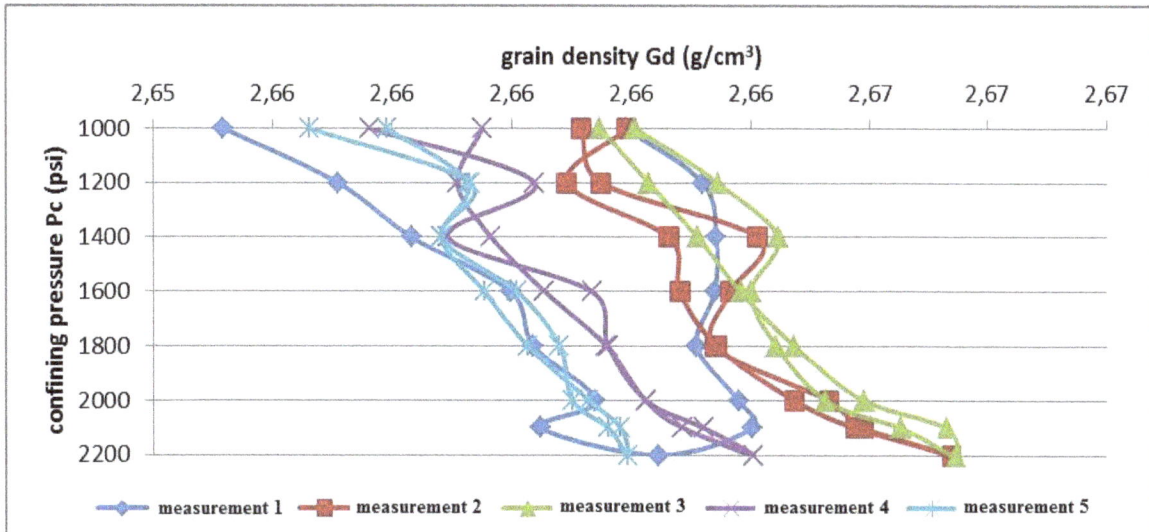

Fig. 4 Hysteresis curves of grain density for different pressure steps

Ad Fig. 4:

Logically, the grain density Gd is higher than the bulk density Bd, due to the low density of fluids filling the pores. Within the whole series of measurements (from 1 to 5), the grain density was changing on a second or third decimal position, therefore the change is negligible. The average grain density of the core sample was found to be 2.6620 g/cm^3. An interesting fact is that the grain density Gd curves shape corresponds with the shape of the grain volume Vg curve and partially with the porosity φ curve.

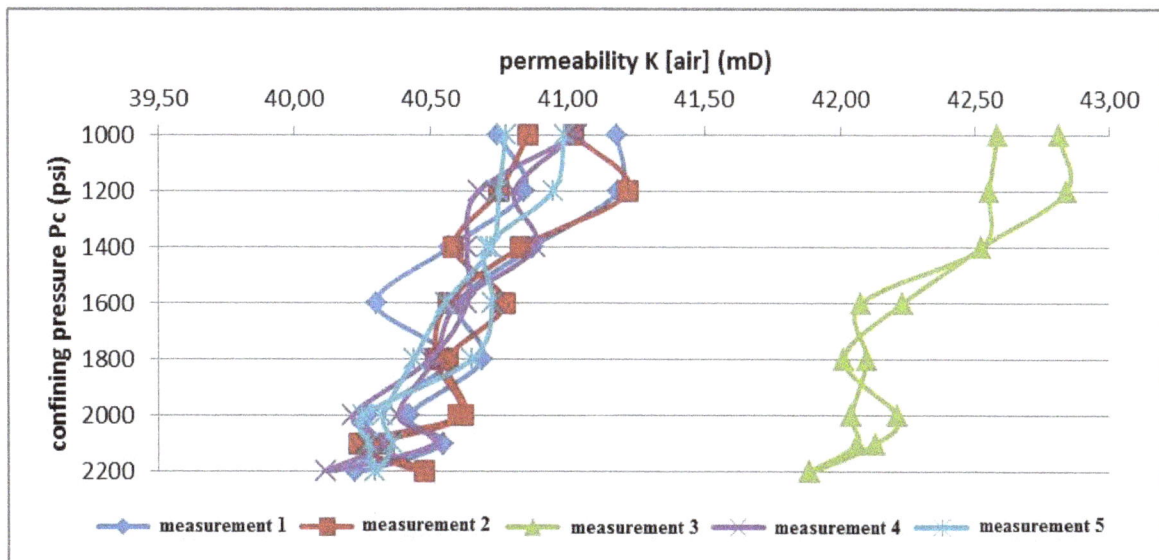

Fig. 5 Hysteresis curves of air (N_2) permeability for different pressure steps

Ad Fig. 5:

The air permeability (medium – N_2) showed itself to be consolidated and, if the measurement no. 3 is not taken into account, it was found to be 40.6139 mD. The measurement no. 3 shows permeability values at least 1 mD higher than the rest of the measurements. It is probably due to the fact that during this measurement, the greatest internal structure changes, which caused the higher permeability, took place. After a partial consolidation the permeability has settled at primeval values.

4 CONCLUSIONS

During the laboratory experiments, a set of cores from a single borehole was tested. Charts 1 and 3 show the most interesting results. And even though, at first glance, these samples from a similar depth look the same, they shown not only different petrophysical results, but also implied possible complications in the internal structure of reservoir rocks considered for the CO_2 storage. For the *porosity*, the difference between the examined cores was *as much as 3 %*, while the difference of the *permeability* was more than *40 mD*. At the same time, the *grain density* was the same for all the cores (*2,64 – 2,66 g/cm³*), it varies on the second or third decimal position. The measurements on these cores indirectly indicated that even though the confining pressure was relatively low (6.895 MPa to 15.168 MPa), the internal structure suffered some damage. It is reflected in hysteresis curves (charts 1 to 3). A "bent V" shape of the curves, with a shape memory, was expected. Only the first measurement approximately resembles this shape, four others are much more chaotic. It is likely that after every pressure step, the deformation of porous space took place, in a way that the effective pore space was squeezed so much that it prevented the flow of measuring medium (N_2). The assumption of internal structure deformation is backed by an evidence of visual reconnaissance of the core surface. The cracks, some more than 0.5 mm deep, appeared on the core's surface. This phenomenon is known from cyclical operations of underground gas storage sites, where due to the thermal and pressure changes, micro-particles of the rock matrix are crumbled away and form a so called "silt cloud". The outcome of the measurement confirms the non-homogeneity of the geological environment. Even though, the samples are from a single borehole from approximately the same depth, the results vary substantially.

The following research works will focus on ascertaining the CO_2 phase permeability in a supercritical state (7.38 MPa and 31.1°C). For this task, another laboratory device of the Laboratory of Wells and Hydrocarbon Deposits Stimulation will be used. It is the BRP 350 multiphasic permeameter made by Vinci Technologies (France).

REFERENCES

[1]   Bethke C.M., 2008, Geochemical and biogeochemical reaction modelling. Cambridge Univ. Press, Cambridge: 1-543.

[2]   Labus, K., Bujok, P.. CO_2 mineral sequestration mechanisms and capacity of saline aquifers of the Upper Silesian Coal Basin (Central Europe) - Modeling and experimental verification. Energy. 2011, vol. 36, issue 8, s. 4974-4982. DOI: 10.1016/j.energy.2011.05.042.

[3]   Palndri J.L., Kharaka Y.K., 2004, A compilation of rate parameters of water-mineral interaction kinetics for application to geochemical modeling. US Geological Survey. Open File Report 2004-1068: 1-64.

[4]   Xu T, Apps JA, Pruess K (2003) Reactive geochemical transport simulation to study mineral trapping for CO_2 disposal in deep Arenaceous Formations. J. Geophys. Res., 108: B2.

[5]   AMERICAN PETROLEUM INSTITUTE. Recommended Practices for Core Analysis [online]. USA: API Publishing Services, 1998 [cit. 2013-05-13]. Recommended practice: 40, 2nd ed. Dostupné z: http://w3.energistics.org/RP40/rp40.pdf

ACKNOWLEDGEMENT

The article has been made in connection with the project of the Institute of Clean Technologies for Mining and Utilization of Raw Materials for Energy Use, no. CZ.1.05/2.1.00/03.0082, supported by the Research and Development for Innovations Operational Programme financed by Structural Founds of the European Union and from the state budget of the Czech Republic.

RESUMÉ

V průběhu laboratorního výzkumu byla otestována řada vzorků vrtných jadérek z jednoho návrtu. Grafy č. 1 až č. 3 ukazují nejzajímavější výsledky. A i když se jednalo o na první pohled stejné vzorky z přibližně stejné hloubky, vykázala testovaná jádra nejenom odlišné fyzikálně petrografické výsledky, ale naznačila i možné komplikace v oblasti vnitřní struktury nádržních hornin pro uskladňování CO_2. U *pórovitosti* je sice rozdíl mezi oběma testovanými jádry *až 3%*, ale při srovnání s *permeabilitou* je rozdíl mezi oběma jádry i více než *40 mD*. Přitom **hustota rostlé části** je pro všechna testovaná jádra takřka stejná (*2,64 – 2,66 g/cm³*), liší se maximálně na druhém až třetím desetinném místě. Měření u tohoto konkrétního jadérka nepřímo naznačila, že i když se jednalo o relativně nízké tlakové hodnoty (6,895 MPa až 15,168 MPa), došlo k poškození vnitřní struktury. Svědčí o tom průběh hysterezních křivek (grafy č. 1 – č. 3). Byl očekáván hladký průběh ve tvaru „vyhnutého V" s předpokládanou tvarovou pamětí. Tohoto stavu pouze přibližně dosáhlo první měření, další čtyři měření již vykázala na první pohled chaotický průběh. Nejspíše při každém tlakovém projevu (Pc) došlo k deformaci vnitřního pórového prostředí, kdy se v určitou chvíli měnilo efektivní pórové prostředí na uzavřené nebo polouzavřené pórové prostředí a tím znemožňovalo nebo výrazně ovlivňovalo průběh prostupu měřícího média (N$_2$). Předpoklad o deformaci vnitřní struktury potvrzuje i vizuální ohledání vnějších stran jadérka. Je patrné poškození vnější strany jádra, kdy některé trhliny vykazují hloubku i více než 0,5 mm. Tento jev je znám z cyklického provozu PZP, kdy vlivem tlakových a teplotních změn dochází k „vydrolování" mikročástic z matrixu horniny a vzniku tzv. siltového mraku. Výsledek měření tak potvrzuje fakt nehomogenity geologického prostředí. I když se jedná o jeden návrt, tedy přibližně stejnou hloubku, jsou výsledky často odlišné a proměnlivé.

Další část výzkumu bude zaměřena na stanovení fázových propustností CO$_2$ za superkritického stavu (tedy tlaku 7,38 MPa a teplotě 31,1°C). K tomu bude využito dalšího laboratorního přístroje Laboratoře stimulace vrtů a ložisek uhlovodíků. Jedná se o fázový permeametr BRP 350 firmy VINCI Technologies (Francie).

13

DETERMINATION OF ELEVATOR SHAFT UPRIGHTNESS APPLYING THE TERRESTRIAL LASER SCANNING METHOD

Ľudovít Kovanič [1]

[1] *Ing., PhD., Institute of Geodesy, Cartography and GIS, BERG Faculty, Technical University of Košice, Letná 9, Košice, Slovak Republic*
e-mail: ludo.kovanic@tuke.sk

Abstract

This paper presents the results obtained from geodetic measurements and processing the data with the objective to determine geometrical parameters of an elevator shaft applying classical as well as modern approaches for obtaining the measured data. The intention was to verify the possibility to apply the terrestrial laser scanning (TLS) method as a suitable, efficient and precise method for collecting spatial data.

Abstrakt

V príspevku sú prezentované výsledky geodetického merania a spracovania údajov na určenie geometrických parametrov výťahovej šachty použitím klasických a tiež moderných prístupov získavania meraných dát. Zámerom je overenie možnosti využitia terestrického laserového skenovania ako vhodnej, efektívnej a detailnej metódy pre zber priestorových údajov.

Key words: elevator shaft, uprightness, terrestrial laser scanning (TLS)

1 INTRODUCTION

The activities to be accomplished by an authorized surveyor and cartographer in the process of building may be divided into several stages as follows:

- stage of background documents and project documentation preparation
- stage of building execution
- stage of building completion, approval and operation

Building objects differ in particular in shape and dimensions, as well as in type of construction. In all construction stages, the stress is placed on the accuracy of building object construction components for the building deviations should be less than the tolerances stated in standards or building project documentation.

In the building execution stage, the important part is to construct such staking grid of the building which meets, by its characteristics of positional and height accuracy, the requirements for the resultant accuracy of the detailed staking. The accuracy of staking alone depends on applied methods and the accuracy of instrumentation.

The requested accuracy of staking works is given as follows:

- directly – by maximum staking deviations of building objects which are given for example by standards,
- indirectly – by building deviations, from which the staking deviations are derived. As a rule, the building deviations are stated in technical standards depending on the type of construction of the building object.

When deriving staking deviations from building deviations, assuming that the effects of errors in staking, building and assembly works are of an equal value, the building deviation U_S is applied, which is considered to be the maximum error for the position of the point m_P. With the applied confidence coefficient t = 2, it holds:

$$U_S = t.m_P = 2.m_P, \tag{1}$$

while:

$$m_P^2 = m_V^2 + m_m^2, \tag{2}$$

where:

m_V - effect of errors in staking,

m_m - effect of errors in building and assembly works.

If $m_v = m_m$ than for the building deviation, it holds:

$$U_S = 2.\sqrt{\left(m_V^2 + m_m^2\right)} = 2.\sqrt{m_P^2} = 2,8.m .$$

(3)

The requested accuracy of staking works is then:

$$m_V = m_m = \frac{U_S}{2,8} = 0,36.U_S ,$$

(4)

It follows from the above that the requested accuracy of staking and building and assembly works should not exceed 36 % of the building deviation. To calculate the accuracy, it is necessary to stress that:

$$m_V^2 = m_1^2 + m_2^2 ,$$

(5)

where:

m_1 - effect of errors in staking grid

m_2 - effect of staking errors, including e.g. the instrument centralisation on site, the effect of measurement of the instrument height error and the height of target in trigonometry measurements of heights and so on.

The accuracy of staking, building and assembly works is based on the equations (1) through (5) and must correspond to the building deviation which is given by the standard STN EN 13670 (732400) – "Execution of concrete structures" of 2010. The numerical value of the building deviation (tolerance) is ±25mm for the position of pillars and walls with regard to the straight line of the secondary top view scheme. Hence the resultant requested accuracy of staking works for the confidence coefficient $t=2$ is about ±5mm.

During the building completion stage, the surveyor's major activities represent in particular checking measurements, and after implementation, setting the meterage for preparing as-built design documents. The objective of the geodetic part of as-built design documents is to determine positioning, shape and dimensions within binding (national) geodetic systems after the execution. The accuracy of measurements for the purpose of as-built design documents corresponds as a rule to the accuracy of staking. In case of the high rise buildings, the geodetic meterage of elevator shaft uprightness is its integral part along with the determination of deviations of the actual building object work in comparison with project documentation, or possibly with a relevant technical standard. To obtain the spatial data on the building object, it is possible to exploit several methods and procedures differing in accuracy, tediousness of measurement and availability of suitable measuring instruments.

2   BUILDING OBJECT, METHODS AND GOALS OF MEASUREMENTS

The object of experimental measurements was SO-03 Dwelling house HB3 at the construction "Košice bývanie - 1. etapa " at Ondavská street in Košice, in the stage of shell construction of monolithic reinforced concrete construction completion (Fig.1).

Fig. 1 SO - 03 Dwelling house HB3 at the construction of "Košice bývanie -1.etapa"

The horizontal dimensions of the shell construction of the dwelling house are 21.70 m x 21.30 m, the construction height of the plate of the first storey is – 0.15 m (239,25 m asl) and that of the eighth storey is + 20.68 m (260.08 m).

The elevator shaft, which was the object of measurements, is located in the object core from the first up to eighth storey. It consists of carrier monolithic reinforced concrete walls with an entry opening at the west side. The construction dimensions of the shaft are 2.650 m x 1.650 m (Fig 2).

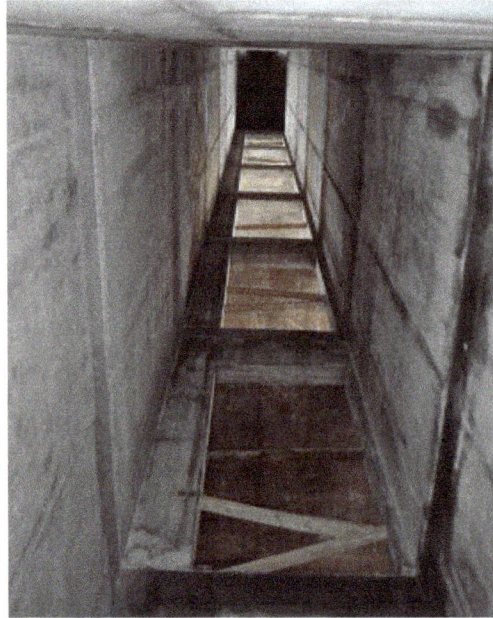

Fig. 2 Object of measurements

At the time of measurements, the elevator shaft was ready to be handed over. It means that all sharp edges were smoothed at the contact between individual sections of casting, and its formwork was completely removed, including the temporary wooden safety and working floors at all the storeys. This fact was decisive for the selection of measuring method.

2.1 Methodology of measurements

The methodology of measurements of elevator shaft uprightness was selected with regard to the requested accuracy and feasibility of measurements under actual conditions.

1. **Method of direct measurements of dimensions and distances of shaft walls from a fixated plummet** as the simplest measuring method. The plummet with a weight of about 1 kg was placed to a shaft wall corner of the elevator shaft on a bar suspension at the eighth storey of the object (Fig .3). It was fixated by immersing it in a tank filled with oil at the level of the first storey. The length of the hanging was about 25.5 m. The dimensions of the elevator shaft and the distance of the walls from the plummet hanging were measured directly using a measuring tape, and in case of non-accessible locations, using the manual laser telemeter Leica Disto A5. The average error of length measurements, declared by the instrument manufacturer, is $m_d = \pm 1,5$mm. The dimensions and the distances were measured at each storey at a height of 1.40 m above the level of the entry plane of the elevator.

Fig. 3 Plummet hanging

2. **Polar method** with exploiting the UMS Leica TCRA 1201+, the average measuring error of vertical and horizontal angles $m_\alpha=m_z=\pm1''$, the mean error of length measurements with a prism $m_d=\pm1mm+1,5ppm$ and without a prism $m_d=\pm2mm+2ppm$. The positions of the instrument were selected for each storey to be possible to measure the maximum number of points along the shaft circumference at a height of 1.40 m above the level of the entry plane of the elevator. During the measurements, the method of subordinate centring on tripods was applied. The determination of coordinates at each storey was executed in the coordinate system of a staking grid of the construction (S-JTSK). For the backward orientation of measurements always the same point at a distance of about 70 mm was used. At each storey, beside the visible edges without a prism, also the shaft walls were measured. The edges, which could not be measured directly from the site as they were in the obstruction of the walls, were constructed in the prolongation of the points on the walls as intersections of approximated lines of the flow-lines of the measured points. For the purpose of checking, also the plummet hanging was positional measured at each storey with the use of a reflection foil (Fig.8) and later on with the use of a 360° mini prism (Fig.9).

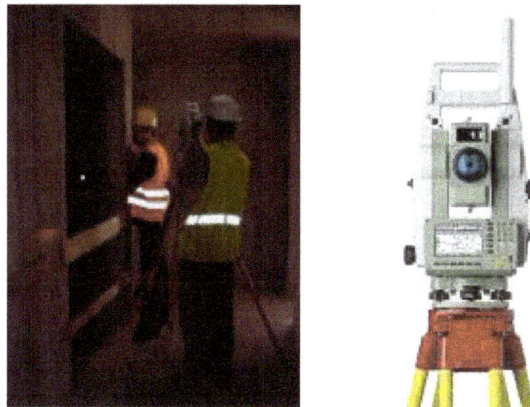

Fig. 4 - TCRA 1201+

3. **Method of terrestrial laser scanning** with the use of TLS Leica ScanStation C10 (Fig.5). The mean error of determination of an individual point position $m_p=\pm6mm$, the mean error of length measurements $m_d=\pm4mm$ (for the lengths up to 50 m), the mean error of measurements of horizontal and vertical angles $m_\alpha=m_z=\pm12''$, the accuracy of the modelled surface $m_m=\pm2mm$. The instrument is furnished with a two-axis compensator with the accuracy of $\pm1,5''$ and a range of $\pm5'$ (Leica ScanStation C10 – data sheet). The measurements were conducted at the same time from one position at the lowest storey of the building object at the bottom of the elevator shaft. The horizontal and vertical density of the measured points was adjusted to 5mm per each 25m of length. The TLS method is an up to date and efficient method for collecting spatial data on objects (Gašinec et al., 2011, Gašinec et al., 2012).

Fig. 5 - TLS Leica ScanStation C10

2.2 Definition of objectives

The measurements of uprightness of an elevator shaft can be executed applying various methods, depending on available instrumentation, configuration and object accessibility, as well as requested accuracy (Erdélyi, et al., 2011, Erdélyi, et al., 2012).

The main objective was to confirm the method of terrestrial laser scanning as a suitable method from the point of accuracy and efficiency in comparison with the conventional measurement methods described in chapter 2.1.

The a priori estimation of the accuracy of this method results from the accuracy parameters of the applied instrument, when based on the expected accuracy of the modelled surface, the dimensions at a level of $m_m = \pm 2mm$ were determined.

The partial tasks, with regard to the fact that the measurements were conducted from the bottom of the shaft whose height exceeded 20 m, were as follows:

1. Verify the accuracy effect of the compensator used for determining the uprightness

2. Verify the effect of the measured ray incidence under an acute angle at top storeys to determine the uprightness and the dimensions of the shaft compared to the checking direct measurements.

3   MEASUREMENTS AND PROCESSING THE RESULTS

The numerical and graphical processing of results for individual methods was carried out individually. **The direct measurement of dimensions and distances with a fixated plummet** was evaluated for each storey individually; the elevator shaft horizontal sections were drawn, and the measured values were inserted. The example of numerical and graphical evaluation for the first storey is given in Fig. 6

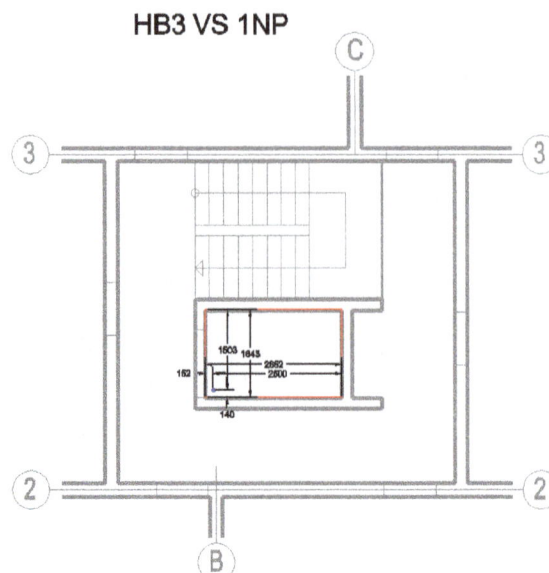

Fig. 6 Numerical and graphical evaluation of the results from direct measurements using the hanging plummet

The processing of measurements, applying the polar method using UMS Leica 1201+, was conducted through the comparison of the measured rectangular coordinates within the coordinate system of the staking grid of the building (S-JTSK), applying the method of laser scanning in a local coordinate system. The coordinates of inside points of the elevator shaft 1 through 4 were compared (Fig.7). These were numerically distinguished according to individual storeys, for example 101-104 for the first storey, 201 - 204 for the second storey and so on.

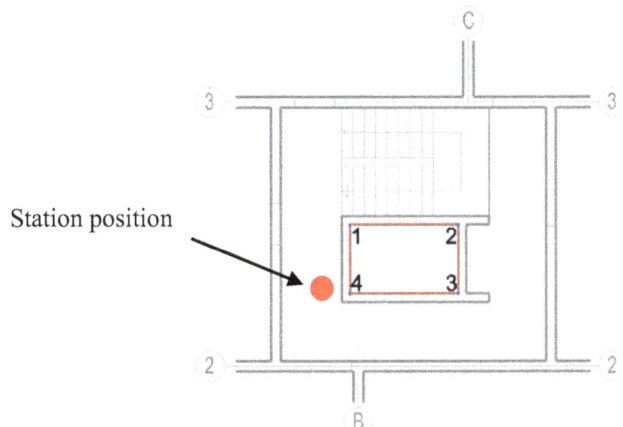

Fig. 7 Designation of observed points

For the purpose of measurements, applying the **polar method using UMS Leica 1201+**, about 15 points were measured at each storey on inside walls of the shaft. The observed corners 2 and 3 were measured directly and the points 1 and 4, with regard to the position of the site outside the elevator shaft, were not measurable directly (Fig.7). These were constructed as intersections of lines led through the points measured on the inside walls of the shaft at a height of 1.40 m above the level of individual storeys. The plummet hanging was multiply focused at each storey as the controlling element to disclose the measurement errors. To determine its coordinates, measuring aids, inspired by the facility for the centric positioning of reflective prism on the hanging wire of the plummet, were experimentally exploited during connecting and regulation measurements under mining conditions exploited by the authors (Černota et al., 2011, Černota et al., 2012), as follows:

- The foil reflector with dimensions of 3 x 3 cm (Fig. 8) fixed on the plummet hanging by an adhesive tape. During the measurements, the adverse effect of its rotation along with the plummet hanging was observed, resulting in the prolonged period needed to its stabilisation when manually rotated into the direction perpendicular to the object,

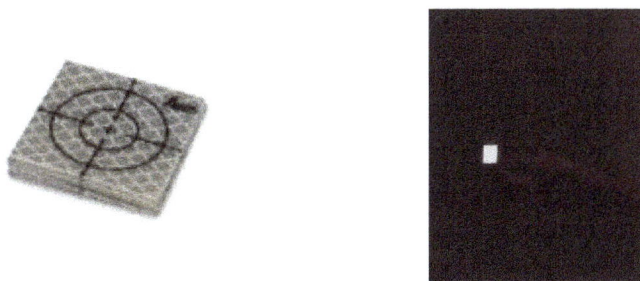

Fig. 8 Foil reflector fixed on plummet hanging

- The instrument 360° mini-reflector Leica GRZ 101 (Fig.9). Its advantage is the possibility of its centric positioning on the plummet hanging; movement to other height levels without the need to be disconnected using fixing screws; as well as the elimination of its rotation effect on the possibility to execute the measurements. The system ATR for automatic targeting the prism was exploited.

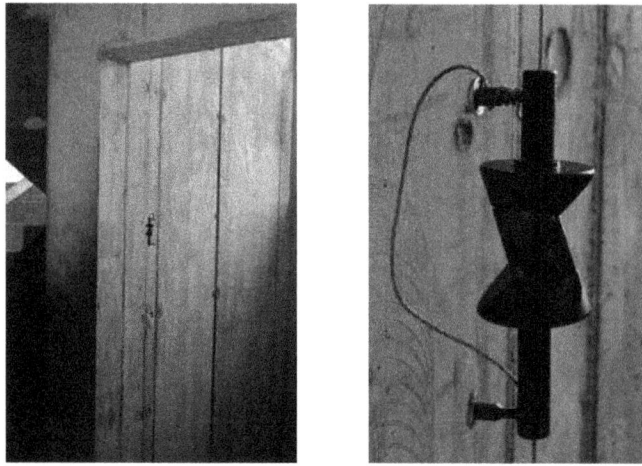

Fig. 9 360° mini-reflector Leica GRZ 101 fixed on plummet hanging

The evaluation of the relationship may be carried out based on the simple difference in coordinates of individual corners of the elevator shaft on condition that the directions of its walls are parallel with the direction of the applied coordinate system (most frequently the local one). In case of other coordinate system (S-JTSK), it is suitable to apply the analytical solution for determining the horizontal distance of the observed point from the vertical plane incorporating two points in 3D space. In this case, the points were located on the first storey of the object – the flow-line of the points 101 and 102 for the coordinate X, the flow-line of the points 102 and 103 for the coordinate Y, and so on.

3.1 The analytical solution for determining the horizontal distance of an observed point from the vertical plane incorporating two points in 3D space

The calculation of the distance between the point and the plane in analytical geometry consists of two steps:

- determine the general equation of the plane

- calculate the distance between the point and the plane

The plane in 3D space is expressed by the equation:

$$ax + by + cz + d = 0 \qquad (6),$$

For the purpose of determining a deviation from the vertical plane, the reference plane is defined, which incorporates two points A, B and is upright. The vector $\vec{u} = A - B$ is determined from the difference in coordinates of the above points. The vector $v = (0,0,1)$ is located on the vertical plane; it means that it is parallel with the axis Z. It is possible to define the reference plane in the direction of any coordinate axis or a general plane, when altering parameters of the vector. The general equation of a plane is unambiguously defined by one point located in the plane and a normal vector. The normal vector of the sought plane ρ may be determined by the vector product of the vectors:

$$\vec{u} \times \vec{v} = \vec{n} \qquad (7),$$

The coordinates of this vector are calculated from:

$$n_1 = u_2 v_3 - u_3 v_2$$

$$n_2 = u_3 v_1 - u_1 v_3 \qquad (8),$$

$$n_3 = u_1 v_2 - u_2 v_1$$

Then the coefficient d is calculated from the equation (6).

All tasks for the calculation of the distance between points, straight lines and planes are transferred into the task to determine the distance between two points. In case of the point and the plane, the formula for such distance calculation is derived. The plane ρ is specified by the general equation (6) and $A[x_0, y_0, z_0]$ which is not located within the plane ρ. The distance between the point A and the plane ρ is equal to the distance between the point A and the foot of the perpendicular led from the point 1 to the plane ρ.

The vector $\vec{u} = (a, b, c)$ is perpendicular to the plane ρ and parallel with the straight line k.

The parametric equations of the straight line k are as follows:

$$x = x_0 + ta$$

$$y = x_0 + tb \qquad (9),$$

$$z = z_0 + tc$$

It holds for the foot of the perpendicular:

$$a(x_0 + ta) + b(y_0 + tb) + c(z_0 + tc) + d = 0 \qquad (10),$$

after the modification the following is valid:

$$ax_0 + by_0 + cz_0 + t(a^2 + b^2 + c^2) + d = 0 \qquad (11),$$

because $a^2 + b^2 + c^2 \neq 0$, it can be written:

$$t = -\frac{ax_0 + by_0 + cz_0 + d}{a^2 + b^2 + c^2} \qquad (12),$$

After the substitution for t, the coordinates of the foot of the perpendicular are obtained:

$$x = x_0 - \frac{a(ax_0 + by_0 + cz_0 + d)}{a^2 + b^2 + c^2}$$

$$y = y_0 - \frac{b(ax_0 + by_0 + cz_0 + d)}{a^2 + b^2 + c^2} \qquad (13),$$

$$z = z_0 - \frac{c(ax_0 + by_0 + cz_0 + d)}{a^2 + b^2 + c^2}$$

The distance between the point and the plane is obtained from the difference in coordinates of the measured point and the foot of perpendicular to the plane:

$$v = \sqrt{(x - x_0)^2 + (y - y_0)^2 + (z - z_0)^2} \qquad (14),$$

After the substitution and modifications, the following is valid:

$$v = \frac{ax_0 + by_0 + cz_0 + d}{\sqrt{a^2 + b^2 + c^2}} \qquad (15),$$

The vector contains the distance between the measured point and the vertical plane (Gašinec et al., 2012).

3.2 Processing the data measured by TSL Leica ScanStation C10

The measurements were carried out with the parameters stated in Chap. 2.1. The measured data were in advance adjusted into the text set within the structure Y, X, Z – the spatial rectangular coordinates of measured points; I – the intensity of a reflected signal; R,G,B - the colour scale (Tab.1), and subsequently imported into the selected software application.

Tab.1 The structure of the list of measured points coordinates in a text form

Y	X	Z	I	R	G	B
264253.293	1226666.361	349.990	82	114	107	65

The program Trimble Real Works 6.5 was the major software tool exploited for modelling the walls, in particular with regard to the option of the program to work with a large number of points. The original set was reduced through the order "Spatial sampling" so that the spatial distance between the measured points was about

1cm. It holds mainly for the closest points to the scanner position. Out of this set of points, horizontal sections were created with a width of 10cm at height levels corresponding to the measurements carried out by the previously mentioned methods, i.e. 1.40 m above the level of the entry plane to the elevator on individual storeys. In such created sections, sets of measured points were created separately for each storey of the elevator shaft. The number of points in a section of each storey was about 2000. The fragments of the walls were formed individually by interlining the planes along the segmented points. The accuracy of the plane interlined by points is characterised by the parameter RMSD – a Root-mean-square deviation which is the mean square deviation of the distance between the points and the plane interlined through those. In case of all segments of the planes, its value was up to 1.5 mm. The corner points of the elevator shaft at each storey were constructed as intersections of these planes. The created 3D model was exported to CAD MicroStation V8 when the point markers were allocated to individual points of the shaft to obtain the list of coordinates for further evaluation (Fig.10 and details in Fig.11)

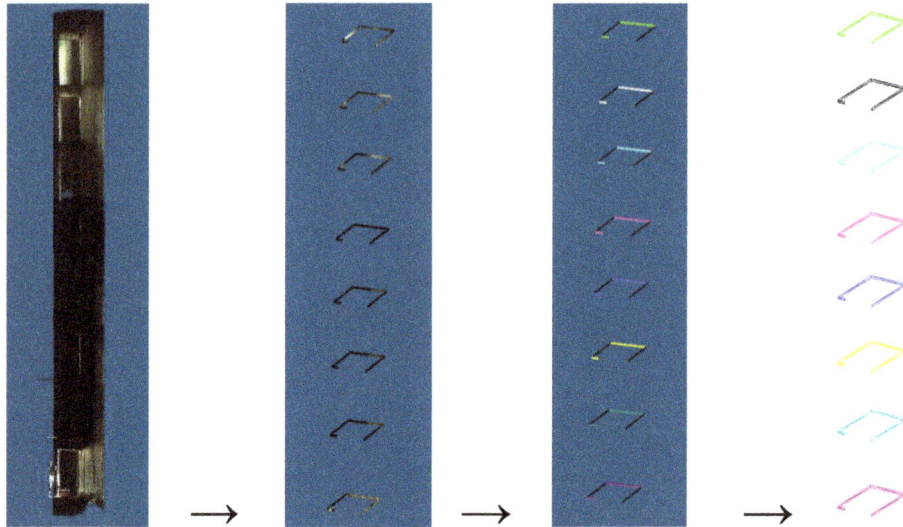

Fig. 10 Processing the data measured by TLS Leica ScanStation C10

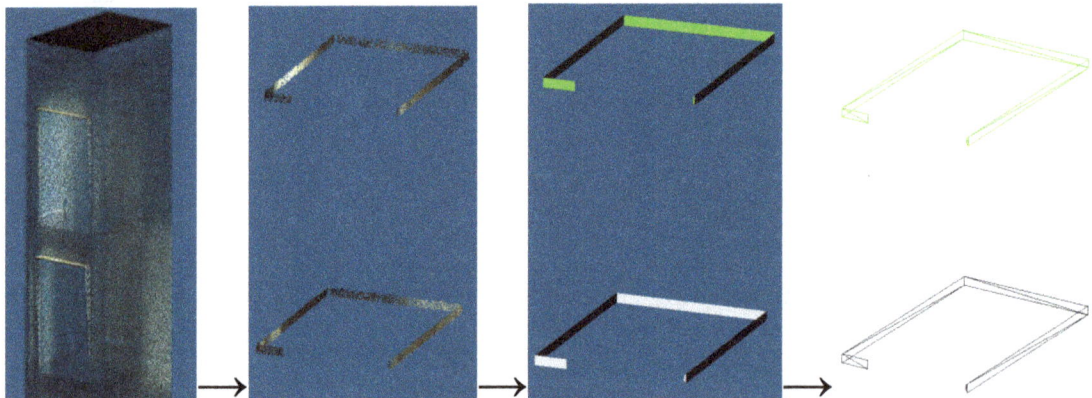

Fig. 11 Processing the data measured by TLS Leica Scan Station C10 - details

3.3 Numerical evaluation

The differences in uprightness for individual methods were estimated from the differences in the coordinates of the corners of the elevator shaft for UMS and TLS, and from the measured length differences for the method of direct measurements. The position of the corners of the elevator shaft on individual storeys in all applied procedures was compared with the position on the first storey of the object, which was selected as the starting one. Tab. 2 provides the numerical comparison of the results of uprightness determination by all methods and the differences in the position coordinates of the individual corners of the elevator shaft.

Tab. 2 The comparison of the results of elevator shaft uprightness determination

	Storey	Elevator shaft uprightness measured with hanging plummet		Elevator shaft uprightness measured with UMS Leica 1201 +		Elevator shaft uprightness measured with TLS Leica ScanStation C10	
		Δy [mm]	Δx [mm]	Δy [mm]	Δx [mm]	Δy [mm]	Δx [mm]
Corner 1	1	0	0	0	0	0	0
	2	4	8	6	13	7	8
	3	7	14	6	17	8	15
	4	11	18	10	21	15	19
	5	12	14	-	-	15	14
	6	10	0	9	5	13	0
	7	7	0	9	2	9	0
	8	7	1	9	3	13	0
Corner 2	1	0	0	0	0	0	0
	2	3	10	1	15	1	14
	3	3	12	0	16	4	14
	4	7	12	8	18	10	14
	5	10	13	-	-	11	15
	6	4	13	6	13	8	12
	7	3	7	4	4	3	6
	8	7	-3	7	-6	10	-6
Corner 3	1	0	0	0	0	0	0
	2	6	15	5	13	6	10
	3	6	8	4	11	3	5
	4	3	8	3	12	2	5
	5	8	10	-	-	7	7
	6	8	7	6	8	8	6
	7	4	-1	5	-2	5	-3
	8	-1	-6	-2	-7	-3	-10
Corner 4	1	0	0	0	0	0	0
	2	4	15	5	18	7	17
	3	4	17	0	20	2	16
	4	3	21	2	23	1	21
	5	3	18	-	-	6	17
	6	2	7	5	7	6	7
	7	3	5	3	5	2	5
	8	-6	4	-2	6	-4	5

With regard to the results from the measurement stated in Tab. 2, it is possible to claim that they differ at the level of measurement accuracy depending on individual methods. The differences in determination of uprightness of the elevator shaft are caused mainly by the accuracy of measurements alone as well as different technology of each method.

3.4 Verifying the effect of TLS compensator error on measurement results

To confirm the definiteness of the results obtained by TLS, further independent measurements with an intentionally deviated compensator of the instrument were executed. When the vertical axis of the instrument is deviated from the vertical, the deformation of the results of uprightness determination is assumed, as shown in Fig. 12.

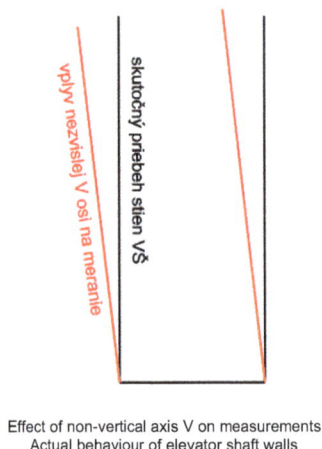

Effect of non-vertical axis V on measurements
Actual behaviour of elevator shaft walls

Fig. 12 The assumed effect of non-verticality of the instrument vertical axis on measurement results

TLS Leica ScanStation C10 is furnished with two axis balance with the accuracy of 1.5″ and a dynamic scope of ±5′. The verification of its function and effect on measurement results was carried out based on the experiment consisting in the execution of two independent scans and their comparison. In case of the first scan of the elevator shaft, the electronic builder's level was balanced as precisely as possible during the entire measurement. In case of the second one, it was declined from the central position to the limit of the compensator scope, i.e. by 5′. The coordinates of elevator shaft corners obtained from these two scans were compared.

The value of the difference in corners uprightness of the elevator shaft between the first and the eighth storey should achieve, with the stated compensator accuracy, the value according to the relationship:

$$l = \Delta h.tg\,\varphi \tag{16},$$

where:

Δh - difference in super elevation between the first and the eighth storey,

φ - angle from the vertical corresponding to the compensator accuracy

If the difference in heights between the first and the eighth storey is about 25m, excluding the consideration of further effects, this value is approximately ±0,2 mm. In case of the value of the instrument vertical axis deviation at the level of the compensator scope when not functioning, the theoretical difference in uprightness is up to 60 mm. The values of the differences in elevator shaft uprightness and the differences in two independent measurements stated experimentally are given in Tab.3. The differences in uprightness determination, found out on the basis of two independent measurements (Tab.3), reach the level of the modelled surface accuracy given by the producer, therefore the influence of the compensator on results at the required level of accuracy was not significant.

Tab. 3 The comparison of the results of uprightness determination from scans using the instrument TLS Leica ScanStation C10

		Elevator shaft uprightness, 1st measurement		Elevator shaft uprightness, 2nd measurement		Measurement differences	
		Δy [mm]	Δx [mm]	Δy [mm]	Δx [mm]	Δy [mm]	Δx
Corner 1	**Storey**						
	1	0	0	0	0	0	0
	2	7	8	6	8	1	0
	3	8	15	7	15	1	0
	4	15	19	14	19	1	0
	5	15	14	13	14	2	0
	6	13	0	12	0	1	0
	7	9	0	8	0	1	0
	8	13	0	12	1	1	-1
Corner 2							
	1	0	0	0	0	0	0
	2	1	14	-1	14	2	0
	3	4	14	2	14	2	0
	4	10	14	8	14	2	0
	5	11	15	9	15	2	0
	6	8	12	6	12	2	0
	7	3	6	2	6	1	0
	8	10	-6	8	-6	2	0
Corner 3							
	1	0	0	0	0	0	0
	2	6	10	5	8	1	2
	3	3	5	2	7	1	-2
	4	2	5	1	7	1	-2
	5	7	7	6	8	1	-1
	6	8	6	6	7	2	-1
	7	5	-3	3	-1	2	-2
	8	-3	-10	-5	-11	2	1
Corner 4							
	1	0	0	0	0	0	0
	2	7	17	5	18	2	-1
	3	2	16	3	16	-1	0
	4	1	21	2	21	-1	0
	5	6	17	6	18	0	-1
	6	6	7	7	8	-1	-1
	7	2	5	3	5	-1	0
	8	-4	5	-3	5	-1	0

3.5 Verifying the effect of the observing ray incidence angle on the determination of uprightness and shaft dimensions

The elevator shaft dimensions were determined based on the model acquired applying the TLS method and the method of direct measurement. The directly measurable dimensions at individual storeys, marked in Fig. 13 as d_1, d_2, d_3, were compared.

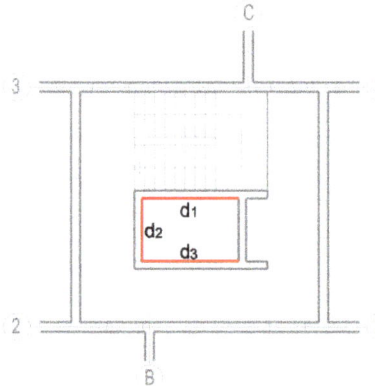

Fig. 13 Designation of the compared dimensions of the elevator shaft

The effect of the staking ray incidence under an acute angle on the measured plane with the resultant deformation of actual shaft dimensions was assumed, as shown in Fig.14.

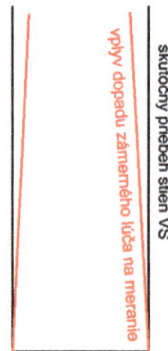

Effect of staking ray incidence on measurements
Actual behaviour of elevator shaft walls

Fig. 14 The assumed effect of the staking ray incidence under an acute angle on measurement results

The internal elevator shaft dimensions, derived from the model obtained applying the TLS method, were compared with the direct measurements at individual storeys. The results are given in Tab.4.

Tab. 4 The comparison of the results obtained from the uprightness determination from two scans applying the instrument TLS Leica ScanStation C10

	TLS			Direct measurement			Dimension difference		
Storey	d_1 [mm]	d_2 [mm]	d_3 [mm]	d_1 [mm]	d_2 [mm]	d_3 [mm]	Δd_1 [mm]	Δd_2 [mm]	Δd_3 [mm]
1	2642	1644	2645	2643	1645	2647	1	1	2
2	2648	1653	2645	2648	1652	2647	0	1	2
3	2647	1647	2645	2646	1649	1646	1	2	1
4	2647	1646	2644	2647	1648	2645	0	2	1
5	2646	1648	2645	2645	1650	2647	1	2	2
6	2647	1651	2644	2646	1649	2646	1	2	2
7	2648	1649	2643	2648	1650	2645	0	1	2
8	2645	1648	2645	2645	1649	2647	0	1	2

It is possible to determine the inclination of partial planes on individual storeys based on the results obtained from processing the 3D model of the elevator shaft and based on the angles to their normal. The elevation angles of the normal to the surfaces are presented in the graph in Fig. 15.

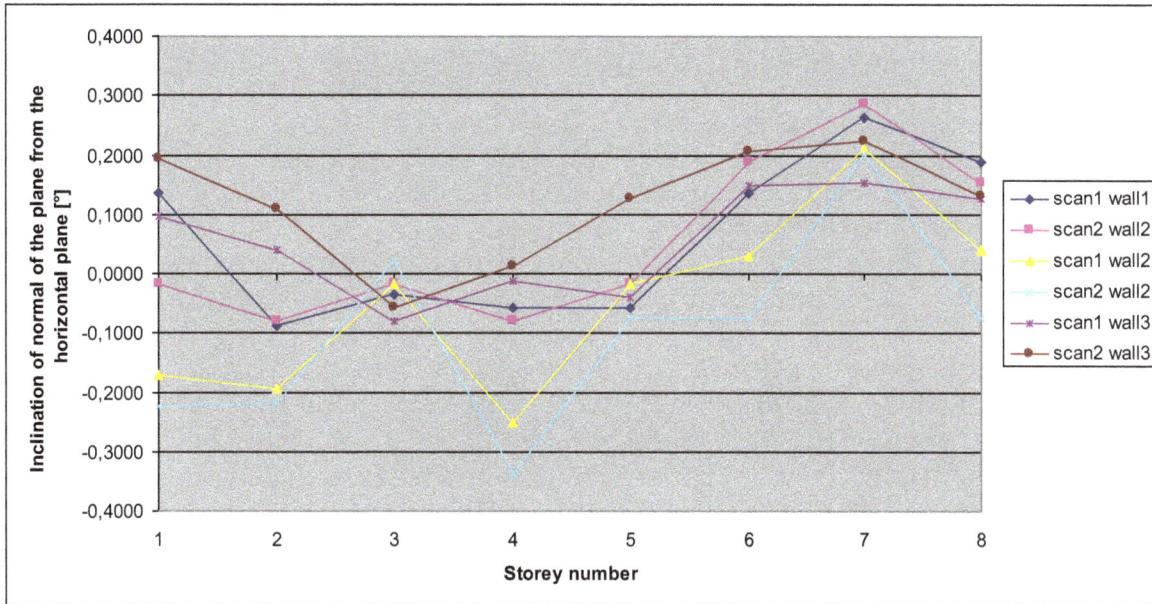

Fig. 15 The elevation angles of the normal to the surfaces from the horizontal plane from the model measured with TLS Leica ScanStation C10

4 CONCLUSIONS

The goal of the experimental measurements was to confirm the method of terrestrial laser scanning (TLS) for its application to determine the uprightness and dimensions of an elevator shaft as a suitable, satisfactory accurate, fast and detailed method as compared with conventional geodetic measuring procedures.

The evaluation of the achieved accuracy of the TLS method may be seen in Tab. 2 where the differences in determined uprightness of individual corners of the elevator shaft, set by three independent methods, are within the scope of the a priori estimated accuracies of the exploited method, the value of which is about ±3mm in all cases, as the outmost differences in uprightness differed by 5mm as maximum.

The definiteness of the results of repeated measurements exploiting the TLS method is apparent from Tab. 3 where the differences in determination of elevator shaft corners uprightness were up to ±2mm. This corresponds to the accuracy declared by the manufacturer of the modelled surfaces. In spite of the fact that in case of the second type of measurements, when an electronic builder's level was applied, intentionally deviated from the central position toward the limit of the compensator scope, i.e. by 5″ with the objective to verify the effect of the unbalanced instrument on results, the determined values of the uprightness differences differed mutually up to ±2mm.

The effect of staking ray incidence under an acute angle on the surface of top storeys on the results of uprightness of elevator shaft measurements is not significant, as can be seen in Tab. 2. According to the values provided in Tab. 3, the angle of incidence of the observing ray has no significant effect on the determination of horizontal dimensions of the elevator shaft. The differences in dimensions determined from the 3D model, obtained when applying the TLS method, are up to ±2mm in comparison with the direct measurement.

While considering the height of a storey to be 3m and the building deviation (tolerance) ±25mm for the positions of pillars and walls on adjacent storeys with regard to the straight line of the secondary top view scheme, the tolerable inclination of the elevator shaft wall from the vertical is ±0.48°. The values of the elevation angle of the inclination of normal to the partial surfaces from the horizontal surface, stated in Tab. 4 and Graph 1, are up to ±0.34° which corresponds to their actual inclination and is within the tolerance scope of the building deviation.

The method of acquiring the data exploiting TLS may be characterised as the fast one with regard to the measuring period of about 30 minutes. However, the administration of the measured data is more demanding as it lasts about 1 day. Its advantage, along with the speed of work in situ, is the high precision and complexity of spatial data as it allows to evaluate the parameters of an elevator shaft at any location of its height, and even to construct its complex 3D model.

REFERENCES

[1]   ČERNOTA, P. STAŇKOVÁ, H., & GAŠINCOVÁ, S., 2011: Indirect Distance Measuring as Applied upon both Connecting Surveys and Orientation One, *Acta Montanistica Slovaca,* Vol. 16 (2011), no. 4, p. 270-275 ISSN 1335-1788.

[2]   ČERNOTA, P., STAŇKOVÁ, H., POSPÍŠIL, J. & MATAS, O., 2012: Nová metoda pozorování kyvů pro určení správné polohy olovnice v tížnici při připojovacím a usmerňovacím měření. In *Geodézia, kartografia a geografické informačné systémy.* Tatranská Lomnica 24. – 25. 10. 2012, Košice, TU, FBERG, ÚGKaGIS, 2012. ISBN 978-80-553-1173-9.

[3]   ERDÉLYI, J., LIPTÁK, I., 2011: Monitoring mostného objektu technológiou TLS. In: 47.*Geodetické informační dny : Sborník přednášek.* Brno,ČR,8.-9.11.2011. – Brno, Český svaz geodetů a kartografů, 2011. - ISBN 978-80-02-02350-0. - p. 58-63

[4]   ERDÉLYI, J., LIPTÁK, I., KYRINOVIČ, P. & KOPÁČIK, A., 2012: Určovanie posunov a pretvorení železobetónových konštrukcií pomocou TLS. In: Geodézia, kartografia a geografické informačné systémy 2012: VII. vedecko-odborná medzinárodná konferencia, Tatranská Lomnica, SR, 24.-25.10.2012. - Košice : Technická univerzita v Košiciach, 2012. – ISBN 978-80-553-1173-9.

[5]   GAŠINEC, J., GAŠINCOVÁ, S., ČERNOTA, P. & STAŇKOVÁ, H., 2012: Zastosowanie naziemnego skaningu laserowego do monitorowania logu gruntowego w Dobszyńskiej Jaskini Lodowej, *Inżinieria Mineralna.* Vol. 13, no. 2 (30) (2012), p. 31-42. ISSN 1640-4920

[6]   GAŠINEC, J., GAŠINCOVÁ, S., 2012: Modelovanie deformácií rovinných plôch. In *Geodézia, kartografia a geografické informačné systémy.* Tatranská Lomnica 24. – 25. 10. 2012, Košice, TU, FBERG, ÚGKaGIS, 2012. ISBN 978-80-553-1173-9.

[7]   GAŠINEC, J., GAŠINCOVÁ, S., ČERNOTA, P. & STAŇKOVÁ, H.: Možnosti použitia terestrického laserového skenovania pri dokumentovaní ľadovej výplne Dobšinskej ľadovej jaskyne a riešenie súvisiacich problémov v programovacom jazyku Python. In *SDMG 2011: sborník referátů 18. konference: Praha, 5.-7. října 2011.* Ostrava : VŠB-TU, 2011 p. 51-59. - ISBN 978-80-248-2489-5.

[8]   Leica   ScanStation   C10   –   technický   list.   Dostupné   na   internete: http://www.geotech.sk/downloads/Laserove-skenery-HDS/Leica_ScanStation_C10_Brochure_sk.pdf

RESUMÉ

Stavebné objekty sa od seba odlišujú najmä ich tvarom a rozmermi a tiež druhom konštrukcie. Vo všetkých etapách výstavby je kladený dôraz na presnosť zhotovenia konštrukčných častí stavebného objektu, tak aby stavebné odchýlky boli menšie ako je ich tolerancia udávaná normami, alebo projektovou dokumentáciou stavby. Vo fáze ukončovania stavby sú hlavnými činnosťami geodeta najmä kontrolné merania a porealizačné zameranie na vyhotovenie dokumentácie skutočného vyhotovenia stavby (DSVS). Pri výškových budovách je jej súčasťou aj geodetické zameranie zvislosti výťahových šácht (VŠ) s určením odchýlok skutočného vyhotovenia stavebného objektu voči projektovej dokumentácii, prípadne príslušnej technickej norme. Na získanie priestorových údajov o stavebnom objekte je možné použiť viac metód a postupov, ktoré sa od seba líšia presnosťou, časovou náročnosťou merania a tiež dostupnosťou vhodných meracích prístrojov. Cieľom experimentálnych meraní prezentovaných v článku bolo potvrdenie metódy terestrického laserového skenovania pre jeho aplikáciu na určenie zvislosti a rozmerov výťahovej šachty ako vhodnej, dostatočne presnej, rýchlej a detailnej pri porovnaní s bežnými geodetickými meračskými postupmi.

14

RISK ASSESSMENT IN MINING-RELATED PROJECT MANAGEMENT

Michal VANĚK [1], Yveta TOMÁŠKOVÁ [2], Alena STRAKOVÁ [3], Kateřina ŠPAKOVSKÁ [4], Petr BORA [5]

[1] *doc. Ing. Ph.D., Institute of Economics and Control Systems, Faculty of Mining and Geology, VŠB – Technical University of Ostrava, 17. listopadu 15/2172, Ostrava, tel. (+420) 59 732 3336 e-mail: michal.vanek @ vsb.cz*

[2] *Ing. Ph.D., Institute of Combined Studies in Most, Faculty of Mining and Geology, VŠB – Technical University of Ostrava, Dělnická 21, Most, tel. (+420) 597 325 702 e-mail: yveta.tomaskova@vsb.cz*

[3] *Ing., Institute of Combined Studies in Most, Faculty of Mining and Geology, VŠB – Technical University of Ostrava, Dělnická 21, Most, tel. (+420) 597 325 703 e-mail: alena.strakova@vsb.cz*

[4] *Ing., Institute of Economics and Control Systems, Faculty of Mining and Geology, VŠB – Technical University of Ostrava, 17. listopadu 15/2172, Ostrava, tel. (+420) 597 324 560 e-mail: katerina.spakovska@vsb.cz*

[5] *Ing., Institute of Economics and Control Systems, Faculty of Mining and Geology, VŠB – Technical University of Ostrava, 17. listopadu 15/2172, Ostrava, tel. (+420) 596 993 860 e-mail: petr.bora@vsb.cz*

Abstract

Risk assessment is an integral part of the assessment of an investment project. Underestimating risks may lead to erroneous conclusions with negative impacts on the economy of the project. With regard to the level of investments and the time factor under mining company conditions, the issue of risk gains importance. The evaluation of existing practical experience shows that managers of mining companies more often approach to the risk assessment based on intuition, than through exact methods. The article is devoted to the risk assessment itself whose procedure is illustrated on a model example of assessing continuous and discontinuous alternatives of exploitation of loose overburden materials during large-scale coal mining operations in progress at pit quarries.

Abstrakt

Hodnocení rizika je nedílnou součástí posouzení investičního projektu. Podcenění rizika může vést k chybným závěrům s negativními dopady na ekonomiku projektu. S ohledem na výši investic a hledisko času v podmínkách těžebního podniku, nabývá problematika rizika na významu. Z hodnocení stávajících praktických zkušeností vyplývá, že manažeři těžebních podniků přistupují ke stanovení rizika častěji na bázi intuice, než s využitím exaktních metod. Článek se věnuje vlastnímu stanovení rizika, jehož postup naznačuje na modelovém příkladu posuzování kontinuální a diskontinuální alternativy exploatace sypkých skrývkových hmot při velkokapacitní těžbě uhlí probíhající na jámových lomech.

Key words: Risk assessment, management, NPV, mining company

1 INTRODUCTION

Managerial decision-making is one of the most important activities performed by management members within their competences (Fotr, 2010). The decision-making is the process of choosing the way which the manager and the organization entrusted to the manager will take. At the end of the decision-making process, there is a specific way, an option. Obviously, the greater the impacts of the decision, the greater the importance of the decision-making process. The complexity and difficulties related to the decision-making function consist mainly in the fact that it is impossible beforehand to verify the impacts that specific managerial decision will bring. In addition, the entire decision-making process takes place under turbulent dynamically changing conditions, and therefore to take decisions free from risks is difficult or quite impossible.

This fact should necessarily lead to the need of the manager to specify, quantify and subsequently control the risks. However, findings from economic studies show the lack of integration of risks and uncertainties into investment decision-making process when any investment project risk analysis does not take place either at all, or in a very simplified form. As a result, wrong investment decisions could be made that threaten, in particular in the case of large-scale investment projects, prosperity and financial stability of firms implementing these projects (Fotr, 2007).

Mining and mineral processing belong to such business sectors which require substantial investments in land, technology, infrastructure and other factors of production, without which miners are not able to exploit the mineral of interest. It is also the sector in which nature is an important influencing factor, because the shape of deposit, storage conditions, the amount and quality of commercial mineral, tectonic disturbances predispose the range of mining, used technology, and thus the amount of necessary investments. In connection with a deposit, it is known that the miner can recognize it only when he mined it out. This is further amplified by the fact that the management of a mining company work with a broader range of threats than the management of a company operating in another industry sector.

The extraction of a raw material from a certain deposit can take tens or hundreds of years. Changing the quarry name does not change the fact that the extraction takes place at the same deposit. The brown coal deposit in the Most Basin, where coal mining began in the 15th century for private purposes only, and since the 19th century the targeted industrial mining of this raw material has taken place here either by underground and surface mining methods, could be an example (Pokorná, 2000).

For quarrying in small quarries, discontinuous mining technology was and still is used. In the Most Basin, a merge of several smaller quarries operating at a single deposit occurred in the early second half of the 20th century, allowing the deployment of more efficient technologies and the gradual commencement of transition from the discontinuous to giant-machine continuous technology. The reason for the massive deployment of this technology was the need of ensuring the mining of large volumes in the 80s of the last century. Other causes can be seen in the fact that the then Czechoslovakia was one of the world-powers in the production of this type of technology, and not least in the fact that the discontinuous technology has not reached the required performance parameters for a long time. Since the discontinuous technology now achieves these parameters, and the continuous technology is costing as for initial investment and overhaul costs, the issue of replacement of existing continuous technological exploitation of loose materials with a discontinuous alternative is discussed for a quite long time (Seidl et al, 2011).

The team of authors of the present article deals with the issue for a long time as well. The starting point of the work became an economic study (Seidl et al, 2011) which analysed the technological options and evaluated them in terms of investment decision-making. From this work, we know that the discontinuous technology is economically feasible. Another study (Vaněk et al , 2012) focused on the identification and evaluation of key threats associated with each of the technologies. It is directly followed by our article, as informal interviews conducted the authors with managers of mining companies indicated that although the managers know about risk management, they usually quantify risk intuitively on the basis of qualified studies.

This article aims at demonstrating the scientific risk assessment approach, and so suggesting the possibilities of its use in practice of managers of mining companies. The focus of the article then can be seen in the assessment of risk of technological alternatives for exploitation of loose overburden materials in the mass production of pit quarries.

2 METHODS AND MATERIALS

The issue of risk management is elaborated in many theoretical works. On conditions of the CR, the work by prof. Fotr and prof. Smejkal can be ranked among fundamental ones. The work of a team of prof. Mařík is important as well.

However, in the professional literature and in business practice there is some confusion about the very concept "risk". Smejkal sais to this: "If someone feels that, when discussing a risk, a confusion of terms occurs, one cannot but agree." It is due to the fact that the term "risk" is promiscuously used for both "a likelihood of event incidence" and "an impact of event incidence on a subject". This promiscuity can be prevented by mentally separating the force, event, activity, or person having an adverse effect on safety (**threat**), and the probability that an adverse event will occur (**risk**) (Smejkal, 2006).

Risk assessment is the result of a comprehensive process which can be aggregated into four basic phases (Smejkal, 2006), as follows:

1. Identification of assets
2. Identification of threats
3. Determination of the significance of threats
4. Determination of risk

Assets are meant to be the values that can be endangered by fulfilling the threats (Smejkal, 2006). When carrying out the identification of threats, such threats are determined that affect the assets. Threats can be of an economic, technological, legislative, political and natural character. Threats caused by human factor are significant as well. Threats may affect the business, organization, project, person, or assets separately or they can interconnect and interact[1] (Vaněk et al, 2012).

To understand the threats and especially their mutual relationships, cognitive maps (CM) can be successfully used. The CM is a representation of the causal relationships that exist among the decision elements of a given object and/or problem, and describe experts' tacit knowledge. CMs are composed of (a) concept nodes (i.e., variables or factors) that represent the factors describing a target problem, (b) arrows that indicate *causal relationships between two concept nodes, and (c) causality coefficients on each arrow that indicate the positive (or negative) strength with which a node affects another node* (Kun, 2009).

The significance of threats is determined essentially in two ways, in particular in an expert manner, or by means of a sensitivity analysis. The expert assessment of the significance of threats consists in their professional evaluation by employees who have the necessary knowledge and experience in the areas to which individual threats fall. In case of the sensitivity analysis, it is determined what change of a criteria indicator will be initiated by a certain change in its direct influencing factor. (Fotr, 2005).

The method of **determining the risk** depends on its nature and the purpose, for which the risk is quantified. Identifying business risks can be performed by using a **modular method** which is a comprehensive method of assessing the monitored sub-risks. The threat identified in previous steps and rated by a risk-level is then transferred to a risk premium. (Ošatka, 2004). This is a method that does not derive the risk, or the risk premium from the capital market, but determines it as the sum of the partial risk premiums being determined for a group of business and financial threats (Mařík, 2011).

In the event that the risk of a project is determined, **statistical characteristics** (variance, standard deviation, coefficient of variation) or **managerial characteristics** (robustness, flexibility) can be successfully used (Fotr, 2005).

By reason that the assessed technological alternatives of exploitation of overburden materials can be seen as projects, the approach of risk quantification through statistical characteristics was chosen.

The prerequisite for the use of statistical characteristics is the knowledge of probability distribution of an evaluation criterion, which may be e.g. a net present value (Fotr, 2005). To determine the probability distribution, scenarios were used while working with the discrete nature of threats.

The scenarios are usually regarded as mutually consistent combinations of values of major threats. Each scenario thus represents a different future development of the evaluation criterion. In the event of two or more threats, probability trees in a form of graphs formed by situational nodes and edges are a useful tool for displaying scenarios. The probability of each scenario is the result of the product of the probabilities of values of considered threats which is based either on the numerical values of the past or on the experience of experts (Fotr, 2005).

The variance is determined by the following relationship (Fotr, 2005):

$$R = \sum_{i=1}^{n}(NPV_i - M)^2 \times p_i \qquad (1)$$

Where:

R – NPV variance,

NPVi – net present value of project of i-th scenario,

M – NPV mean value,

Pi – probability of i-th scenario,

n – number of scenarios.

---

[1] For example, a change of an exchange rate – the threat of an economic nature will cause a threat on the part of the supplier which specifically reveals itself in higher investment costs of the project (in case of foreign supplier).

The mean value of NPV is then:

$$M = \sum_{i=1}^{n} NPV_i \times p_i \tag{2}$$

The coefficient of variation is determined as follows:

$$k = \frac{\partial}{M} \tag{3}$$

Where:

k – coefficient of variation,

δ – standard deviation,

M – NPV mean value,

The determination of threatened assets is one of the starting points of risk quantification. The overview of the assets considered in the model coal pit quarry in prices of the year 2012 is shown in Tab. 1. Since the continuous technology is supplied by domestic manufacturers, the prices are listed only in CZK.

Tab. 1 Threatened assets

Continuous technology		
Costs	**in thous. €**	**in thous. CZK**
Excavator	XX	1,650,000
Back filler	XX	455,000
DPD – long-distance belt haulage	XX	1,230,000
Total	**XX**	**3,335,000**
Discontinuous technology		
Costs	**in thous. €**	**in thous. CZK**
Excavator RH120 - RH120 - BH	3,700	92,944
Excavator RH120 -RH120 - FS	3,700	92,944
Caterpillar Cat 785D Dump Truck (8x)	17,600	442,112
Caterpillar 16M Motor Grader	700	17,584
Caterpillar 730 BWT Ejector	400	10,048
Caterpillar 824 H Wheel Dozer	600	15,072
Caterpillar Cat D8T Dozer	500	12,560
Cat D10T Dozer	820	20, 598
Cat D11TCD Dozer	1,800	45,216
M318D Excavator	200	5,024
Total	**30,020**	**754,102**

Source: Inherent processing

3   RESULTS

These results can be understood as a recommended decision with the consideration of accurate risk assessment.

The study aimed at determining the significance of threats (Vaněk, 2012) showed that in case of continuous technology the following key threats were identified: capital expenditure, energy consumption, repairs and related services, and in case of discontinuous technology: costs of services – repairs, fuel and capital expenditure.

The starting points for creating scenarios are foreseen value statuses of major threats and the probabilities of the occurrence of the statuses. The bases for quantifying were the results of a panel discussion with experts on the issue. The discussion results are summarized in Tabs. 2 and 3.

Tab. 2 Input data for continuous technology

Threat / Status	Change in capital expenditure	P	Change in energy costs	P	Change in costs of repairs and related services	P
1	0%	0.50	0%	0.10	0%	0.10
2	5%	0.40	10%	0.40	2.5%	0.40
3	10%	0.10	20%	0.50	5%	0.50

Source: inherent processing

Tab. 3 Input data for discontinuous technology

Threat / Status	Change in capital expenditure	P	Change in fuel costs	P	Change in costs of services	P
1	5%	0.10	0%	0.10	5%	0.10
2	10%	0.50	10%	0.40	10%	0.40
3	15%	0.40	20%	0.50	15%	0.50

Source: inherent processing

Since the significance of threats was determined by analysing the sensitivity of NPV, the net present value at the same time became the evaluation criterion. In order to determine the probability distribution of NPV, 27 alternative scenarios were created for each technology, resulted from the combination of the statuses of major threats, see Tab. 2 and 3. In total 54 scenarios were created.

When calculating the NPV, the values of discount rate of 8 % were used. In the publication (Fotr, 2005), the value is recommended for the renewal of existing technology (the case of continuous technology). For the introduction of new machinery, which corresponds to the deployment of discontinuous technology instead of continuous one, the publication (Fotr, Souček) recommends to use a rate of 10 % which reflects the increased risk for standard projects. Due to the specific way of spending capital expenditures and non-standard monetary revenues of the surveyed project, the use of the higher discount rate induces a decrease in the resulting net present value, which makes the criterion indicator of the more risky project more advantageous. Therefore, the uniform above mentioned discount rate of 8 % was used. Tab. 4 shows the value of NPV for each scenario of relevant transport technology and the resulting probability of the scenario.

Tab. 4 NPV scenarios and their probability

Scenario	Continuous technology		Discontinuous technology	
	NPV	Probability	NPV	Probability
1	-4,711,906	0.005	-3,964,002	0.001
2	-4,790,486	0.020	-4,026,732	0.004
3	-4,805,198	0.025	-4,089,462	0.005
4	-4,838,544	0.020	-4,077,384	0.004
5	-4,853,256	0.080	-4,140,114	0.016
6	-4,867,968	0.100	-4,202,844	0.020
7	-4,901,314	0.025	-4,190,765	0.005
8	-4,916,026	0.100	-4,253,496	0.020
9	-4,930,738	0.125	-4,316,226	0.025
10	-4,908,925	0.004	-4,019,978	0.005
11	-4,923,637	0.016	-4,082,708	0.020
12	-4,938,349	0.020	-4,145,438	0.025
13	-4,971,696	0.016	-4,133,360	0.020
14	-4,986,407	0.064	-4,196,090	0.080
15	-5,001,119	0.080	-4,324,111	0.100
16	-5,034,466	0.020	-4,312,033	0.025
17	-5,049,178	0.080	-4,374,763	0.100
18	-5,063,889	0.100	-4,437,493	0.125
19	-5,042,077	0.001	-4,178,342	0.004
20	-5,056,789	0.004	-4,241,072	0.016
21	-5,071,500	0.005	-4,303,803	0.020
22	-5,104,847	0.004	-4,291,724	0.016
23	-5,119,559	0.016	-4,354,454	0.064
24	-5,071,500	0.020	-4,417,185	0.080
25	-5,167,617	0.005	-4,405,106	0.020
26	-5,182,329	0.020	-4,467,836	0.080
27	-5,197,041	0.025	-4,530,566	0.100

Source: inherent processing

In order to be able to assess competently the risk of the considered technological alternatives, statistical parameters, the mean value, variance, standard deviation and the coefficient of variation were calculated, see Tab. 5.

Tab. 5 Values of statistical characteristics

Technology	Mean value	Variance	Standard deviation	Coefficient of variation
Continuous	-4,962,565	9,419,620,327	97,055	-0.0196
Discontinuous	**-4,347,135**	**14,848,639,348**	**121,855**	**-0.0280**

Source: inherent processing

Due to the higher difference in values of the calculated variances, the coefficient of variation (5,429,019,021) is crucial for the risk assessment.

The value of the coefficient of variation is higher in the discontinuous technology, but only by 0.0084. This difference is negligible, therefore both technological alternatives can be considered equivalent in terms of risk.

4   DISCUSSION

The authors do not conceal that the result achieved by the risk analysis is a certain surprise for them. However, the result is based on the relevant bases, therefore there is no choice but to accept it.

A certain pitfall of the risk analysis is that the quantification of major threats is usually based on expert estimates, so it is necessary to pay due attention to the selection of experts. This is the only way how to ensure the bases for determining the risk of a project to be valid.

For the management of the mining enterprise that intends to replace the continuous technology with the discontinuous technology, the result is crucial in that the risk does not disqualify any of the alternatives and the management can focus in decision-making entirely on technical and economic issues.

NPV suggests which technology of exploitation of loose overburden materials in the model example of a coal pit quarry is economically advantageous for miners. Since the NPV method was applied to the process that does not generate revenues, the net present value takes negative values. Therefore, it is recommended to implement the one of the two technologies for which the NPV takes a value closer to zero. In our case it is the discontinuous technology.

5   CONCLUSION

The complexity of managerial decision-making lies among other things in that managers are fully responsible for their decisions. It is therefore up to them on which kind of decision-making methods and forms they ground their decisions.

Pitfalls of decision-making that is based solely on intuition and experience can be seen in a minor strength of arguments. However, if the manager will rely on exact and heuristic approaches, not only the persuasiveness of his arguments will increase, but also the quality of the resulting decisions. This quality can be crucial for achievements of the manager in the management of a project, a company (institution), or its organizational units.

In the comprehensive assessment of an investment plan, it is also necessary to evaluate the risks associated with the project. Especially, when the manager (management) decides on a project that requires high investment costs and potential long life.

To these criteria, a series of projects implemented in the extraction and processing of raw materials corresponds as well. These include, among others, the restoration project of technological equipment deployed in the large-scale mining of loose overburden materials taking place at coal pit mines. The project became a model example of the application of risk management in terms of a mining company, within which the risks of existing continuous and discontinuous solution alternatives were assessed.

The discontinuous technological solution requires lower initial investments. It is also flexible in time. It would thus appear that this alternative is less risky as well. However, it is apparent from the results of the risk analysis presented in this paper that both assessed technologies are in principle equivalent in terms of the level of risk. Thus, the investment decision made on the basis of intuitive risk assessment can be misleading, if not completely wrong. With regard to economic aspects and the NPV value, the management of a (model) mining company could be recommended to change to the discontinuous technology as a better alternative.

REFERENCES

[1]   FOTR, Jiří; SOUČEK, Ivan. Podnikatelský záměr a investiční rozhodování. Business plans and investment decisions. 1st edition. Prague: Grada Publishing, 2005. 356 p. ISBN 80-247-0939-2.

[2]   FOTR, Jiří; ŠVECOVÁ, Lenka et al. Manažerské rozhodování. [Managerial decision-making]. Postupy, metody a nástroje. The procedures, methods and instruments. 2nd revised edition. Prague: Ekopress, s.r.o., 2010. 474 p. ISBN 978-80-86929-59-0.

[3]   FOTR, Jiří; ŠEVCOVÁ, Lenka. Probabilistic approaches to investment decisions and their implementation. Ekonomika a management. Economics and Management. VŠE 2007, 16 pp.

[4]   KUN, Chang Lee; NAMHO, Lee; HONGLE, Li. A Particle Swarm Optimization-Driven Cognitive Map Approach to Analyzing Information Systems Project RiskJournal of the American Society for Information Science & Technology, Jun2009, Vol. 60 Issue 6, p1208-1221, 14p.

[5]   MAŘÍK, Miloš et al. Metody oceňování podniku : Proces ocenění základní metody a postupy. Company valuation methods: The process of fundamental valuation methods and procedures. 3 The revised and expanded edition. Prague: Ekopress, s. r. o., 2011. 494 p. ISBN 978-80-86929-67-5

[6]   OŠATKA, Jiří. Náklady na vlastní a cizí kapitál. The costs of own and foreign capital. In Juniorstav 2004, Brno : 4 – 5. 2 2004 Brno : Faculty of Civil Engineering University teaching of Technology, [online] 2004.   [cit.   02/02/2013].   Available   from   WWW: <http://www.fce.vutbr.cz/veda/dk2004texty/pdf/07_Soudni%20inzenyrstvi/7_01_Soudni%20inzenyrstvi/Oscatka_Jiri.pdf >.

[7]   POKORNÁ, Libuše. Kniha o Mostecku. 1st edition. Litvínov: Dialog, 2000. 464 p. ISBN 80-85843-80-3.

[8]   SEIDL, Miroslav; TOMÁŠKOVÁ, Yveta; KOLMAN, Petr. Alternative Options For Deploying Extraction Equipment At Large Pit Quarries. In *SGEM 2011 The International Multidisciplinary Scientific GeoConference*. Sofia: STEF92 Technology Ltd., 2011. pp 677-684. ISSN 1314-2704.

[9]   SMEJKAL, Vladimír; Rais, Karel. Řízení rizik ve firmách a jiných organicích. Risk management in companies and other organizations. 2nd and expanded edition. Prague: Grada Publishing, 2006. 296 p. ISBN 80-247-1667-4.

[10]   VANĚK, Michal; TOMÁŠKOVÁ, Yveta; SEIDL, Miroslav; KOLMAN, Petr. Identification of the Threats, and Determining Their Significance, in Stripped Overburden Transport. In 12th International Multidisciplinary Scientific GeoConference SGEM 2012, Albena, Bulgaria, June 2012. Sofia : STEF92 Technology Ltd., 2012. Conference Proceedings, Volume 1, Geology, Exploration and Mining, pp 611 – 618.

RESUMÉ

Těžba a zpracování nerostných surovin patří k podnikatelským odvětvím, které vyžadují nemalé investice do výrobních činitelů zajišťující samotnou exploataci a následnou úpravu zájmové nerostné suroviny. Tyto investice mají charakter projektů a management těžebního podniku tak musí (v souvislosti s těmito projekty) přijmout investiční rozhodnutí a také rozhodnout o způsobu jejich financování.

Nedílnou součástí posuzování investičních projektů je problematika zhodnocení rizik, jejíž akcentování managementem může eliminovat chybné závěry se všemi dopady na ekonomiku projektu.

Jednou z aktuálně diskutovaných oblastí je nahrazení kontinuální technologie přepravy sypkých skrývkových hmot při velkokapacitní těžbě na jámovém lomu technologií diskontinuální. Článek stanovuje rizika obou technologií (projektů), přičemž navazuje na předchozí studii zaměřující se na určení významnosti hrozeb. Posouzení rizika obou dopravních alternativ je provedeno prostřednictvím statistických charakteristik rozdělení pravděpodobnosti hodnotícího kritéria. Tímto kritériem je NPV projektu. Ke stanovení rozdělení pravděpodobnosti bylo využito metodologie scénářů, přičemž se pracovalo s diskrétním charakterem hrozeb. Pro každou technologickou alternativu bylo vytvořeno 27 scénářů, které vznikly kombinací stavů významných hrozeb. Celkem tak bylo vytvořeno 54 scénářů.

Vzhledem k vyššímu rozdílu hodnot vypočtených rozptylů je pro posouzení rizika rozhodující variační koeficient. Hodnota variačního koeficientu je sice vyšší u diskontinuální technologie, avšak pouze o 0,0084. Tento rozdíl je zanedbatelný, a proto lze obě technologické alternativy považovat z hlediska rizika za rovnocenné.

Poněvadž metoda NPV byla aplikována na proces, který negeneruje příjmy, nabývá čistá současná hodnota záporných hodnot. Proto se doporučí ta technologie, která nabývá hodnoty bližší nule. V našem případě se jedná o diskontinuální technologii.

15

OCCUPATIONAL COMPETENCE FOR IMPROVING INDUSTRIAL ENTERPRISE COMPETITIVE STANDARDS

Lucie KRČMARSKÁ [1], Igor ČERNÝ [2], Michal VANĚK [3]

[1] *Ing. Ph.D., Institute of Economics and Control Systems, Faculty of Mining and Geology, VŠB – Technical University of Ostrava 17. Listopadu 15, Ostrava, tel. (+420) 59 732 4530*
e-mail: lucie.krcmarska@vsb.cz

[2] *Ing. Ph.D., Institute of Economics and Control Systems, Faculty of Mining and Geology, VŠB – Technical University of Ostrava 17. Listopadu 15, Ostrava, tel. (+420) 59 732 33240*
e-mail: igor.cerny@vsb.cz

[3] *doc. Ing. Ph.D., Institute of Economics and Control Systems, Faculty of Mining and Geology, VŠB – Technical University of Ostrava 17. Listopadu 15, Ostrava, tel. (+420) 59 732 3336*

e-mail: michal.vanek@vsb.cz

Abstract

Success of enterprises is usually measured by their achievements. This success results from the competence of people working for the company. The employee competence can be understood as the sum of their actual performance and their latent abilities. Both is essential and provides for the competence completeness. Missing or inadequate occupational competences cause problems in running industrial enterprises, and thus a continuous improvement process of competence levels is a must. If a company is aware of the importance of people competences in relation to success or failures and strives to develop them, it is able not only solve current problems, but also remove causes of future problem incidences. This paper focuses on the assessment of the occupational competences and competence models as an important tool in the management of human resources.

Abstrakt

Úspěchy firem jsou většinou poměřovány jejich dosaženými výsledky. Za základ těchto úspěchů lze považovat úroveň kompetencí lidí, kteří pro firmu pracují. Kompetence lidí chápeme jako souhrn dosahovaného výkonu a přinášeného potenciálu. Jestliže chybí jedno, pak chybí i kompetence jako celek. Problémy, které vznikají ve firmách můžete tedy převést na chybějící nebo nedostačující kompetence a je tedy nutné doplnit to, co v oblasti kompetencí chybí. Uvědomí-li si firma význam kompetencí lidí ve vztahu k úspěchům nebo neúspěchům a zaměří-li se na jejich rozvoj, odstraňuje tím příčiny vznikajících problémů, ne jen jejich viditelné problémy. Článek je zaměřen na posouzení kompetencí a kompetenčních modelů jako významného nástroje v oblasti řízení lidských zdrojů.

Key words: competence, competence model, job market, industrial enterprise, investigation.

1 INTRODUCTION

The key managerial tasks of a company – its size makes no difference – consist in collecting, combining, and utilizing financial, material, and human resources. As the human resources activate all other resources, they represent a decisive factor of entrepreneurial prosperity and competitiveness. [3]

Whereas material, financial, and information resources can be easily emulated (for example by utilizing the same or better technologies), the competence levels are not that easy to balance and level off. As such, the occupational competence is a key competitive edge factor, and its systematic monitoring should be a matter of routine.

All companies need competent employees whose performance standards exceed average levels. Preconditions of such performance are not only professional knowledge and training, but also their personal and social characteristics.

Job application requirements should be the focus points of the selection procedure. A competence model might be an appropriate tool for assessing the occupational competence. The model could provide for the selection of applicants whose abilities suit best the job and its related position requirements. However, competence models are usually not too general and universally oriented. Only the models oriented by specific competence requirements can be of major assistance as regards the selection of appropriate candidates for specific job positions.

Only some competences can be regarded as relevant to all job position needs. They are called key competences. If more specific competence assessments are needed, differences between individual professions should be taken into account, and specific models developed.

The authors of this paper focused on the occupational competence as a major HR management tool. The related research provided for the distinguishing of individual competence relevance as seen from three vantage points, namely:

- Competence relevance to discharge of office,

- Significance of incorporating development of competences into HE study programmes,

- Competence categories and their levels anticipated by selection procedures.

2   MEANING AND CATEGORIES OF COMPETENCES

If practical relevance of competence model utilization is scrutinized, some clarification of the meaning of competence as such is needed.

The concept of competence has basically two meanings – legal and technical. For one thing, it means an official right of direct or delegated control – power, authority, jurisdiction. For another, it means skills that are needed to perform a particular job or a particular task. The lawful idea of competence is of a French-German origin (competénce, die Kompetenz). The technically oriented idea of competence is rather common for the Anglo-Saxon environment. For example, the Penguin Dictionary of Psychology, quote: competence 1. Generally, ability to perform some task or accomplish something.

This paper deals with the competence as suggesting ability to perform actions to be able to do something.

There are many approaches to competence categorization and it is difficult to cover the subject in its entirety.

The competences can be basically classified as generic and specific or cognitive, social, key, and functional competences. [5]

A clear, 'traditional' arrangement of the competence is shown in Fig. 1. [7]

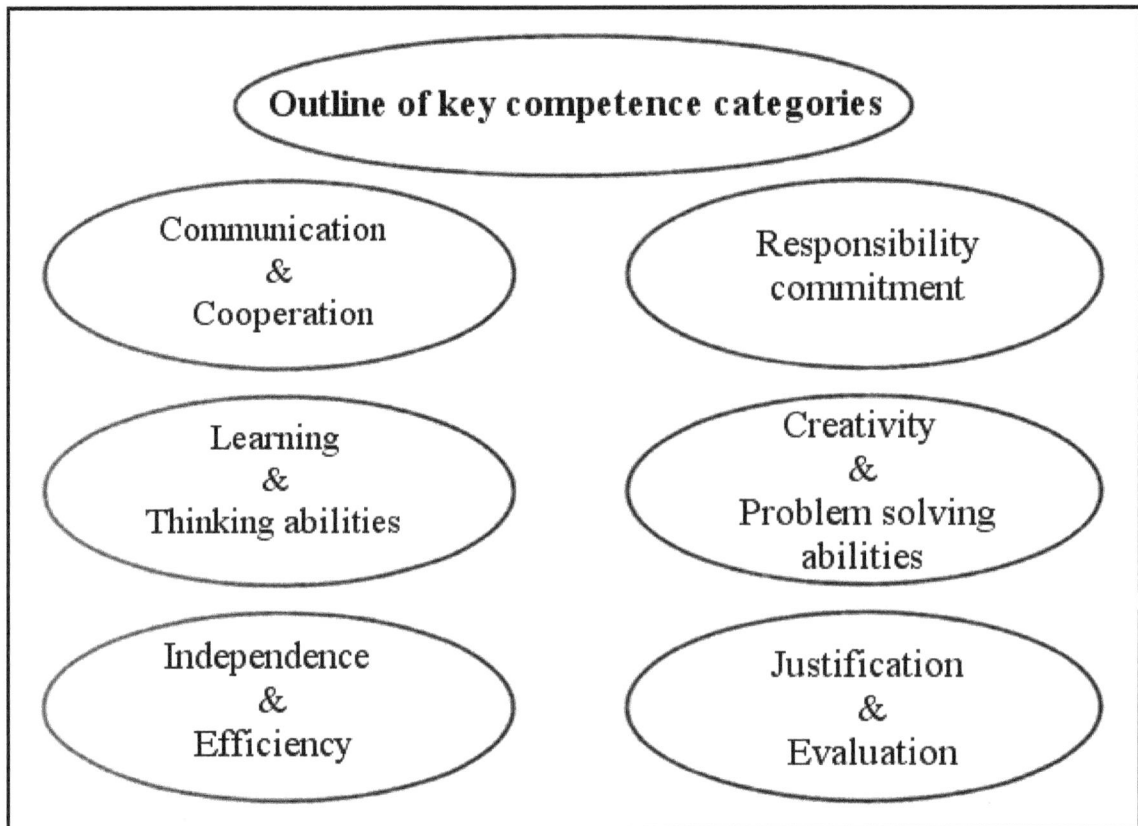

Fig. 1 Traditional arrangement of key competence categories (Veteška, Tureckiová, 2008)

The competence categorization based on the work performance and its related skills and knowledge was used by Tyron whose classification includes the following competence categories: managerial, interpersonal, and technical competences. Carrol and McCrackin classify the competences as key, team, functional, leadership, and managerial competences. [4]

Armstrong's categorization of competences is also quite well known:

- Soft skills,
- Hard skills,
- Generic, basic, specific competences. [5]

As such, there are many approaches to the classification of competences which results in a wide spectrum of individual competence categories.

3   COMPETENCE MODELS

Competence models identify the knowledge, skills, personal, and professional abilities that are necessary to successfully perform critical work functions in an industry or occupation. They provide for a HR tool of major importance.

The competence model development is influenced by many factors. It is important to be aware of the individual model function and utilization. In the framework of personal activities, it is especially about solving problems of employee selection or planning educational programmes. The models can be also instrumental in the performance evaluation of employees and their career planning.

The competence model can embody a competence common to the staff of a whole organization - the so called core competence. Another model can focus on specific competence categories of working for a particular company.

A company can develop its own competence model or can have it tailor made. Another option is in customizing a generic competence model in offer. [7]

The question is whether it is of advantage to develop models at generic levels or specify an individual competence more closely, concentrating on differences between specific professions (branches, functions, management hierarchy, etc.). The research conducted by Tett et al. [6] pointed to the conclusion of the competence model increased specificity. They consider the existing models to be too generic and unable to provide for appropriate selection procedures, career development, and assessment of employees.

The importance of competence models becomes clear if we want to answer the question which competence categories of individual employees are needed to ensure the existence of a sustained competitive advantage for a company.

4   INVESTIGATION OF SOME COMPETENCE CATEGORIES

As mentioned above, the competence model forte is in facilitating the selection, career development, and performance assessment of company employees. The true is that the staff selection procedures can be decisive factors of competitiveness and that an appropriate attention should be paid to the preparation and running of the employee selection procedures.

The authors of this paper researched the field of the technical competence. It was a two-stage investigation. In the framework of these two stages, the respondents answered questions about the competence categories as follows:

Professional competence: Fundamental knowledge of statistics and company financial analysis, computer literacy, knowledge of company functioning, role of modern machinery, and state-of-the-art technology knowledge.

Language competence: Knowledge of foreign languages (this competence was subject to a detailed analysis within the second stage of investigation), Czech language correct communication, i.e. grammar and vocabulary.

Personal competence:

- Intrapersonal: Self-confidence, self-control and time management, stress management, change process flexibility, target orientation, initiative, independence;
- Cognitive: Conceptual thinking, analytical thinking;
- Interpersonal: Task commissioning, staff motivation, briefing, coaching, dialog performance, resolving disagreements, reaching agreements, stimulating competition, empathy, presentation abilities, managerial ethics, deportment and etiquette.

The choice of the specific competence categories was made so that specific features of managerial work were best reflected, having regarded the intermediate management level of industrial enterprises.

For the competence assessment, a five-grade scale was employed that provided for sufficient differentiation. Without distinguishing individual features of observation, the results of the questionnaire investigation served as the input data for the calculation of competence significance values as arithmetic averages for all respondents who had participated in the questionnaire action. The values of the aggregated competence were always calculated as an arithmetic average of partial, further indivisible competences. The MS Access 2007 provided for the needed calculations which were adapted and presented in the form of figures and tables by the MS Excel 2007.

In the framework of the first stage of investigation, an interrogatory action was performed among students of combined courses of the Faculty of Mining and Geology, i.e. it concerned students that had some previous practical experience, and as such were able to assess significance of some specific competences in practice.

In total, 157 students were addressed. In view of the fact that the questionnaire investigative action took place during the students' regular teaching hours, the rate of return was 100 %.

The respondents' task was to assess some competence categories as regarded their significance for performing a job position and for developing them by HE teaching courses. The significance assessment for the job performance competence is given in Tab. 1.

Tab. 1 Significance of competence for job performance

Competence	Significance
Personal competence	**3.76**
Thereof:	
Intrapersonal	4. 05
Cognitive	3.85
Interpersonal	3.50
Language	**3.72**
Professional	**3.16**

Tab. 2 shows the significance assessment of competence as developed by the HE programmes of study.

Tab. 2 Competence significance as regards development by HE study programmes

Competence	Significance
Language competence	**3.85**
Personal competence	**3.66**
Thereof:	
Cognitive	3.98
Intrapersonal	3.78
Interpersonal	3.46
Professional	**3.39**

Fig. 2 provides for significance comparison from both perspectives.

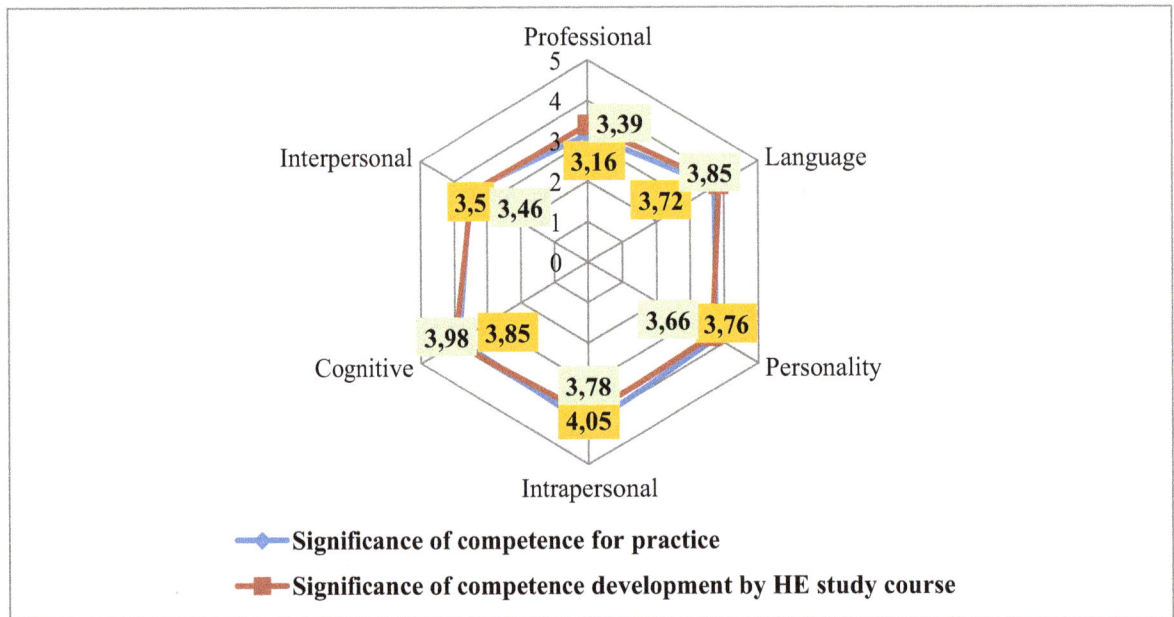

Fig. 2 Competence significance assessment from the perspectives of practice and HE study programme incorporation and development.

The comparison of the significance assessments of competences for practice and HE development yielded these results:

The assessment rate for all the investigated competence categories is in excess of 2.5 which means that all respondents considered the specific competence significant not only for the job performance, but also as a subject for development by HE courses of study.

In contrast to others, the professional competence got the lowest assessment rates, namely job performance, 3.16; HE study programme incorporation, 3.39. The computer literacy was rated rather highly; job performance, 4.00; HE development, 3.99 which suggests that computer the literacy belongs to basic qualifications of job applicants.

The language competence was rated high as regards its HE development significance, namely 3.85, which can be interpreted as respondents' expectations of coping successfully with needs of practical foreign language communication after graduation.

The personal competence was also rated quite high from both perspectives, i.e. job performance, 3.76; HE development, 3.66. The intrapersonal, cognitive, and interpersonal competence categories were assessed individually in the framework of the personal competence.

As regards the intrapersonal competence, the independence was rated as the most significant from both perspectives (job performance, 4.36; HE development, 4.09. Concerning the cognitive competence, the conceptual thinking received the highest rating, namely job performance, 3.80; HE development, 3.93. The interpersonal competence significance was rated the same from both perspectives where the presentation ability took up the first place (job performance, 3.87; HE development, 4.03). [2]

The questionnaire action evaluation was also performed regarding respondent individual characteristics. The women-to-men ratio was almost even, namely 82 men (52 %), and 75 women (48 %). The age group, 31-50 years, was the biggest, 104 respondents (66 %). The age group, 18-30 years, comprised 53 respondents (34 %). There were no respondents in the age category of 51 years plus.

The respondent education structure was as follows: secondary, 76 respondents (48 %), higher professional, 9 respondents (6 %) and HE, 72 respondents (46 %). The respondents usually worked in the tertiary sector (101 respondents, i.e. 64 %). Concerning the primary sector, there were 16 respondents (10 %), and 40 respondents (26 %) from the secondary sector. The assessment of competence significance, as regards the individual respondent characteristics, is illustrated in Figs. 3-6.

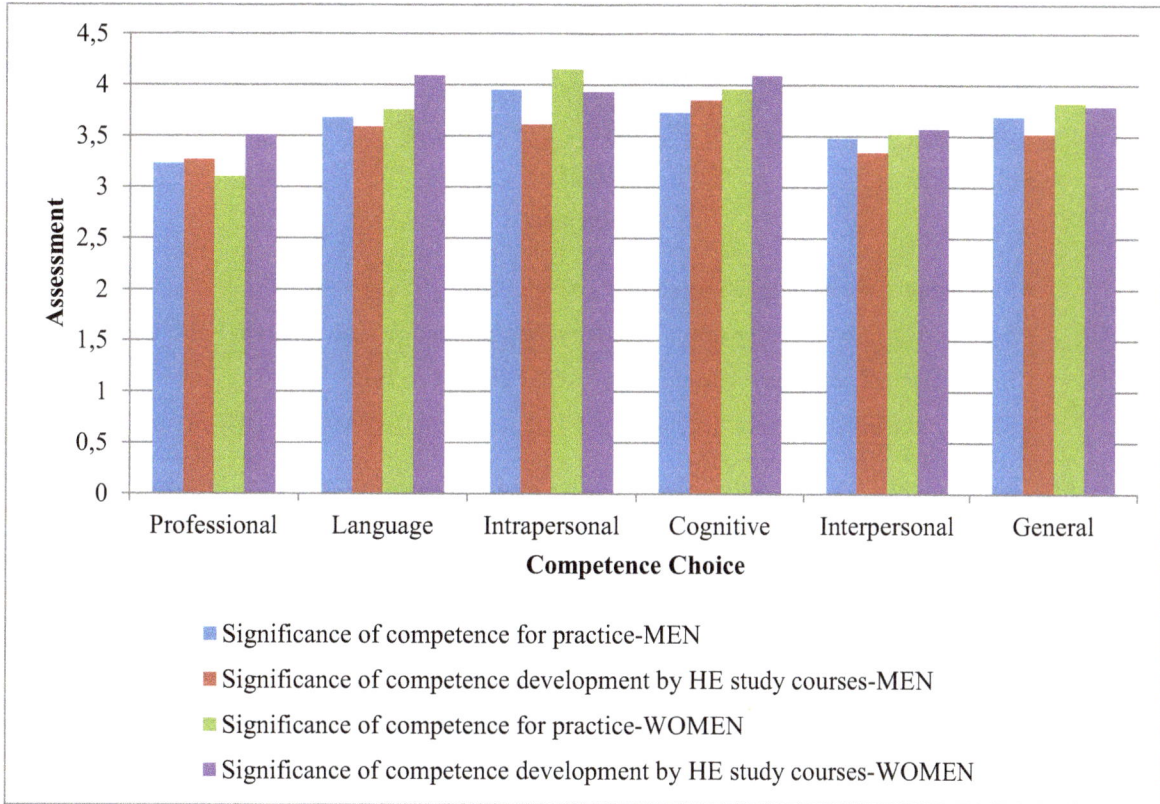

Fig. 3 Competence significance according to sex of respondents

The assessment of competence as sex dependent evidenced that women considered some competence categories more important than men did, which was valid for both the aspect of job performance and the aspect of competence development by HE study programmes. The only opposite view concerned the professional competence needed to meet requirements for a specific job position.

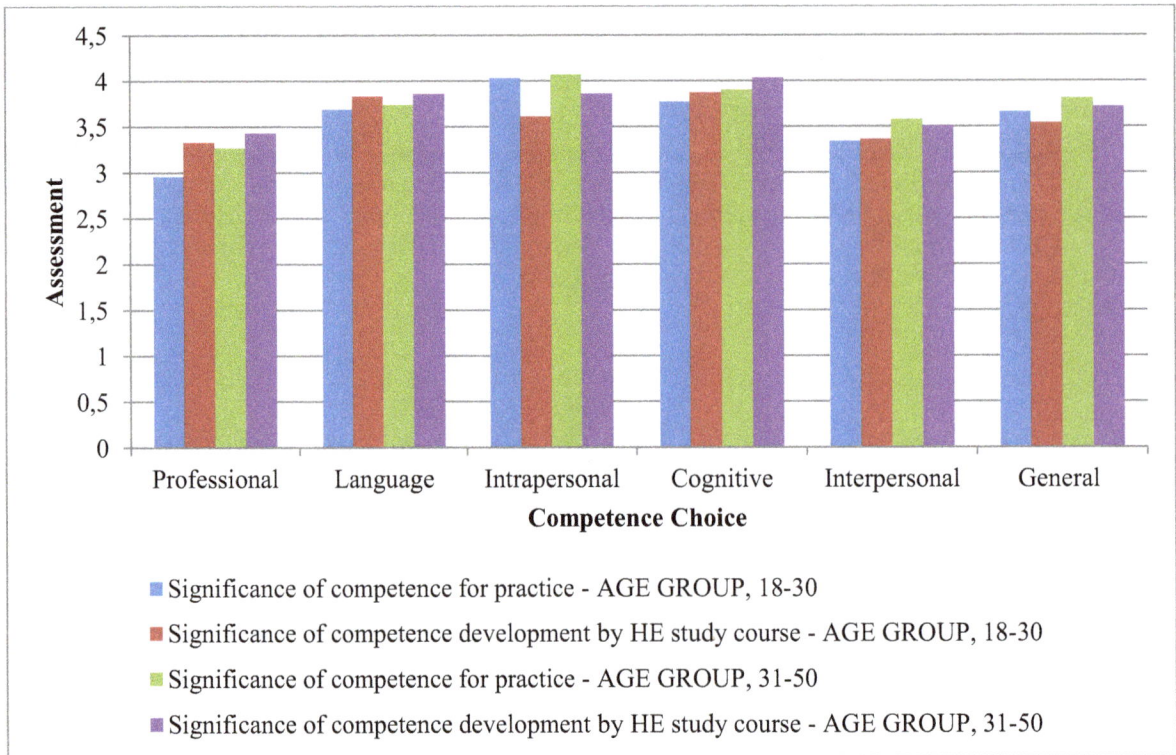

Fig. 4 Competence significance related to respondents' age

The age related responses testified increased the significance of the language and cognitive competences from the perspective of developing them by HE study programmes. In contrast to this view, the intrapersonal competence was considered more important for meeting job performance requirements which was also the case of the interpersonal competence as regarded the respondent age group, 31-50 years. The respondent age group, 18-30 years, attached almost the same importance to the competence from both perspectives.

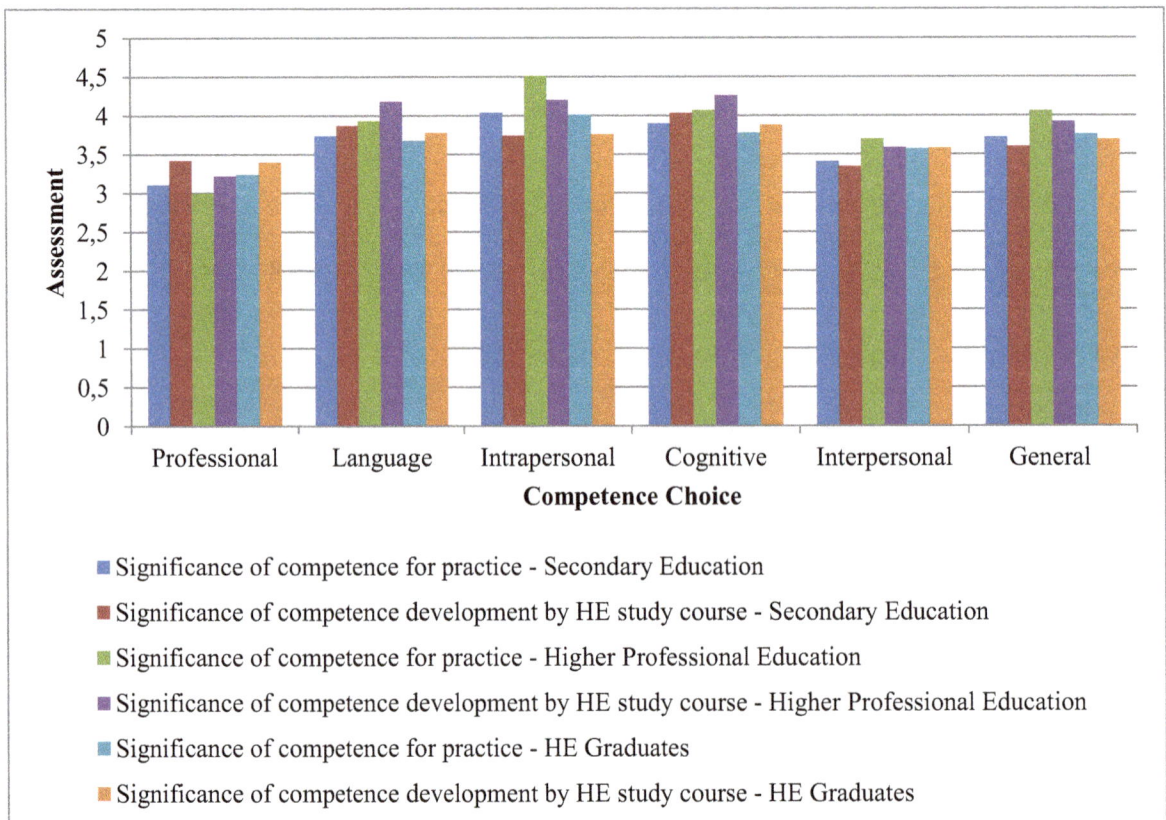

Fig. 5 Competence significance related to respondents' education level

The education level of respondents influenced their assessments of competence significance in analogy to the age dependence. The stress was given to professional, language, and cognitive competences as regarded their HE study programme development. An opposing view concerned the intra- and interpersonal competences.

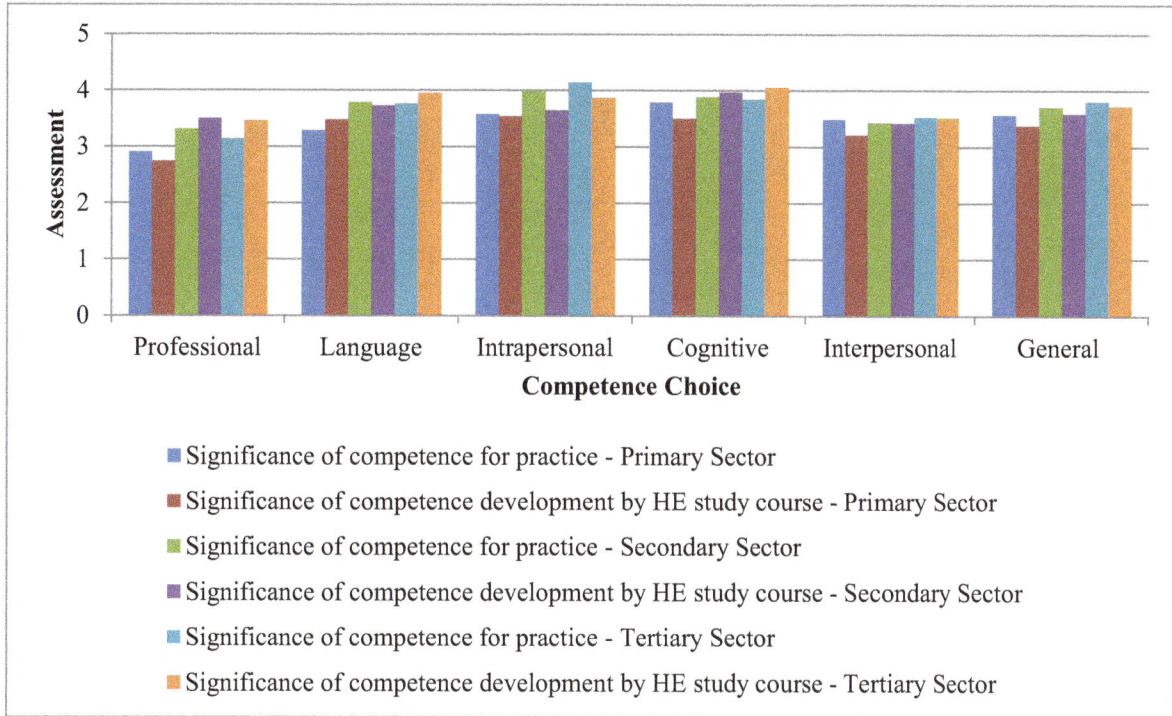

Fig. 6 Sector oriented assessment of competence significance

The sector (primary, secondary or tertiary) oriented assessment of the competence significance was the last item for the respondents to decide. This was to the same effect as was the case for the intrapersonal and interpersonal competences. The employees of individual work sectors put a special stress on competences related to job performance. Nevertheless, they consider the competence development by HE study programmes necessary.

The second part of the investigation concentrated on the assessment of competence by mining enterprises. The objective was to find out the level of competence that the HE graduates should have as job applicants. The respondents also voiced their opinion to the issue of hiring fresh HE graduates. The total of 15 mining companies was addressed.

As it is obvious from Fig. 7, not even a single respondent preferred the hiring of fresh HE graduates. A positive fact is that 46 % of respondents did not demanded previous practical experience which represents a chance for new graduates to succeed as job applicants.

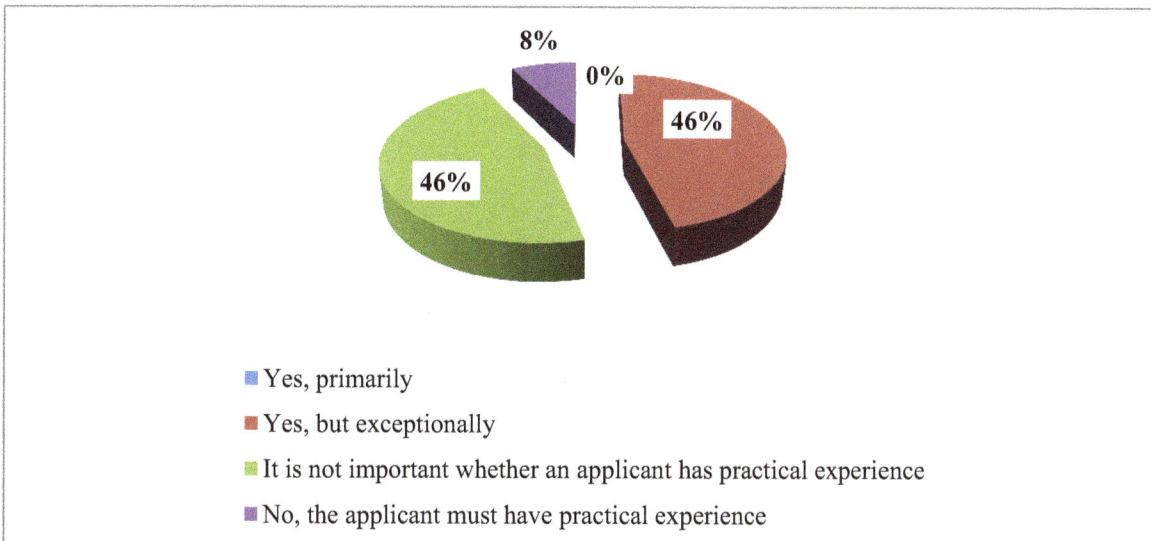

Fig. 7 Hiring selection procedures - applicants' preferences

Tab. 3 Expected competence level

Competence	Significance
Language competence	**3.92**
Personal competence	**3.40**
Thereof:	
Intrapersonal	3.82
Cognitive	3.69
Interpersonal	3.38
Professional	**3.23**

The both perspective investigations provided for the following:

- The evaluation of the specific competence from both the perspectives of job performance and HE development was always in excess of 2.5.
- The mining companies' assessment of the expected competence levels again always resulted in the rates above 2.5.
- The results of both investigations indicate the necessity of the competence model provisions as regards selective procedures for hiring.

The evaluation results for some competence categories studied in the framework of both investigative efforts are given in Tabs. 4-6.

Tab. 4 Assessment of professional competence

Competence	Significance for job performance	Significance for HE study programme development	Expected level
Professional competence	3.16	3.39	3.42
Fundamental knowledge of company statistics and financial analysis	2.62	3.16	2.67
Computer literacy	4.00	3.99	4.00
Knowledge of enterprise functioning	3.2	3.29	3.33
Knowledge of state-of-the-art technologies and equipment	2.83	3.13	3.67

Tab. 5 Assessment of language competence

Competence	Significance for job performance	Significance for HE study programme development	Expected level
Language competence	3.72	3.85	3.89
Knowledge of foreign languages	3.22	3.83	5.00
Grammar (Czech language)	3.98	3.85	3.78
Vocabulary (Czech language)	3.97	3.87	4

Tab. 6 Assessment of personal competence

Competence	Significance for job performance	Significance for HE study programme development	Expected level
Personal competence	3.76	3.66	3.40
Intrapersonal:	4.05	3.78	3.82
Self-confidence	3.83	3.61	3.38
Self- and time management	4.1	3.73	3.69
Coping with stress	4.11	3.79	3.69
Competence	Significance for job performance	Significance for HE study programme development	Expected level
Flexibility vs. change	4.08	3.70	4.15
Target orientation	3.92	3.83	3.85
Initiative	3.97	3.68	415
Independence	4.36	4.09	3.85
Cognitive:	3.85	3.98	3.69
Conceptual thinking	3.8	3.93	3.77
Analytical thinking	3.78	3.97	3.62
Interpersonal:	3.5	3.46	3.38
Task commissioning	3.41	3.3	3.23
Motivating people	3.32	3.46	3.62
Briefing	3.02	3.29	3.08
Dialog management	3.71	3.64	3.54
Conflict management	3.7	3.25	3.46
Stimulating competition	2.86	3.01	3.23
Understanding needs of others, empathy	3.78	3.43	3.15
Managerial ethics, social behaviour	3.82	3.75	3.54
Presentation abilities	3.87	4.03	3.62

5 CONCLUSIONS

The competence models have their important role not only in the field of human resource management, but also as regards their development by HE institutions. The competence developed by education becomes an advantage for job applicants, and the employers can better choose their future employees if they have their demands on competence specified. The competence models can provide for quality selection procedures of the staff whose performance is becoming decisive for companies' competitiveness.

This paper investigation has been facilitated by the financial support of the project, MŠMT SP 2012/8 „*Vytvoření kompetenčního modelu pro studenty HGF VŠB – TU Ostrava z pohledu těžebních společností*" (Competence Model Development for Students of the Faculty of Mining and Geology, VSB-Technical University of Ostrava, from the Perspective of the Mining Companies).

REFERENCES

[1] HRONÍK, F. Hodnocení pracovníků. 1. vydání. Praha : Grada Publishing, 2006. 128 s. ISBN 978-80-247-1458-5.

[2] KRČMARSKÁ, Lucie; ČERNÝ, Igor; ROLČÍKOVÁ, Markéta; VANĚK, Michal; MAGNUSKOVÁ, Jana. The Role of Competence Development in the Process of Education. In: 12th International Multidisciplinary Scientific GeoConference: SGEM2012 Conference Proceedings: June 17-23, 2012, Albena, Bulgaria. Albena: 1215 - 1222 pp. ISSN 1314-2704.

[3]   KOUBEK, J. Řízení lidských zdrojů. 2. vydání, Praha, Management Press, 2000, 350 s. ISBN 80-85943-51-4

[4]   KUBEŠ, M.; SPILLEROVÁ, D.; KURNICKÝ, R. manažerské kompetence. Způsobilosti výjimečných manažerů. 1. vydání. Praha. Grada Publishing, 2004, 184 s. ISBN 80-247-06398-9

[5]   SMEJKAL, J. Kompetenční vybavenost absolventů VŠ. 1. vydání. Ústí nad Labem : Univerzita Jana Evangelisty Purkyně v Ústí nad Labem, 2008. 110 s. ISBN 978-80-7471-708-9

[6]   TETT, R. P.; HAL, G. A, ; BLEIR, A.; Murphy, M. J. Development and Content Validation of a Hyperdimensional Taxonomy of Managerial Competence. In: *Human Performance* 13, no. 3 (2000): 205-25.[online]   [cit.   2012-12-05].   Dostupný   z   WWW: <http://condor.depaul.edu/profpjm/Tett%20et%20al.%20(2000).pdf>

[7]   VETEŠKA, J.; TURECKIOVÁ, M. Kompetence ve vzdělávání. 1. vydání. Praha : Grada Publishing, 2008. 160. ISBN 978-247-1770-8

RESUMÉ

Článek je zaměřen na problematiku významu kompetencí jako nástroje pro zvyšování konkurenceschopnosti průmyslových podniků.. Za cíl si autoři stanovili posoudit význam jednotlivých kompetencí z pohledu studentů kombinované formy studia a z pohledu zaměstnavatelů, konkrétně z pohledu těžebních podniků.

Pro pochopení dané problematiky obsahuje článek stručnou charakteristiku kompetencí a kompetenčního modelu. Stěžejní částí je pak průzkum kompetencí, který byl proveden ve dvou etapách. Jednalo se o posouzení významu kompetencí z pohledu studentů kombinované formy studia na VŠB-TU Ostrava, HGF. V rámci tohoto průzkumu byla hodnocena významnost jednotlivých kompetencí pro výkon funkce a významnost začlenění rozvoje kompetencí do studijních programů VŠ.

Druhá etapa průzkumu byla zaměřena na posouzení očekávané úrovně kompetencí absolventů VŠ z pohledu těžebních podniků.

16

HOMOGENEOUS MAGNETIC FIELD SOURCE FOR ATTENUATED TOTAL REFLECTION

Doc. Dr. Ing. Michal Lesňák[1]; RNDr. František Staněk, Ph.D.[1], Prof. Ing. Jaromír Pištora, CSc.[2], Ing. Jan Procházka[1]

[1]*VŠB - TU Ostrava, Institute of Physics, Faculty of Mining and Geology, VŠB –Technical University of Ostrava, 17. listopadu 15/2171, 708 33 Ostrava – Poruba, e-mail: michal.lesnak@vsb.cz*

[2]*Nanotechnology Centre, VŠB –Technical University of Ostrava, 17. listopadu 15/2171, 708 33 Ostrava – Poruba, e-mail: jaromir.pistora@vsb.cz*

Abstract

The paper is focused on the study of two-dimensional magnetic field distribution used for an analysis of samples containing magnetically active films by means of the Attenuated Total Reflection (ATR) method. The design of a proposed electromagnet and the magnetic field model computation are presented together with the results obtained from magnetic field distribution measurement. The ATR method can provide information about a thin film thickness, refractive index, and attenuation in addition to the perfunctory coupling of an optical wave into and off a waveguide [1, 2]. The prism coupling conditions are determined for magnetic structures with induced anisotropy.

The prism – a film coupler is located in the central cavity of a magnetic yoke. By current switching in the coils, we can change the amplitude and magnetic field direction in order to modulate the induced anisotropy in a thin film with magnetic ordering. By the in-plane modulation of the magnetization direction in the samples, we can change the rotation and elasticity of outgoing light.

Keywords: Attenuated Total Reflection (ATR), magnetic field, prism coupler

1 INTRODUCTION

Currently, magnetic materials are of great interest because they have found wide application in data storage, sensing, and integrated optical devices. An optical waveguide consisting of a magneto-optical (MO) layer can be advantageously used as light nonreciprocal isolators, filters and modulators driven modes. Magneto-optical waveguides are typically designed as an iron garnet layer sandwich or thin ferromagnetic films.

The theory of electromagnetic wave propagation in layered media has been described in literature – see for example [1-3]. Magneto-optical experiments can be divided into two groups according to the orientation of the external magnetic field with respect to the surface of the sample. In one case, an external magnetic field vector is perpendicular to the sample surface, in the latter case it lies in the surface of the sample. This is called a plane configuration.

The Dark Mode Spectroscopy (DMS) method, for which the magnetic field is determined, belongs to the family of ATR methods. The DMS is a technique based on the excitation of guided modes in a planar structure with a prism coupler. The electromagnetic wave interaction of magnetic anisotropic layered structures is carefully described by the 4 x 4 Yeh's matrix algebra [7].

As to the study of in-plane magnetic field influence on light guided in a thin film system using the DMS [4, 5], the generator of the appropriate magnetic field is an essential part of the measuring system. It has to allow changing both the orientation and the magnitude of magnetic induction of the external field in the space occupied by the studied sample [6, 7]. At the same time, it is crucial to keep the field homogenous.

2 WORK AIM

The objective was to create a homogeneous magnetic field with variable vector orientation of magnetic induction. The vector variable orientation of magnetic field induction can be obtained between the pole pieces of magnets which are perpendicular to each other, because the change in the orientation vector of the magnetic field is proportional to the change of the direction of the flow of electric current in the coils. Because the space for the measurement samples is large, the magnetic field homogeneity has to be measured. It results in a map of intensity (homogeneous areas), which gives us the necessary position of the measuring points of the sample.

3 WORK PROCEDURE

The measuring setup is composed of two units – the optical part and the magnetic part. The optical part is schematically represented in Fig. 1 where the pass of the light beam is represented schematically.

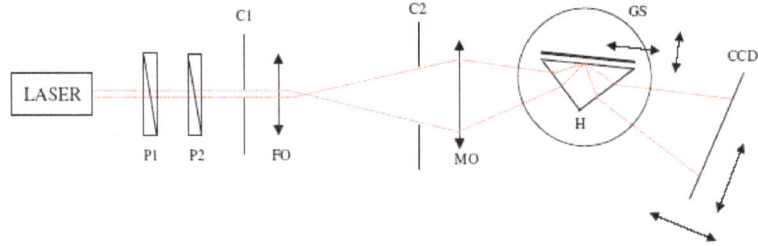

Fig. 1 Measuring apparatus for Dark Mode Spectroscopy. P1, P2 – polarizers, C1; C2 - shutters, FO - lens, MO - microscope objective lens, H – coupling prism, GS - rotation table, CCD camera.

A magnetic field generator for the study of light propagation in a magnetic garnet thin film is shown in Fig. 2. It consists of a pair of mutually perpendicular pole extensions and appropriate excitation coils. The intensity and orientation of a resulting field is controlled via the coil currents. The magnetic field intensity distribution is required to be constant in the planes perpendicular to the pole extensions faces, at least in the central part of the measuring area. The detail of the setup can be found in Fig. 3 where the sample is clearly seen together with the excitation prism and launching optics.

In order to fulfil all the method's requests, a special electromagnet had to be designed. The design of the magnetic field source had to respect the size of the studied samples (20 x 20 x 8 mm). Because the field homogeneity had to be kept in a perpendicular direction as well, the pole extension thickness was proposed to be 8 mm. The result is the magnetic circuit with a 20 x 8 mm size cross section, made from a soft magnetic material (AREMA steel). Two coils of the horizontal magnetic circuit (707+707 windings) as well as one coil of the vertical magnetic circuit are wound from a copper enamelled wire with a 0.67 mm diameter with total number of 1414 windings for each magnet circuit. The coils are designed for a standard maximal current density of $4A/mm^2$ in the used conductors.

The actual study of the magnetic field was realized in two ways. Partly with the help of a supported computational model and partly by the measuring of real magnetic field distribution by means of LOHET II connected to a digital voltmeter and with the METRA gaussmeter (made by METRA Blansko).

Fig. 2 Magnetic field source setup

Fig. 3 Detail of pole extensions of magnetic field source for DMS.

3.1 Magnetic Field Mapping

The main task of the experimental measurement work was to determine various parameters of the proposed magnetic field source. The first three measurements were performed using the LOHET II probe connected to the digital voltmeter. The detection range of the probe was ±0.05 T (for detailed description of the probe parameters see [10]). The advantage of using a simple probe was an easy data acquisition and computer processing of obtained values. On the other hand, the magnetic field intensity excited by the required current I=2.5 A exceeded the measuring range of the probe. That is why the METRA gaussmeter (measurements ranges ±0.2 T, ±0.5 T and ±2 T) had to be used in subsequent measurement, even if the data had to be recorded manually (no digital interface).

The experimental results obtained using the LOHET II probe are depicted in Figs. 5a and 5b. It is clearly seen that the magnetic field distribution in Fig. 5a looks quite 'rough'. Based on the preliminary results, it was decided to anneal the core of the magnet. The procedure was the following: a two-hour ramp from 20 ° to 800 °C, two-hour annealing at 800 °C and an eight-hour linear ramp cooling from 800 °C back to a room temperature. The magnetic field of the magnetic circuit is not quite homogeneous (findings of previous measurements). When annealing changes, the internal structure of the material changes as well, and it affects the homogeneity of the magnetic field [11].

Magneto-optic in-plane experiments can be roughly divided in two categories: pure configurations (longitudinal or transversal), or mixed configurations where the external magnetic field is arbitrarily oriented in the plane of the sample. Considering the first category, only one electromagnet connected to the power source is needed during the measurement. The question is how much the excited magnetic field is influenced by the presence of the pole extensions of the other electromagnet. That is why the next measurements were oriented on the influence of the pole extension remanence. All measurements were performed using the METRA gaussmeter. The vertical electromagnet was used for the excitation of the magnetic field, whereas the horizontal one was switched off. Two cases were considered: in the first case, the power leads of the unused coil were just disconnected, in the second one, the coil was short-circuited. The results of the experiments led to the conclusion that the influence of the residual magnetic field could be neglected and confirmed the suitability of the used soft magnetic material.

3.2 Study of Magnetic Field by Modelling Supported by ANSYS Program

A mathematical model of an electromagnet results from Maxwell's equations [7]. The solution of the equation system can be simplified by introducing new values, a vector potential $\mathbf{A}$ and a scalar potential φ. The magnetic flux density vector can be expressed as a vector potential rotor:

$$\mathbf{B} = \nabla \times \mathbf{A}.$$

$$(3.1)$$

For a solution of 2D propositions, it is advantageous to use the vector potential, whereas for 3D propositions, the scalar potential is better. The ANSYS program applies the transformation for the solution [8]

$$\mathbf{B} = \nabla \times \left(\mathbf{N_A}\right)^{\mathbf{T}} \mathbf{A}_e,$$ (3.2)

$$\nabla^{T} = gradient\ operator = \left[\frac{\delta}{\delta x} \frac{\delta}{\delta y} \frac{\delta}{\delta z}\right]$$

where $\mathbf{N}_A$ is a shape function; $\mathbf{A}_e$ defines a magnetic vector potential and T is a hash function.

4 RESULTS AND DISCUSSION

4.1 Magnetic Field Mapping

The result of the annealing can be seen in Fig. 4b. The annealing of the yoke helped greatly to improve the homogeneity of the magnetic field. The results were: the in-plane magnetic field maps for various positions of the perpendicular coordinate (see Fig. 6 for orientation). The magnetic field probes were mounted on the computer-controlled step-motor driven x-y translation stages with calibrated actuators. The perpendicular coordinate was measured using the dial gauge.

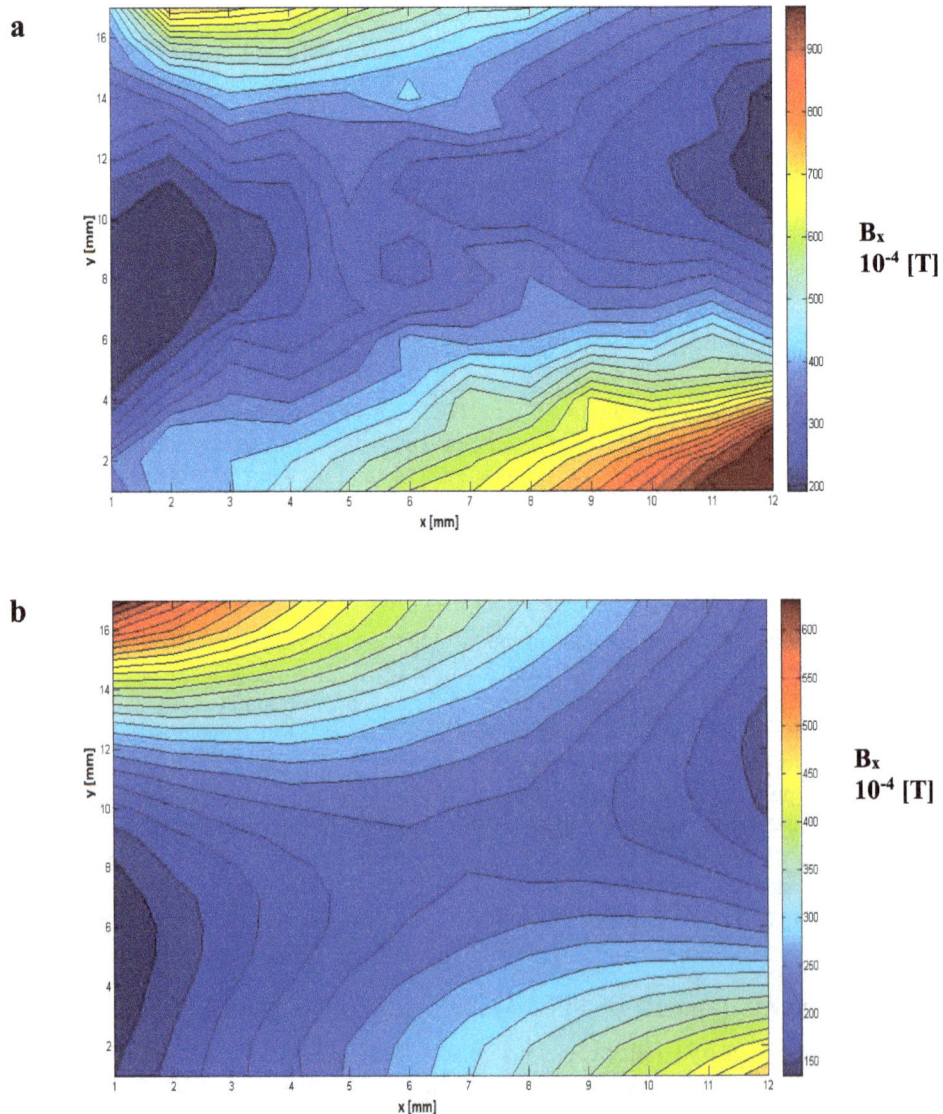

Fig. 4 Distribution of B_x in the measured sample area: (a) before annealing, (b) after annealing. (I$_1$=-0.11A, U$_1$=0.9V vertical coil; I$_2$= -0.9A, U$_2$=-0.8V horizontal coils), (z = 3 mm).

The main task of the experimental work was to determine the in-plane magnetic field distribution, which is crucial for the proper understanding of magneto-optic DMS experimental results. The example of the results can be seen in Fig. 5 where the components B_x (Fig. 5a) and B_y (Fig. 5b) are depicted.

In order to get the information about the perpendicular component of the magnetic field as well, the field mapping was performed for various depths measured with respect to the front plane of the pole extensions. The in-plane magnetic induction components were measured with a 1 mm probe movement step in both directions and both coils were fed by the dc current of 2.5 A.

a

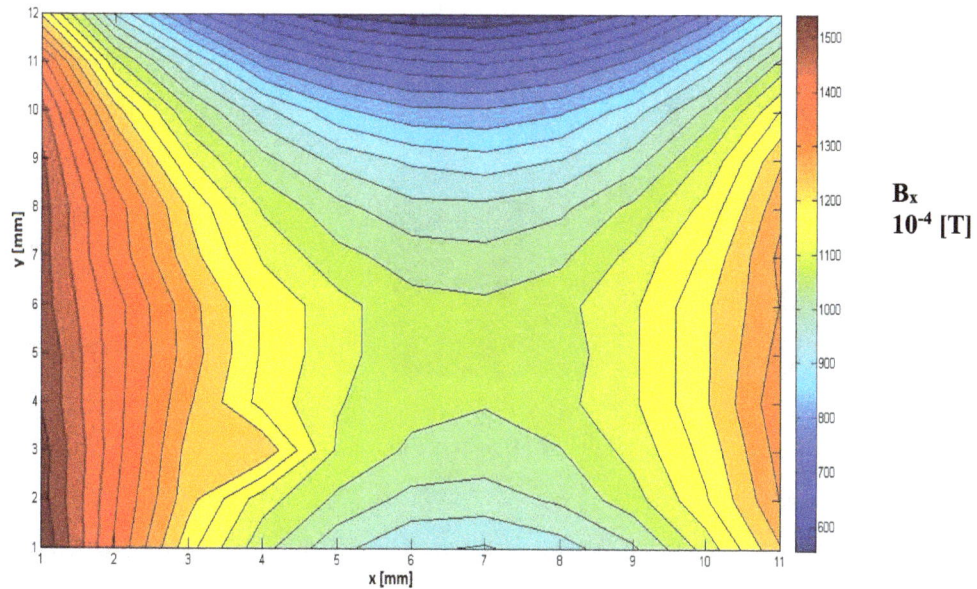

b

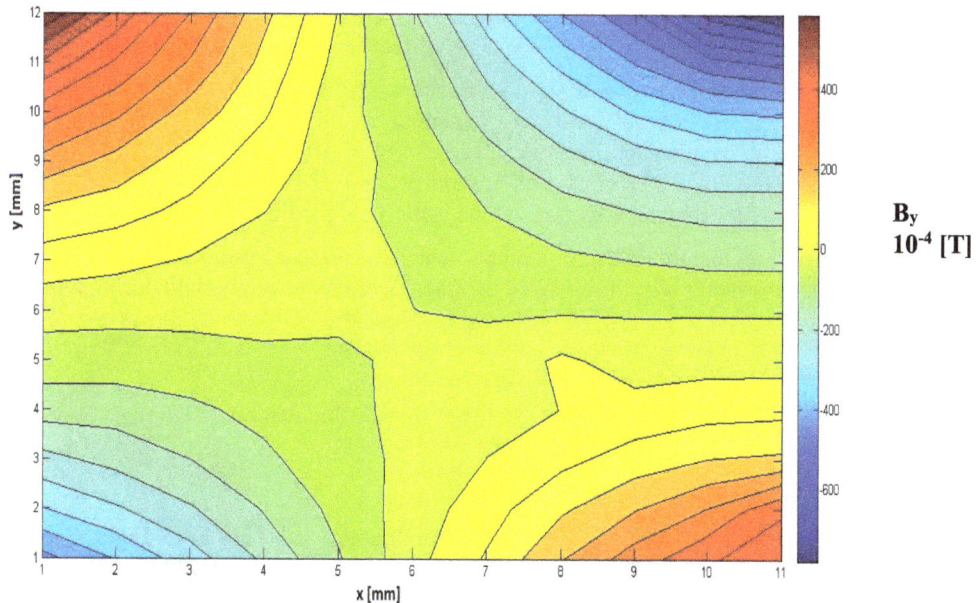

Fig. 5 (a) The B_x component of magnetic induction in detail in the measured area in a depth z = 3mm at 2.5 A current in coils, the measure detector is horizontal; **(b)** the B_y component of magnetic induction in detail in the measured area in a depth z = 3mm at 2.5 A current in coils, the measure detector is horizontal.

4.2 Study of Magnetic Field by Modelling Supported by ANSYS Program

While designing the model of magnetic field sources, at first the calculation with the help of the FEM – Finite Element Method was performed. The ANSYS program for solving the FEM on an IBM SP/2 computer was used for the calculation. The model of the 3D scheme of the magnetic field source for the DMS is shown in Fig. 6.

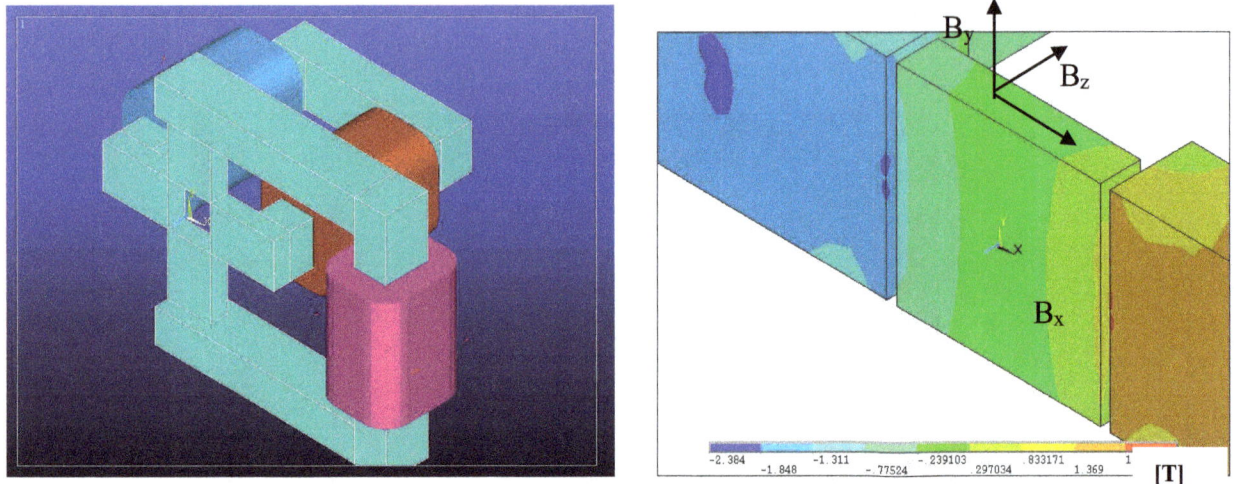

Fig. 6. Model of 3D scheme of magnetic field source.

5 CONCLUSION

On the basis of the created mathematical model, it is evident that the magnetic field distribution in the whole space between the pole extensions is not homogenous. The values of the magnetic field obtained from the measurement differ from the values obtained by modelling within an order. The differences are partly due to the parameters used in the model. Nevertheless, the magnetic field can be considered to be homogenous around the geometric centre of the measured area between the pole extensions in an area of 5x5 mm and 4 mm along the central axis parallel to z axis.

As to the temperature influence, a significant change of magnetic induction for the given value of the excitation current was not observed. Anyway, it is essential to note that at higher current loading it is necessary to apply forced cooling or to interrupt the work. Overheating of magnets does not influence magnetic induction for current loading of the coils. The construction modification of the extension (the recess for the launching microscope lens) has no fundamental effect on the magnetic field parameters in the sample area.

Positioning of a miniature measuring probe near the measured point (on the thrust tip) appears to be optimal for real time observations of the required parameters of the magnetic field during DMS experiments.

Acknowledgements

This work has been partially supported by the IT4Innovations National Supercomputing Center (#CZ.1.05/1.1.00/02.0070).

REFERENCES

[1]   Tien, P. K., Ulrich, R.: Theory of Prism-Film Coupler and Thin-Film Light Guides, Journal of the Optical Society of America, Vol. 60, No. 10, October 1970, pp. 1325-1337.

[2]   Ulrich, R.: Theory of Prism-Film Coupler by Plane-Wave Analysis, Journal of the Optical Society of America, Vol. 60, No. 10, October 1970, pp. 1337-1350.

[3]   Višňovský, Š., Yamaguchi, T., Pištora, J. and Postava, K., Beauvillain, P. and Gogol, P.: Unidirectional propagation in planar optical waveguides. Schenk, 2006, ISBN 80-248-1186-3, s.82.

[4]   Bárta, O., Pištora, J., Staněk, F., Postava, K.: Magnetic permeability effect on light propagation in planar structures. Proceedings of SPIE, 2000, Vol. 4239, pp. 204-209.

[5]   Lesňák, M.: The sours of magnetic fields for magneto optic measurement (in Czech). Jemná Mechanika a optika, 3/2002, s. 79-80.

[6]   Bárta, O.: Study of guiding modes in anisotropic andabsorbing planar structures. Ph.D. thesis (in Czech). VŠB – Technical University of Ostrava, 2006

[7]   Yeh, P.: Optics of anisotropic layered media: a new 4 x 4 algebram, Surf. Sci., 96, 1980 s. 41-53.

[8]   ANSYS, Inc. Theory Reference, www.ansys.com.

[9]   Kvasnica, J.: Electromagnetic field theory (in Czech). Praha Academia 1985.

[10]   http://uk.rs-online.com/web/search/searchBrowseAction.html?method=getProduct &R=17856

[11]   B.D. Cullity,: Introduction to Magnetic Materials, Massachusetts, Addison-Wesley, 1972.

ACIDIFICATION PROCESS IN THE AREA OF THE ABANDONED ĽUBIETOVÁ - PODLIPA CU-DEPOSIT, SLOVAKIA

Vojtech DIRNER[1], Jozef KRNÁČ[2], Lenka ČMIELOVÁ[3], Eva LACKOVÁ[4], Peter ANDRÁŠ[5]

[1] *Prof., Ing.,CSc. Institute of Environmental Engineering, Faculty of Mining and Geology, VŠB-Technical University of Ostrava, 17.listopadu 15, Ostrava, tel. (+420) 597 324 168*
e-mail vojtech.dirner@vsb.cz

[2] *RNDr., Ph.D. Department of Environmental Management, Faculty of Natural Sciences, Matej Bel University, Tajovského 52, 974 01 Banská Bystrica, Slovensko, tel. (+421) 484 465 809*
e-mail Jozef.Krnac2@umb.sk

[3] *Ing., Ph.D. Institute of Environmental Engineering, Faculty of Mining and Geology, VŠB-Technical University of Ostrava, 17.listopadu 15, Ostrava, tel. (+420) 597 325 585*
e-mail lenka.cmielova@vsb.cz

[4] *Ing., Ph.D.Institute of Environmental Engineering, Faculty of Mining and Geology, VŠB-Technical University of Ostrava, 17.listopadu 15, Ostrava, tel. (+420) 597 324 168*
e-mail eva.lackova@vsb.cz

[5] *Prof., RNDr., CSc. Geological Institute, Slovak Academy of Sciences, Ďumbierska 1, 974 01 Banská Bystrica, Slovensko, tel. (+421) 484 465 809*
e-mail andras@savbb.sk

Abstract

Acidity in surroundings of the abandoned Ľubietová-Podlipa Cu-deposit depends predominantly on the geochemical behaviour (weathering) of particular minerals (mainly pyrite). The article presents the results of measurements of basic physicochemical parameters – pH and Eh in technogenic sediments and in surface and drainage water of a dump-field. Although the dump material shows a significant amount of mobility-able metals and thus also a potential to form acidity, a massive AMD formation in future is not probable.

Abstrakt

Acidita v okolí opusteného Cu-ložiska Ľubietová - Podlipa závisí predovšetkým od geochemického správania sa (zvetrávania) niektorých minerálov (hlavne pyritu). V článku sú prezentované výsledky meraní základných fyzikálno-chemických parametrov – pH a Eh v technogénnych sedimentoch a v povrchovej a drenážnej vode haldového poľa. Haldový materiál síce stále vykazuje značné množstvo mobilizovateľných kovov a teda aj istý potenciál tvoriť kyslosť, podľa výsledkov výpočtu rizika, k masívnejšej tvorbe AMD nebude v budúcnosti pravdepodobne dochádzať.

Key words: dump-field, technogenic sediments, soil, acidity, neutralisation potential

1 INTRODUCTION

Surroundings of mine workings in Ľubietová represents an area having been changed by historical exploitation of copper ores [3]. Changes in pH and Eh in technogenic sediments cause a release of heavy metals from the solid phase, where they are in a form of heavier soluble minerals or a sorption complex, to groundwater and surface water [1,10]. The mobility of heavy metals in solutions and complex compounds is shown by the fact that the contents of many of heavy metals in technogenic sediments and soils and in products of oxidation of sulphides in the area of dumps is sometimes lower than the contents in soils under a dump-field, as well as the formation of numerous secondary minerals (mainly copper minerals) carbonates, phosphates, sulphates and oxides [4]. These secondary mineral phases originated mainly in the process of precipitation from solutions circulating in technogenic sediments and soils, but also due to the oxidation of primary minerals. Their formation controls and slows down migration of heavy metals, due to their stability under surface conditions [2].

Important soil properties include soil reaction which is used to indicate the acid-base reactions in soils. We distinguish between active and changeable forms of soil reaction. The active soil reaction is determined by oxonium cations and hydroxide anions present in a soil solution. It is determined from a H_2O soil solution. The exchangeable soil reaction is given except for free H^+ and OH^- ions also by H^+ and Al^{3+} ions adsorbed by a soil colloidal complex, which are released into the soil solution by the action of hydrolytic neutral salts (NaCl, KCl, $CaCl_2$). Generally, it is determined in 1M of the KCl solution [20].

Numerous chemical and biochemical reactions take place only under certain specific conditions of soil reaction. These chemical and biochemical reactions affect the decomposition of mineral and organic substances, formation of clay minerals, solubility (mobility) of substances and hence their bioavailability for living organisms, availability of nutrients, adsorption and desorption of cations, biochemical reactions, soil structure and physical properties [9]. The production of H^+ ions in soils is affected by acid rain and degradation of sulphides. Oxidation of sulphide minerals represents a complex of biogeochemical processes [15]. The rate of acidification depends on several factors such as the effect of bacteria, air O_2, the presence of water, etc. Signs of acidification are usually subdued by buffering and neutralizing ability of surrounding rocks (carbonates, clay minerals and organic matter), but beyond the buffering and neutralizing capacity of the environment, acute acidification begins to manifest [17].

Nand and Verloo [14] characterized the mobile fraction of metals as the sum of the soluble portion of metal in the liquid phase and the portion which indeed remains in the solid phase, but can gradually move into the soil solution. Rieuwerts et al. [16] emphasizes the importance of the mobility of metals for the possibility to estimate their concentrations in the soil solution, surface, drainage and groundwater. The total metal concentration depends on the content of metal in the solution, which in turn depends on the sorption to natural sorbents (such as clayey minerals, hydrogoethit, zeolites, ...) and the metal release into the soil solution [21].

The acidification is significantly influenced by the rock composition of the surrounding terrain and especially the material of dump-fields. The vast majority of rocks in the area of the studied dump-field is created by rocks of a terrigenous crystalline complex of Permian age [8]. Main representations of them are greywackes and arkoses, colourful slate and conglomerates. These rocks have only a minimum neutralizing potential. Carbonates, which may enter the reaction as a neutralizing agent, are represented only by calcite veins, ankerite and rare siderite.

The determination of the risk of acidification is a guide to determine the degree of risk of environmental contamination of the country by heavy metals. The aim of this study is to determine the risk of occurrence of acid mine water (AMD) whose production by the material of mine dumps could lead to a release of ore components into percolating water, an increase of the degree of their bioavailability and thus the acidification and contamination of landscape components by heavy metals.

2 METHODOLOGY

From the surface of the dump-field, 90 samples of technogenic sediments numbered 1-89 and 12 samples numbered A-1 through A-12 were taken across the surface in a regular network. Each sample incurred by the homogenization of 8-10 samples among which each have a weight of approximately 2 kg. The sample A-12 represents the reference surface with no mineralization outside the dump-field (Fig. 1). The resulting sample had a mass of about 3 kg. The heavy fraction of the samples was studied by the scanning microscope JEOL JSM-840 at the State Geological Institute Dionysus Stur, Bratislava.

The samples of surface and drainage water (sample V-1 and V-13) were collected in four periods: after rains (in June 2006 – samples identified with the index "a" and in March 2008 – samples identified with the index "c") and in dry seasons (in February 2007 – samples identified with the index "b", in May 2008 – samples identified with the index "d"). The surface water samples were collected from the mountain creeks and marshes in the vicinity of mining dumps as well as from the creek flowing through the municipality of Ľubietová. Water sampling sites are indicated in Fig. 2.

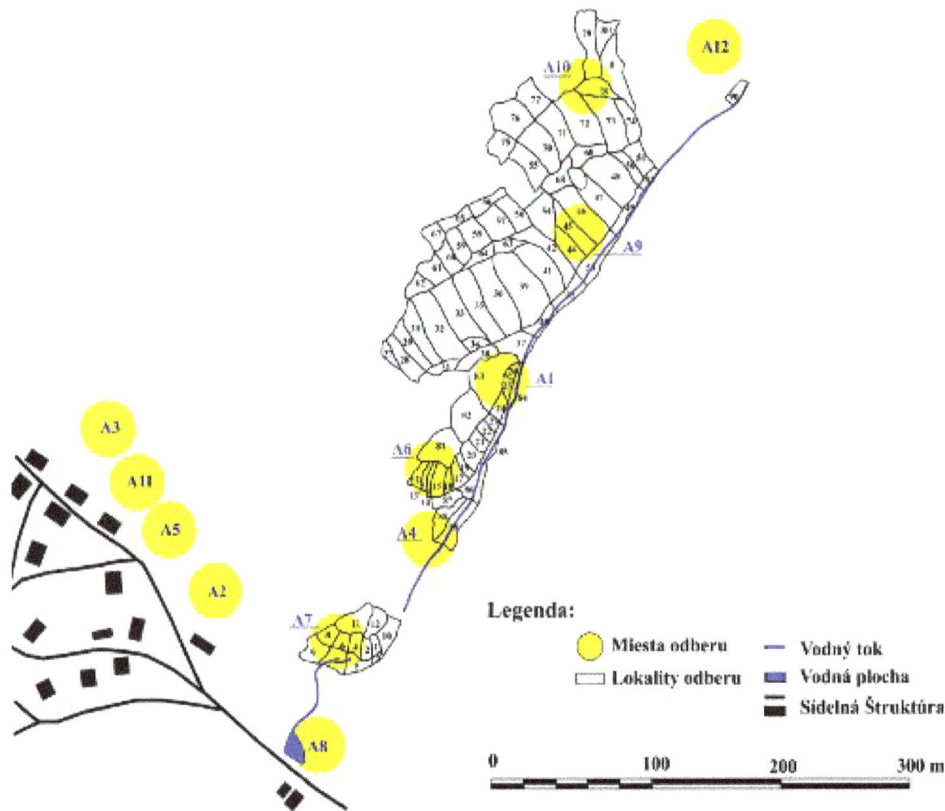

Fig. 1 Location of sampling technogenic sediments and soils from the site Ľubietová - Podlipa.

Active and changeable pH was also determined in surface water of a broader area of the dump-field as well as in drainage water percolating through dump sediments, and was measured in the suspension of 20 g of the sample and 50 ml of distilled water, or 1M KCl, according to the methodology by Van Reeuwijk [22] used for soils using the pH meter EcoScan pH 5/6 by the companuy EUTECH Instruments. The monitoring points are indicated in Fig. 2.

The measurement of the oxidation-reduction potential (Eh) was carried out by means of the in situ measuring instrument WTW Multi 3420 with the combined redox-electrode SenTix ORP with the reference system Ag/AgCl, containing the 3M KCl electrolyte solution. The Eh reading was converted to the standard hydrogen electrode.

The measurement of pH and Eh of water (Samples V-1 to V-13) was carried out "in situ" directly while taking water samples. The Eh measurement of sediment samples (samples 1 to 90, A1 to A12) was carried out in an aqueous suspension, and the working procedure was as follows:

The samples of soils and technogenic sediments were homogenized. A 20g weighed portion of the samples of soil and technogenic sediments was poured with 50 ml of distilled water in a 200ml flask. After 20 min of shaking out, the contents of the flask was poured into a 200ml beaker and while stirring constantly the suspension, pH and Eh were determined.

For the determination of the total acidity (AP) and neutralization potential (NP), it was necessary to know pH and Eh of sediments, contents of sulphur and carbon. For the prediction of acid mine products in mine waste, the U.S. EPA methodology was used: Acid Mine Drainage Prediction EPA530-R-94-036 from 1994, which was applied for the needs of research in the Western Carpathians by Lintnerová and Majerčík [11].

The analysis of total (S_{tot}), sulphidic (S_s) and sulphate sulphur (S_{SO4}) was performed from a 1g weighed portion of the samples in the Geoecological Laboratories of the Geological Survey, Spišská Nová Ves. Carbon (total carbon – $C_{tot.}$, organic carbon – $C_{org.}$ and inorganic carbon – $C_{inorg.}$) was determined in the laboratories of the Geological Institute of the Slovak Academy of Sciences with IR-spectroscopy of carbon using the device Ströhlein C-MAT 5500. The values of contents of CO_2 and $CaCO_3$ were determined by the recalculation from the values of $C_{tot.}$, $C_{org.}$ and $C_{inorg.}$ The analytical data was processed by the GIS system.

Σ Fe and Fe^{2+} were determined in the Testing Laboratories GEL, Turčianske Teplice by titration.

The samples of surface and drainage water were stabilized with 10 ml.l^{-1} HNO$_3$ and analyzed in the Water Research Institute, Bratislava.

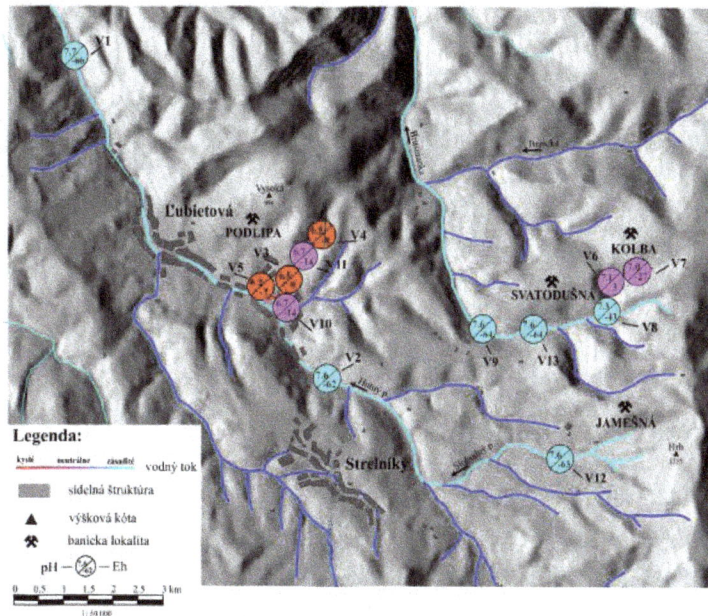

Fig. 2 Location of monitoring sites of samples of surface and drainage water around Ľubietová indicating measured pH values.

Processing the map image attachments were made using Grass GIS, a modular tool for processing spatial data. Plotting the data was carried out with the v.surf.rst module ensuring spatial interpolation of data (regularized spline with tension, RST).

3 Results

The active pH value in samples from the dump-field Podlipa (and a reference surface) varies within the range of 3.75 – 6.32 and the changeable pH value from 3.58 to 6.48 (Tab. 1, 3). The average of the active pH value of the reference area is 4.93 and the average of the changeable pH value is 3.47 (Tab. 1). The acidity of the dump-field on the basis of the leachates in distilled water and in the 1M KCl solution is shown in Fig. 3 a, b.

The chemical reactions of soil are as follows: a) acid – pH 4 or less, b) neutral – pH 6.5 to 7.4 and c) alkaline – pH 7.5 and above [7]. Under this classification, the acidic soil reaction was (with minor exceptions) confirmed throughout the entire study area, especially on landfill platforms of terraces (compare Fig. 1; lower acidity of slopes is probably due to a greater leaching of material with rainfall water

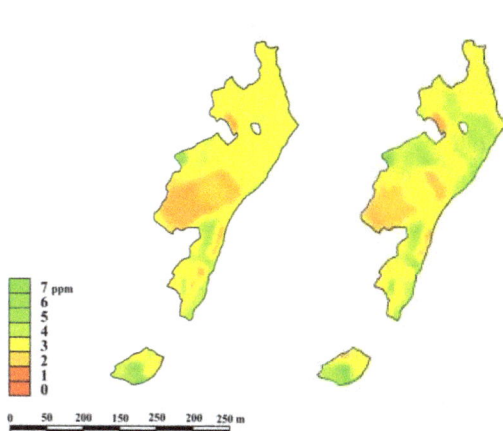

Fig. 3 Acidity of technogenic sediments in the dump-field Podlipa;figures indicate pH values in the a) water leachate, b) 1M KCl leachate.

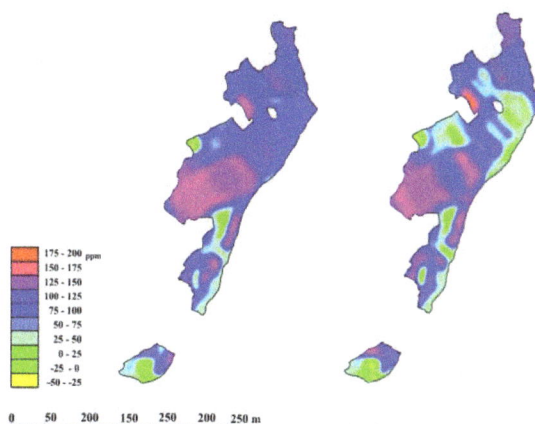

Fig. 4 Values of Eh of technogenic sediments and soil in the dump-field Podlipa;figures indicate Eh values in the a) water leachate, b) 1M KCl leachate.

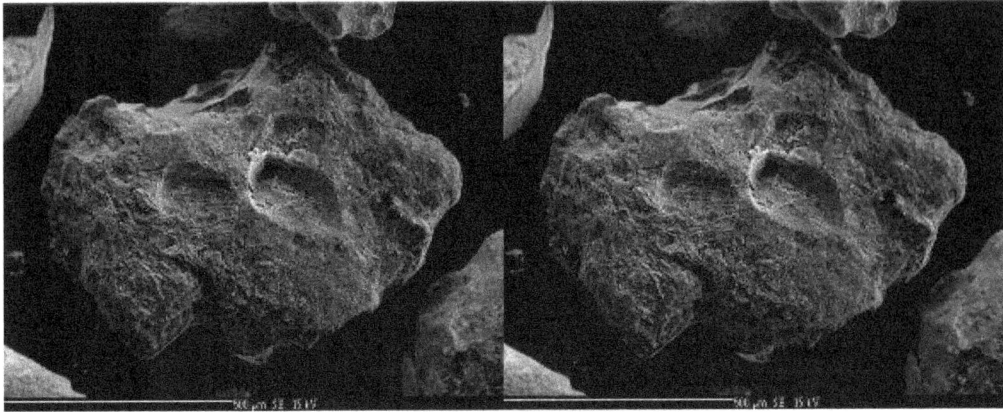

Fig. 5 Porous pure copper
Fig. 6 Pyritic pentagon dodekaeder – FeS$_2$

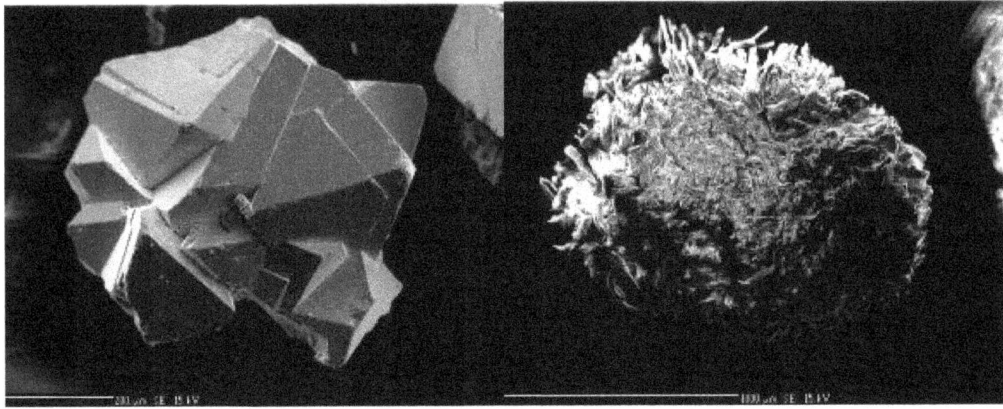

Fig. 7 Aggregate of fused crystals of euchroite - Cu$_2$(AsO$_4$)(OH)·3H$_2$O
Fig. 8 Aggregate of lepidocrocite – FeOOH

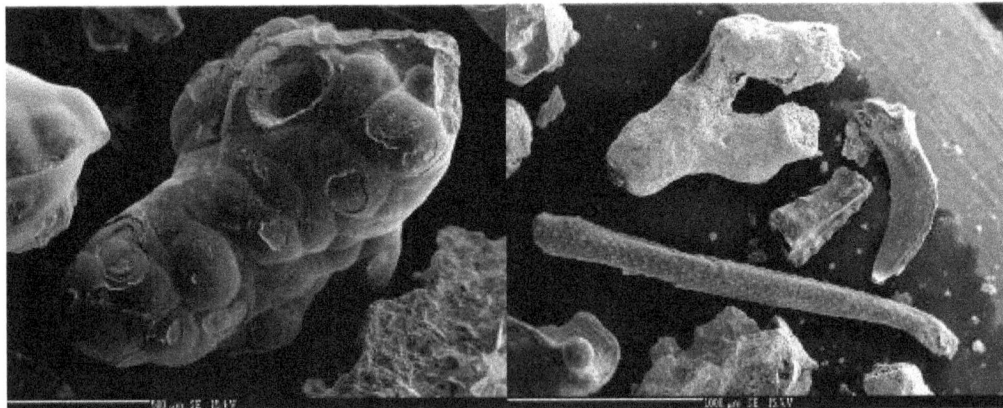

Fig. 9 Nodules of malachite
Fig. 10 Organic remains of animal and plant origin in sediments of landfill material

The oxidation-reduction potential (redox potential - Eh) is a parameter that allows you to define aerobic (oxidation) and anaerobic (reduction) processes in soil. The most important from the factors that determine the conditions for such processes are soil moisture, soil reaction (pH), O$_2$ content in the soil air and solution, organic matter content, presence of elements – Fe, Mn, N, S, and activity of micro-organisms [7]. The lower the value Er, the more intense the reduction processes taking place in the soil and vice versa. The values higher than 600 mV indicate soil oxidation processes; the values lower than 200 mV indicate the dominance reduction processes in soil.

The Eh measured value is not stable, it fluctuates at a very small distances and varies depending on the amount of rainfall. The Eh values in the water leachate vary in a wide range from 5 to 156 and in the 1M KCl leachate from 16 to 174 (Tab. 1).

The most oxidizing conditions indicate the measured Eh values (up to 133 mV in the water leachate and 173 mV in the 1M KCl leachate) on the platforms and top parts of the dump-field (Tab. 1; Fig. 4). The highest values of Eh are in the sample from the reference area A-12. The lowest Eh values (-84 in the water leachate and -58 in the 1M KCl leachate) were measured in the waterlogged lower zone of heaps (Tab. 1; Fig. 4).

In the dump material, the following were identified: silicate minerals, oxides, as well as pure metals (Fig. 5), oxidized and non-oxidized sulphide minerals (Fig. 6), arsenates (Fig. 7), hydroxides (Figure 8), secondary sulphides, secondary Cu carbonates (Fig. 9) and residues of organic matter (Fig. 10).

Weathering kinetics of these minerals depends on pH, temperature and mainly on the reaction surface. The weathering of carbonates and certain silicates is able to deliver neutralizing agents to the system, which are able to consume the acid released from the oxidation of sulphides and thus buffer the system to the point so that the resulting pH is close to neutral values [12].

Tab. 1 Characteristics of samples A-1 to A-12 of technogenic sediments from the dump-field.

Sample	H_2O		1M KCl		%							
	pH	$Eh_{(mV)}$	pH	$Eh_{(mV)}$	S_{tot}	S_{SO4}	S_s	$C_{tot.}$	$C_{org.}$	$C_{inorg.}$	CO_2	$CaCO_3$
A-1	5.14	77	4.61	109	0.25	0.10	0.15	0.74	0.20	0.54	1.97	4.48
A-2	5.89	34	5.40	63	0.02	0.01	0.01	0.86	0.38	0.48	1.75	3.99
A-3	4.87	94	4.21	131	0.10	0.03	0.07	0.62	0.34	0.28	1.02	2.32
A-4	5.46	59	5.33	66	0.33	0.13	0.01	0.34	0.26	0.08	0.29	0.66
A-5	5.77	42	5.37	64	0.05	0.01	0.05	0.78	0.35	0.43	1.57	3.57
A-6	5.17	74	5.06	83	0.42	0.15	0.27	0.40	0.27	0.13	0.47	1.08
A-7	7.93	-84	7.34	-58	0.03	0.02	0.01	1.63	0.10	1.53	5.61	12.71
A-8	5.42	36	5.22	42	0.01	0.01	0.01	0.45	0.13	0.32	1.17	2.66
A-9	5.03	83	5.01	85	0.03	0.03	0.01	0.40	0.37	tr.	tr.	tr.
A-10	5.25	71	5.14	78	0.04	0.02	0.02	0.48	0.46	tr.	tr.	tr.
A-11	6.11	22	5.95	30	0.11	0.04	0.07	4.31	4.18	0.13	0.47	1.08
A-12	4.21	133	3.47	173	0.02	0.01	0.02	4.05	4.03	tr.	tr.	tr.

It is interesting that the lowest pH value was measured in the area of the reference surface – sample A12. This is probably due to the fact that despite the absence of sulphides there are no carbonates there; inorganic carbon contents ($C_{inrg.}$) are below the limit of quantification (Tab. 1). The highest contents of carbon within the dump-field were found in the sample A-7 ($C_{tot.}$ 1.63 %, recalculated to $CaCO_3$ up to 12.71 %; Tab. 1), where also the highest pH_{H2O} value of 7.93 was recorded.

Tab. 2 Calculation of reliability of the value R^2.

	Measurement		Calculation	
	H_2O	1M KCl	H_2O	1M KCl
R^2	0.9968	0.9992	0.9945	0.9959

The pH-Eh diagrams of the stability of compouds with the contents of Fe in the system $Fe^{3+} - SO_4^{2-} - H_2O$ in Figs. 11 and 12 show the form of occurrence of Fe in the dump material from the Podlipa and Reiner deposits. The diagram shows a strong correlation between pH and Eh. High values of R^2 reliability equations (Tab. 2, Fig. 11) show a very accurate determination, especially the determination of pH and Eh in an leachate.

The activity diagram of the components in the system $Fe^{+3} - SO_4^{-2} - H_2O$ describes the stability of solid phase for active and changeable pH values. Based on in situ measurements of physico-chemical parameters (pH and Eh) and by calculating Eh [18, 19] it was found out that the main form of iron occurrence in sediments and soils of the dump-fields Podlipa and Reiner is the sulphate form – $FeSO_4$ (Fig.12). This data is inconsistent with the data indicating that a high incidence of sulphide sulphur was confirmed in the material of dump-fields [5]. This discrepancy can be explained by the fact that Fe occurs in sulphide form only in fresh rock fragments that do not enter the reaction during the measurement of pH and Eh due to the time factor (60 minutes). The possibility of Fe occurrence also in sulphide form is indicated by the calculated values of R^2 which are unlike those obtained by in situ measurements of pH and Eh at the boundary of stability of fields $FeSO_4$ and FeS_2 (Fig. 12).

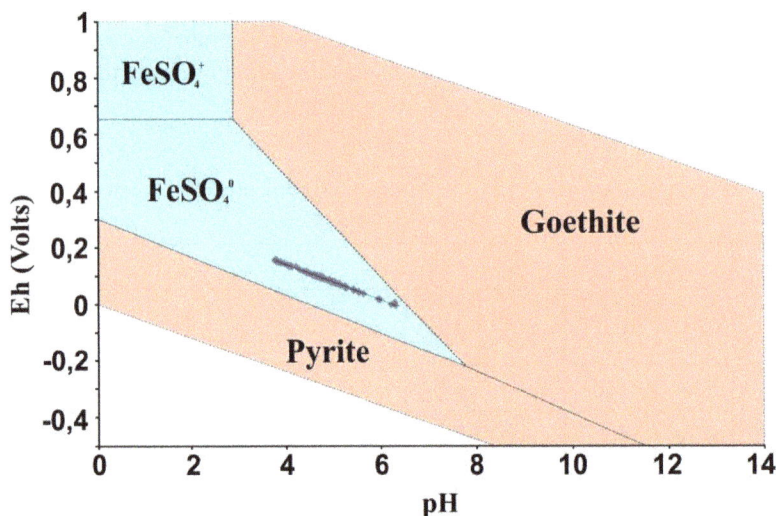

Fig. 11 Activity diagram of components in the systeme $Fe^{3+} - SO_4^{2-} - H_2O$, describing the stability of solid phases of activities $[Fe^{3+}] = 1.9136e-02$ and $[SO_4^{2-}] = 2,5259e-02$ [6] for samples of technogenic sediments from the dump-fields in Ľubietová.

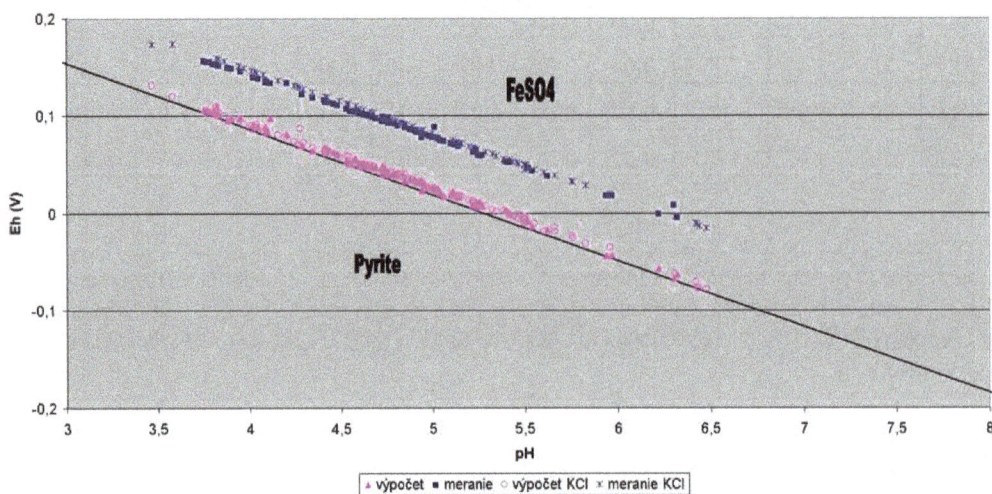

Fig. 12 Activity diagram of components in the system $Fe^{3+} - SO_4^{2-} - H_2O$, for active pH ($pH_{H2O}$) and for changeable pH (pH_{KCl}) of the samples of the dump material from Podlipa and Reiner.

The pH value of surface water around the village of Ľubietová approaches usually neutral values. Nevertheless there are some differences. Even though both surface water flows and mine water in the area of the Svätodušná dump-field show probably slightly alkaline reaction due to the different lithology (especially the higher contents of dolomite in the rock complex), but also due to the different composition of the primary ores (e.g., a lower representation of pyrite), surface and drainage waters from the Podlipa area are slightly acidic in nature (Tab. 3, Fig. 2), primarily resulting from weathering of ore minerals (mainly pyrite, but also chalcopyrite) in dump sediments and soils. This characteristic was also confirmed by multiple measurements: the data collected during the dry season and the rainy season differs substantially rarely only.

Tab. 3 Values of pH and Eh in samples of surface and mining water.

Sample	pH	Eh (mV)	Sample	pH	Eh (mV)
V-1a	7.7	-69	V-7a	7.0	-27
V-1c	7.7	-66	V-7b	7.1	-29
V-2a	7.6	-62	V-8a	7.7	-66
V-3a	6.7	-12	V-8b	7.3	-43
V-3b	6.2	14	V-9a	7.6	-64
V-3c	6.5	0	V-9a	7.7	-67
V-4a	6.5	-6	V-10a	6.7	-14
V-4b	7.5	-58	V-10b	6.2	14
V-4c	6.54	-8	V-11b	6.7	-14
V-5a	6.2	-11	V-11c	6.9	-21
V5b	6.1	-8	V-12a	7.6	-63
V-6a	6.4	1	V-13a	7.6	-64
V-6d	7.1	3	V-13b	7.4	-60

To determine the risk of acidity, the set of samples A-1 to A-12 is used, which represents the entire dump-field. The pH value in sediments and soil of these samples set out in the leachate by distilled water varies between 4.21 and 7.93 (Tab. 1). Sulphide sulphur is in most samples in a higher amount than sulphur sulphate, indicating a relatively high content of not yet oxidised primary sulphides.

Tab. 4 Values of total created acidity (AP), neutralization potential (NP) and net neutralization potential (NNP).

Sample	AP	NP	NNP
A-1	7.81	44.8	37.0
A-2	0.62	39.9	39.3
A-3	3.12	23.2	20.1
A-4	10.31	6.6	-3.7
A-5	1.56	35.7	34.1
A-6	13.12	10.8	-2.3
A-7	0.93	127.1	126.2
A-8	0.31	26.6	26.3
A-9	0.93	0	-0.9
A-10	1.25	0	-1.3
A-11	3.43	10.8	7.4
A-12	0.62	0	0.6
Average	3.7	27.1	23.5

The total acidity formation (AP) corresponds to the amount of acid that can potentially be produced by the dump material. At the Podlipa site, AP ranges from 0.3125 to 13.125 (at average 3.7, Tab. 4). The value of neutralization potential (NP) indicating the content of neutralizing substances in a dumping site, capable to neutralize the acidity produced by the dump material is different in each section of the dumping site (0 to 127.1, $\bar{x}$ = 27.1; Tab. 3) and in the negative correlation to the AP. A high NP (127.1) is only in the sample A-7 (Tab. 4), in which the highest content of $C_{tot.}$ (which after recalculation corresponds to 12.71 kg.t^{-1} of $CaCO_3$; Tab. 1) was determined.

The net neutralization potential (NNP) corresponds to the amount of a neutralizing substance that must be added to neutralize the acidity produced by the dumping site (NNP = NP – AP). The NNP values in the dump-field Podlipa (Tab. 2) show that the neutralization of mine tailings requires as much neutralizing agent as it corresponds, on an average, to 23.5 kg of $CaCO_3$ per 1 ton of the dump material. The best NNP value shows the risk of the occurrence of acid mine water (AMD). If the value is close to 1, the risk of the AMD formation is high. Where the ratio is equal to or greater than 3, the risk of formation of AMD is negligible [20] (Sobek et al. 1978). The average value of the NNP at the Podlipa dump-field is 23.5 (if we exclude the extreme value of 126.08, the average value will change to 14.62; Tab. 4), so the risk of the AMD formation can be ruled out.

4 DISCUSSION

Spatial distribution of heavy metals in the dump-field is a reflection of their geochemical properties: content, solubility, acidity (migration potential) and sorption properties. Lack of carbonates (as the primary natural neutralizing agent) causes that in 5 from 12 samples (A-1 to A-12), corresponding to average samples from 12 parts of the dump-field, the NPP values are negative (neutralizing agents are missing at all), and only two values (from the samples A-3 and A-11) are very low (7.4 and 20.1). Unlike Sobek et al. [20] , who states that if the NNP values are greater than 3, the risk of AMD formation is negligible in terms of the US EPA methodology [13], the NNP values from -20 to 20 (kg of $CaCO_3.t^{-1}$ of the dump material) in terms of the potential formation of acidic substances classified as a „*range of uncertainty*“ as it is impossible to make it clear whether or not the formation of AMD occur. The pH value in leachates in distilled water usually fluctuates around 5.3. pH <5 which indicates that the sample contains potential acidity, while the values measured in carbonates generally oscillate between 8 – 10. The values above 10 can be considered to be alkaline ones [20].

5 CONCLUSIONS

The ongoing process of oxidation of primary sulphide minerals at the deposit is indicated by coatings of secondary copper minerals (carbonates and oxides) and confirmed by the observed values of pH and Eh in leachates from technogenic sediments. The dump material still shows a considerable amount of mobilized metals and a certain potential to generate acidity. As a result, the surface water in the area of the dump-field Podlipa is slightly acidic, while the water from the nearby dump-field Svätodušná shows a slightly alkaline character due to the presence of carbonates (calcite, ankerite and siderite).

The dump material still shows a considerable amount of mobilized metals and a certain potential to generate acidity.

Although a relatively high level of total sulphide sulphur in technogenic sediments indicates a considerable content yet unoxidised primary sulphides, the overall result of the study of acidity formation suggests that the formation of AMD is not likely to occur in the future.

Acknowledgement The work was supported by the grants APVV-0663-10, APVV-51-015605 and in the framework of the project Opportunity for young researchers, reg. no. CZ.1.07/2.3.00/30.0016, supported by Operational Programme Education for Competitiveness and co-financed by the European Social Fund and the state budget of the Czech Republic.

REFERENCES

[1] ALLOWAY, B. J.Soil processes and the behaviour of metals. In: B. J. Alloway (ed.) Heavy Metals in Soils. 1995, Glasgow, Blackie, 214 pp.

[2] ASHLEY, P. M., CRAW, D., GRAHAM, B. P., CHAPPEL, D. A. Environmental mobility of antimony around mesothermal stibnite deposits, New South Wales Australia and southern New Zealand. Journal of Geochemical Exploration. 2003, 77,pp 1-14.

[3] ANDRÁŠ, P., JELEŇ, S., KRIŽÁNI, I. Cementačný účinok drenážnej vody z haldového poľa Ľubietová-Podlipa. Mineralia Slovaca. 2007, 39, 4, pp 303-308.

[4] ANDRÁŠ, P., LICHÝ, A., RUSKOVÁ, J., MATÚŠKOVÁ, L. Meavy metal contamination of the ladscape at the Ľubietová deposit (Slovakia). Proceedings of World Academy of Science, Engineering and Technology. 2008. 34, ISSN 2070 – 3740, Venice, Italy, pp 97-100.

[5] ANDRÁŠ, P., LICHÝ, A., KRIŽÁNI, I., RUSKOVÁ, J. Heavy metals and their impact on environment at the dump-field Ľubietová-Podlipa (Slovakia). In: Advanced Technologies. Ed.: Jayanthakumaran, K. In-Tech, Olajnica. 2009, 19/2, 32000, ISBN: 978-953-307-009-4, pp 163-185.

[6] BETHKE, C. The Geochemist's Workbench software package. 2000. University of Illinois, Urbana-Champaign, 220 p.

[7] ČURLÍK, J., BEDRNA, Z., HANES, J., HOLOBRADÝ, K., HRTÁNEK, B., KOTVAS, F., MASARYK, Š., PAULEN, J. Pôdna reakcia a jej úprava, 2003 Bratislava, ISBN 80-967696-1-8, 249 p.

[8] EBNER, F.; PAMIČ, J.; KOVÁCS, S.; SZEDERKÉNYI, T.; VAI, G. B.; VENTURINI, C.; KRÄUTNER, H. G.; KARAMATA, S.; KRSTIČ, B.; SUDAR, M.; VOZÁR, J.; VOZÁROVÁ, A., MIOČ, P. Variscan Preflysch (Devonian-Early Carboniferous) environments 1 : 2 500 000: Tectonostratigraphic terrane and paleoenvironment maps of the Circum-Pan-nonian region. Budapest: Geological Institute of Hungary. , 2004. 63-671-245X CM, 125 p.

[9] GOULD, W. D., BÉCHARD, G., LORTIE, L. The nature and role of microorganisms in the tailings environment. In: Jambor, J. L., and Blowes D. W., (eds.), Environmental Geochemistry of Sulphide Mine-Wastes, Mineralogical Assoc. of Canada Short Course Handbook. 1994, pp 185-200.

[10] LEE, J., CHON, H., KIM, J. Human risk assessment of As, Cd, Cu, and Zn in the abandoned metal mine site. Environ. Geochem. Health. 2005, 27, pp 185–191.

[11] LINTNEROVÁ, O., MAJERČÍK, R. Neutralizačný potenciál sulfidického odkaliska Lintich pri Banskej Štiavnici – metodika a predbežné hodnotenie. Mineralia Slovaca. 2005, 37, 4,pp 517 – 528.

[12] LINTNEROVÁ, O., ŠOLTÉS, S., ŠOTNÍK, P. Environmentálne riziká tvorby kyslých banských vôd na opustenom ložisku Smolník. UK Bratislava. 2010, ISBN 978-80-223-2764-0, 157 p.

[13] MISSANA, T., GARCIA-GUTTIEREZ, M., ALONSO, U. Sorption of strontium onto illite/smectite mixed clays. Physics and Chemistry of the Earth. 2008, 33, Supl. 1.,pp 156-162.

[14] NAND R., VERLOO, M. Effect of various organic materials on the mobility of heavy metals in soil. Environ. Pollution (B). 1985, 10, pp 241–248.

[15] NORDSTROM, D. K. Aqueous pyrite oxidation and the consequent formation of secondary iron minerals, Kittrick, J.A., Fanning, D.S. and Hossner, L.R., eds. Acid sulfate weathering. Soil Science Society of America. , 1982, pp 37-63.

[16] RIEUWERTS, J. S., THORNTON, I. FARAGO, M. E., ASHMORE, M. R. Factors influencing metal bioavailability in soils:preliminary investigations for the development of a critical loads approach for metals. Chem. Spec. Bioavail. 1998, 10(2), pp 61–75.

[17] RIMSTIDT J. D., VAUGHAN D. J. Pyrite oxidation: A state-of-the-art assessment of the reaction mechanism. Geochimica Cosmochimica Acta. 2003, 67, pp 873-880.

[18] SATO, M. Oxidation of sulfide ore bodies, I. Geochemical environments in therms of Eh and pH., Economic Geology. 1960, 55, pp 928-961.

[19] SATO, M., Oxidation of sulfide ore bodies, II. Oxidation mechanisms of sulfide minerals at 25°C, Economic Geology. 1960a, 55, pp 1202-1231.

[20] SOBEK, A. A., SCHULLER, W. A., FREEMAN, J. R., SMITH, R. M. Field and laboratory methods applicable to overburden and minesoils, U. S. Environmental Protection Agency, Environmental Protection Technology. 1978, EPA 600/2-78-054, Cineti. OH. 203 p.

[21] STERCKEMAN, T., DOUAY, F., PROIX, N., FOURRIER, H. Vertical distribution of Cd, Pb and Zn in soils near smelters in North of France. Environ. Pollution. , 2000, 107,pp 377–389.

[22] VAN REEUWIJK, L. P. Procedures for soil analysis. International soil reference and information centre (ISRIC) a FAO OSN. Technical report, 9. 1995, Zubkova, N. V., Pushcharovsky, 87 p.

RESUMÉ

V článku sú prezentované výsledky meraní základných fyzikálno-chemických parametrov – pH a Eh v technogénnych sedimentoch a v povrchovej a drenážnej vode haldového poľa opusteného Cu-ložiska Ľubietová – Podlipa. Výsledky výpočtu rizika potvrdili, že acidita okolia zkúmanej plochy závisí predovšetkým od geochemického správania sa (zvetrávania) niektorých minerálov (hlavne pyritu), a že haldový materiál stále vykazuje značné množstvo mobilizovateľných kovov a teda aj istý potenciál tvoriť kyslosť, ale ďalej výsledky poukazujú, že k masívnejšej tvorbe AMD nebude v budúcnosti pravdepodobne dochádzať.

CONTRIBUTION OF ELECTRICAL RESISTIVITY TOMOGRAPHY APPLIED TO THE SLOPE DEFORMATION SURVEY IN LIDEČKO

Bladimir CERVANTES[1], Aleš POLÁČEK[2], Jaroslav RYŠÁVKA[3]

[1]Ing., Institute of Geological Engineering. Faculty of Mining and Geology,
VŠB – Technical University of Ostrava, tř. 17. listopadu 15/2172, 708 33 Ostrava-Poruba
E-mail: bladimir.cervantes@vsb.cz

[2]Ing., Institute of Geological Engineering. Faculty of Mining and Geology,
VŠB – Technical University of Ostrava, tř. 17. listopadu 15/2172, 708 33 Ostrava-Poruba
E-mail: ales.polacek@vsb.cz

[3]   Ing., PhD., Unigeo a.s., Místecká 258, Ostrava-Hrabová,
E-mail: rysavka@unigeo.cz

Abstract

In the last years, electrical resistivity tomography (ERT) has been increasingly used to solve various types of problems in engineering geological survey, geotechnical investigations, etc. It gradually replaces a traditional combination of methods of resistivity profiling (RP) and vertical electrical sounding (VES). This paper provides selected results obtained from the survey of a slope deformation in Lidečko. It brings also some new details about its construction and results of monitoring carried out in the year 2011. The largest landslide hazards result from its position over a water pipeline line, where there is a real risk of a massive landslide of the slope ending in the valley of the Senice River. It is an old landslide reactivated during the floods in the years 1997 and 2006.

Abstrakt

Elektrická rezistivitní tomografie (ERT) je v současné době stále více používána k řešení různých problémů v oblasti inženýrsko-geologického průzkumu, geotechnických výzkumech atd. Postupně nahrazuje tradiční kombinaci metod odporového profilování (OP) a vertikálního elektrického sondování (VES) při průzkumu svahových deformací. V tomto příspěvku jsou uvedeny vybrané výsledky získané při průzkumu svahové deformace u obce Lidečko. Přináší některé nové poznatky o jeho stavbě a výsledky monitoringu provedeného v roce 2011. Největší nebezpečí sesuvu vyplývá z jeho polohy nad vodovodním přivaděčem, kde existuje reálné riziko mohutného sesuvu svahu končícího v údolí řeky Senice. Jde o starý sesuv oživený při povodních v roce 1997 a 2006.

Key words: electrical resistivity tomography, slope deformation, engineering geological survey.

1   INTRODUCTION

In the last ten years, we have encountered the presentation of results of using a relatively new method, or measurement methodology, which uses a large number of connected electrodes – 25 and more (Loke 2001), which is most commonly known as resistive tomography or shortly called as "Multi-cable", etc., however, the most suitable indication is electrical resistivity tomography (ERT). The principle of measurement, as well as the processing of measured data are already well known, e.g. Loke (1996), so the attention is paid to the use of the ERT method in the exploration of a hazardous slope deformation near the town of Lidečko. Currently, the landslide front is located about 17 meters from the main water pipeline, which supplies with water the citizens of Horní Lidečko, Valašské Klobouky, Slavičín and Luhačovice Regions.

Recently initial stages of remediation work have been made on this landslide, (Ryšávka, Skopal 2008), which were based mainly on the results of existing reconnaissance of the slope deformation and also the results of geophysical survey carried out by the firm KOLEJ CONSULT & servis spol. s r.o., Brno, in the years

2007 – 2008. The first stage of redevelopment work includes drainage of surface water from the areas of the greatest subsidies to the landslide body.

2 LOCATION, GEOLOGICAL AND GEOMORPHOLOGICAL CONDITIONS

The massive landslide is located in forest stands east of the spot elevation of the peak called "Kopce" (699 asl), 1 km north-west of the town of Lidečko in the Vsetín Region (Fig. 1). The affected area is located on an old landslide, where the active part of the landslide develops in the space above a fossil landslide. This area is heavily violated. The landslide is classified as a current block slide with a thickness of up to 30 m (KOLEJ CONSULT & servis s.r.o., in Ryšávka, Skopal 2008).

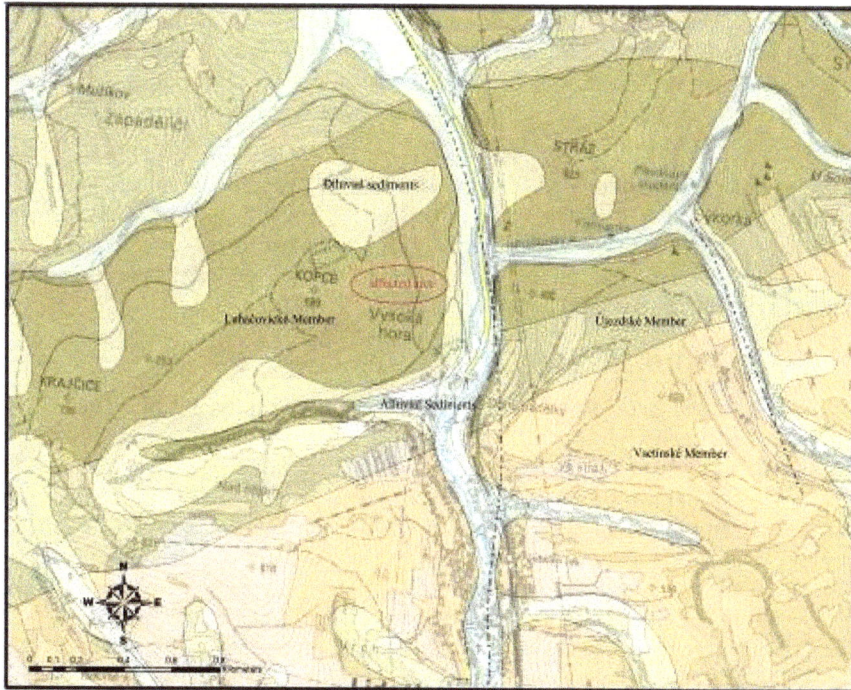

Fig. 1 Situation map of the Lidečko locality and cut of the geological map with description of basic rock complexes.

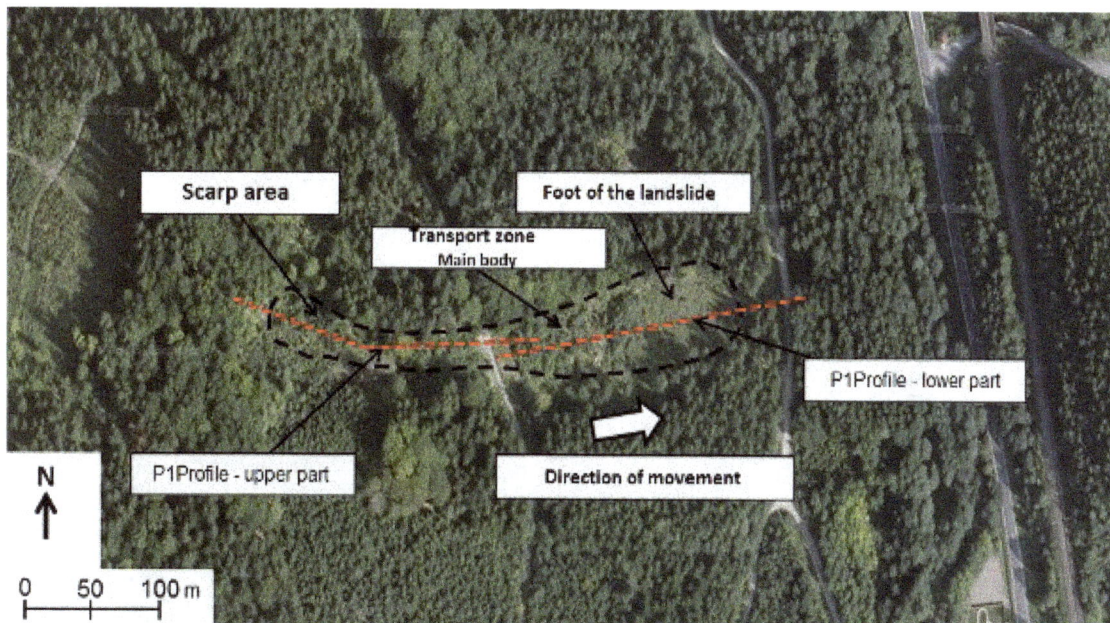

Fig. 2 View of individual parts of the landslide and presentation of profiles along the slide axis - satellite image (www.mapy.cz)

From the geomorphological point of view, the area is located in the subprovince of the Outer Western Carpathians, in the area of the Slovak-Moravian Carpathians, the unit of the Vizovice Highlands, the subunit of Komonec Upland. The landslide terrain is manifested by significant degrees of slope and laterally elongated depressions. Blocks of sandstone are separated along a series of cracks in the ENE-WSW direction, perpendicular to the direction of movement (Baroň 2004). The average slope inclination is between 25 to 30°, locally up to 40°. The presence of relatively thick resistant sandstones and their tectonic disturbance cause a relative difference in elevation up to 260 m along the length of 500 m (Baroň 2004).

From the regional and geological points of view, the locality is situated in the territory belonging to the Rača Unit of Magura Paleogene. Pre-Quaternary bedrock of the locality of interest is built up by Palaeoceneous to Eoceneous Soláň Formation - arkose of a Luhačovice menber. The sandstones are light grey to tan, mostly medium to coarse grained, often slightly conglomeratic with calcareous or clay cement (Baroň 2004). They are thin-to-thick tabular and in different parts of their surface affected by weathering processes. The Quaternary cover is in the area of interest represented by loose diluvial sediments that are highly heterogeneous, somewhere having a character of sands, at the base gravels. Somewhere else it is rocky and boulder deluvium with sandy and loamy sediments. The complex of Quaternary sediments ends close to a special-purpose forest road with a layer of anthropogenic made-up ground of a gravel character (Merta 2006).

From the hydrogeological point of view, fissure permeability is applied in the pre-Quartenary formations, which is predominantly bound to near-surface eroded parts of the Soláň Formation. Underground water is also bound to the quaternary diluvial sediments of a character of clastic sediments (gravel, sand) – collectors with intrinsic permeability. Diluvial cohesive soil (clay) probably behaves in the water-bearing systems as insulators or as semi-insulators (Ryšávka, Skopal 2008).

3   BASIC CHARACTERISTICS OF LANDSLIDE

The landslide has an elongated shape in the W-E direction with a length of about 350 and an average width of 50 m. In the accumulation area of the landslide (foot of the landslide), the width exceeds 70 m and in the upper half, it is about 30 m. In height the landslide is bordered with the spot elevations – 465 m asl and 615 m asl. The landslide can be divided into three parts: scarp area (main scarp and minor scarp), transport zone (main body) and foot of landslide (Fig. 2). The scarp area is bordered by a distinct scarp line with a spot elevation of 615 m asl (Fig. 3) and the lower part by a forest path with a spot elevation of 545 m asl. This is the longest part of the landslide with a length of approximately 160 m. This scarp area has a concave shape. Earth and rock material is moved into the transport zone and foot of landslide. In the scarp area, a large sandstone sheet can be seen, on which the rock material rolled off. This area represents a part of the left side border of the landslide scarp area. The transport zone is located below the forest road level at 542 m asl. In this part, the slope terrain is chaotically covered with rock blocks together with the remains after uprooted trees. The transport zone passes to the foot of the landslide at a level of spot elevation of 525 m asl. The surface of the foot area is similar to that in the transport zone. The foot area is made up of accumulated rock sandstone blocks, to a lesser extent conglomerates and uprooted trees (Fig. 4). Rock blocks in the foot area reach sizes up to several meters. The massive sandstone blocks show a grossly low to moderate level of weathering.

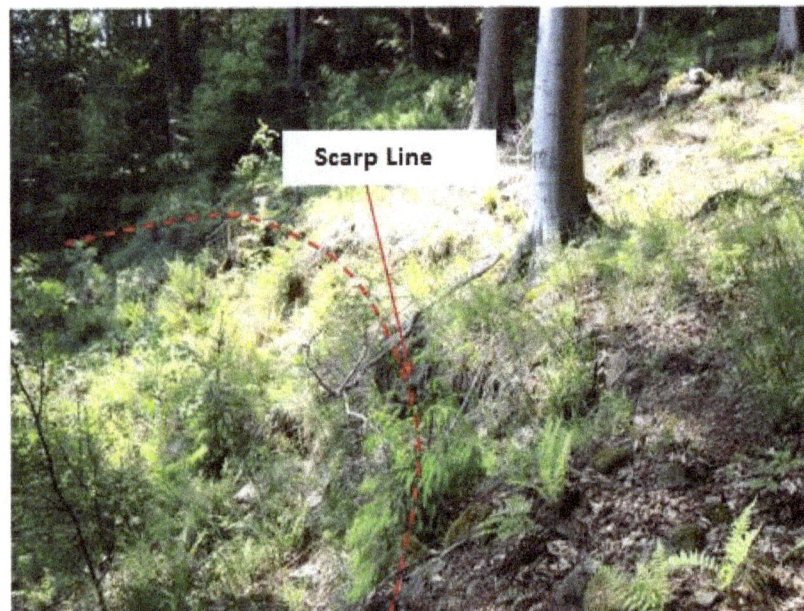

Fig. 3 Illustration of the scarp line of the landslide (photo: A. Poláček, 2011)

Fig. 4 The accumulation area and the foot of the landslide (photo: A. Poláček, 2011)

According to the determined structure of the landslide, geological, hydrogeological conditions and geotechnical parameters (Mack 2008) it can be assumed that the triggering mechanism of landslide activation was extreme precipitation in July 1997. In this period, the maximum saturation of permeable sandstones and near-surface layers occurred. Deposit layers and their tectonic fracturing (Merta 2006) cause the creation of slickensides (discontinuities and jointing) along which landslide movements take place.

4   FIELD WORK AND MEASUREMENT METHODOLOGY

Geophysical works were carried out using the method of electrical resistivity tomography (hereinafter referred to as ERT) which followed previous geophysical measurements, where the methods of ground penetrating radar (hereinafter referred to as GPR), refraction seismic and the VES method (in two cases) were applied. A detailed assessment is provided in a report (KOLEJ CONSULT & servis s.r.o., in Ryšávka, Skopal 2008).

Geoelectric measurements in the area of interest took place in the period March – November 2011. The measurements were performed at a total of eight profile lines along the axis of the landslide body and transverses to it (Fig. 5). In the axis of the landslide, the main profile was carried out marked as P 1, which consists of two parts with lengths of 254 m and 258 m. Further measurements took place at six cross sections Pp0 - Pp5. The cross sections Pp0 and Pp5 are outside the landslide body itself. The cross sections were 156 m long. The behaviour of all profiles is indicated in Fig. 5. The total length of the geoelectric profiles was 1448 m.

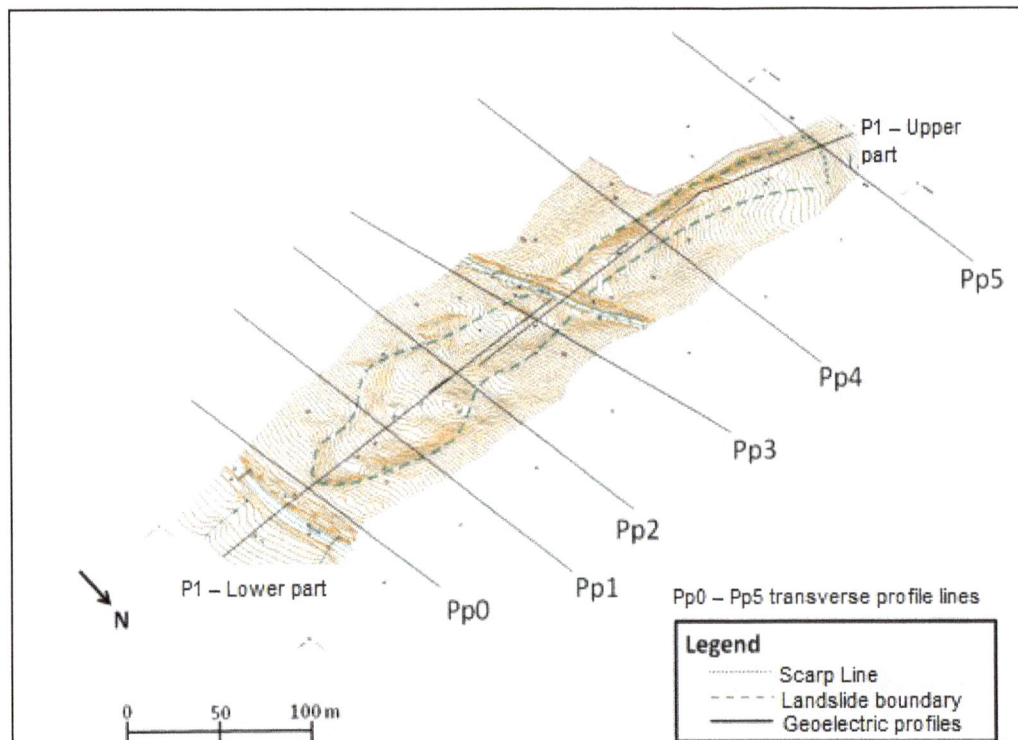

Fig. 5 Schematic representation of all geoelectric profiles

In these measurements, a Wenner-Schlumberger array was used. Under this arrangement, it was possible to detect horizontal and quasi-horizontal structures of larger sizes, different shapes and orientations, to a lesser extent tectonic zones or faults, contacts of layers with high different specific resistivity, etc. The real depth measurement range reached about 1/5 of the maximum distance between the first and the last electrode at the profile (Cervantes, Poláček 2011).

The output data was processed using the algorithm of inverse task by means of the RES2DINV software. This computer program automatically determines two-dimensional models (2D) of the base resistance for the data obtained by the ERT method. The program uses an inverse model, which consists of a series of rectangular pseudo-blocks (Loke 2001). Using this program, basic ideas of physical inhomogeneity of investigated environment using appropriate measurement methodology can be obtained faster than ever before.

5 ASSESSMENT OF GEOELECTRIC MEASUREMENT RESULTS

5.1 P 1 Profile

The results of geophysical measurement at the main – longitudinal profile P 1 consisting of two parts, are shown in Fig. 6. The division of the profile into two parts is performed mainly due to its large total length. It was not possible to perform measurements within one day with regard to quite difficult conditions for movement on the landslide surface as it was covered with dense vegetation of self-seeding trees. Considering the purpose of measurements, the attention is given in the next text especially to the transport and accumulation parts of the P 1 profile, not to the scar one; thus to that landslide part where it was possible to assume changes in the internal construction of landslide and a possible increase of the risk of further movement. The complete graphical illustration of the interpreted vertical resistivity cut at the main profile is shown in the work by Ryšávka, Poláček, Cervantes (2011). In Fig. 6, next to the slickensides, the points are indicated through which cross sections were led. It should be noted that the choice of their direction and position in relation to the results obtained at the longitudinal profile was largely influenced by the terrain possibility and viability even outside, thus in peripheral parts of the landslide.

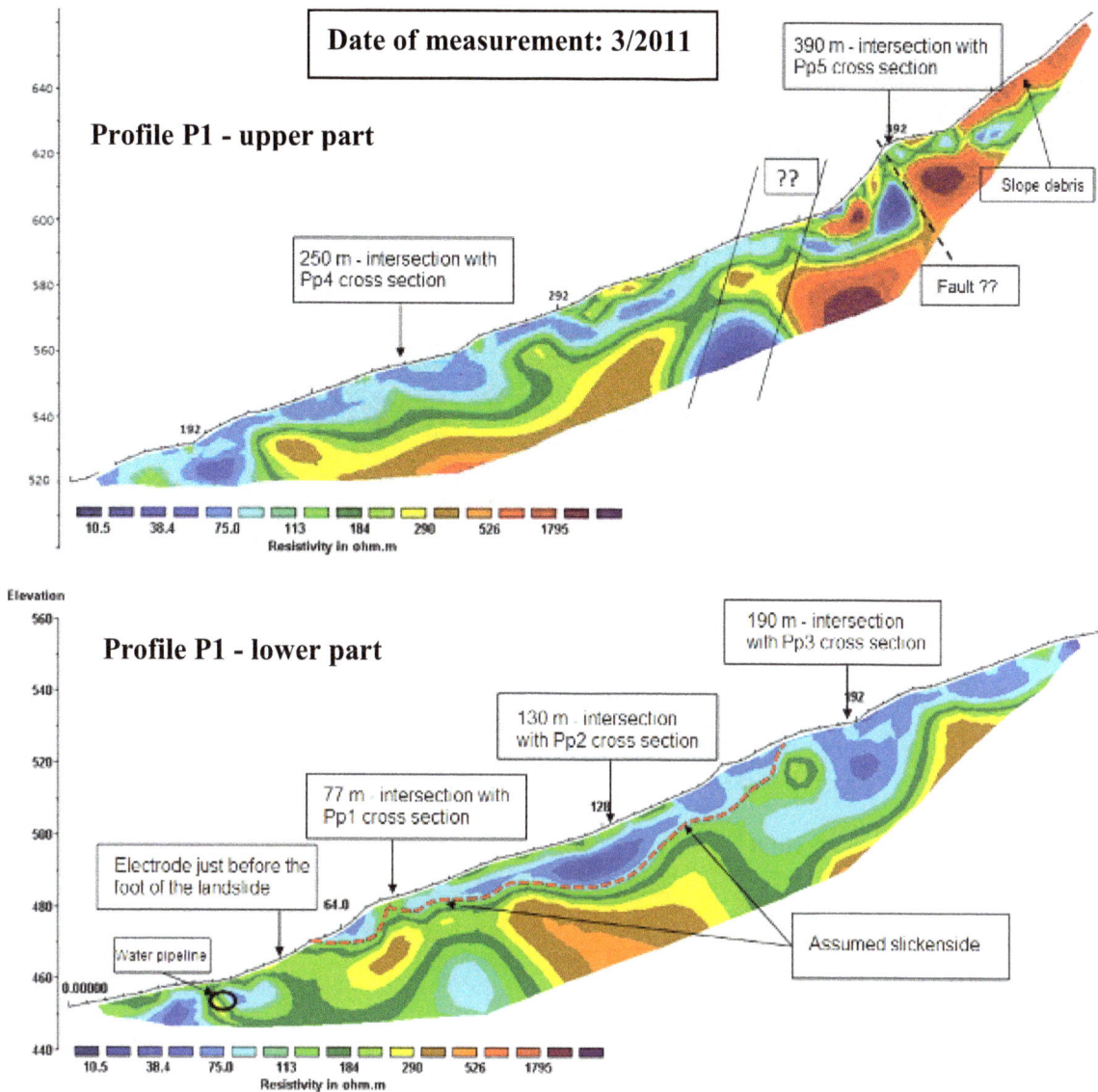

Fig. 6 Longitudinal resistive cut P1 and basic representation of slickensides.

5.2 Pp1, Pp2 and Pp3 cross sections

Fig. 7 shows the interpreted vertical resistivity cuts at the Pp1 to Pp3 (cross sections led through the accumulation (bottom) part of the slope deformation and for comparison then the Pp0 cross section led outside the landslide in the bottom forest path. From the comparison of the resistivity images at the profiles led through the landslide body with the behaviour of the resistivity at the profile Pp0 it is evident that the values of the specific resistivity at the cross sections are similar to each other. On the basis of the results, the Pp1 to Pp3 cross sections differ from the Pp0 cross section by both resistivity values and its overall image. Taking into account the work (Baroň 2004), there is then indicated the lithological interpretation of main petrographic types of rocks, resolution of sandstone and claystone positions as well as slope debris, including the indication of their borders.

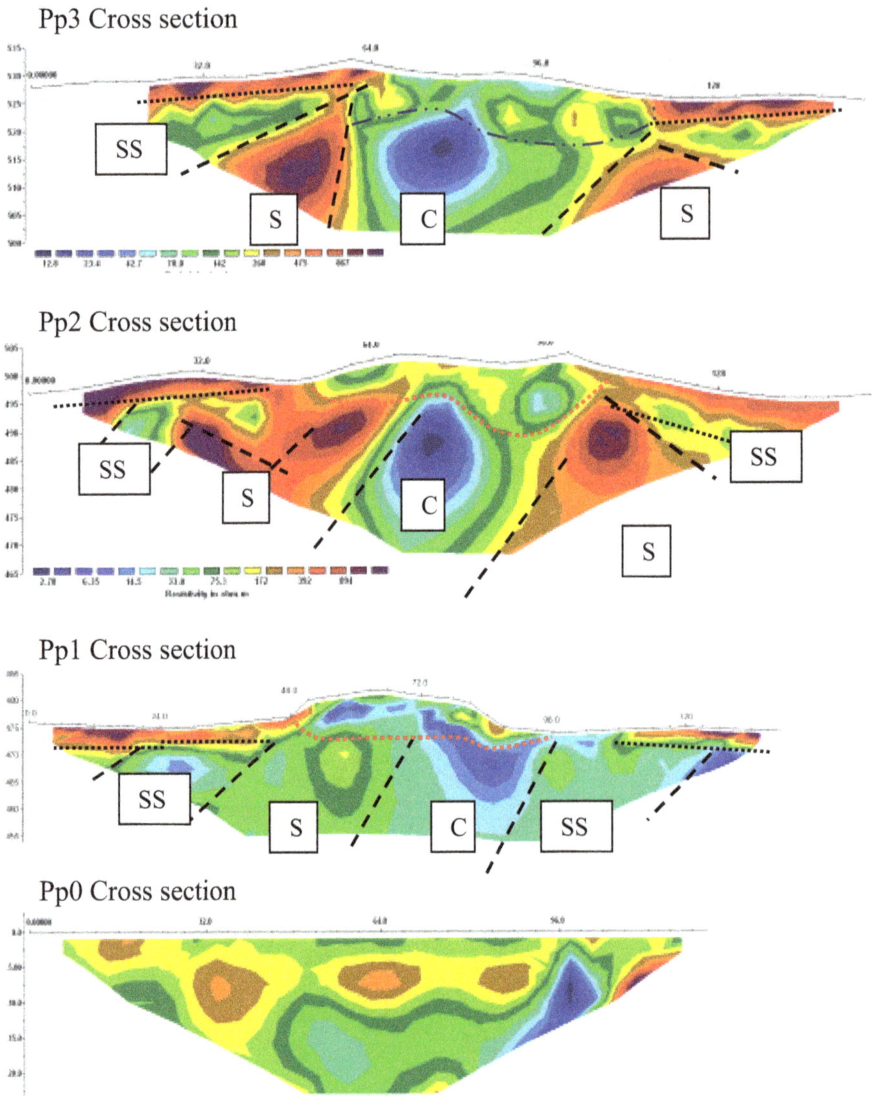

Fig. 7 Interpreted vertical resistivity cuts at the Pp1, Pp2 and Pp3 cross sections at the bottom of the P 1 main profile (the Pp0 cross section was led outside the landslide body).

Legend:

S	– sandstone positions
C	– claystone positions
SS	– saturated sandstones
– – – – –	– interface between rock types
– · · – · – · · ·	– old slope accumulations with a possibility of partial reactivation
· · · · · · · · · · · · · · ·	– slope debris
– – – – –	– active part of landslide

5.3 Comparison of measurement results at the bottom of the P 1 profile in March and November 2011.

With regard to the fact that based on visual reconnaissance a movement of the slope deformation was found out, when the foot of the landslide shifted by about 1 m towards the forest path and thus also towards the water pipeline, repeated geophysical measurements were carried out in this part. The results obtained are shown in Fig. 8. Comparing the obtained resistivity cuts from measurements made at an interval of six months it was found out that adverse development of the slope deformation occurred in this period. A significant completion of the slickenside at a depth of about 8 to 15 m occurred. A loosen, partially consolidated part of the landslide moved slightly across the slickenside. The estimated initiation of this movement is related to the rainfall in this period.

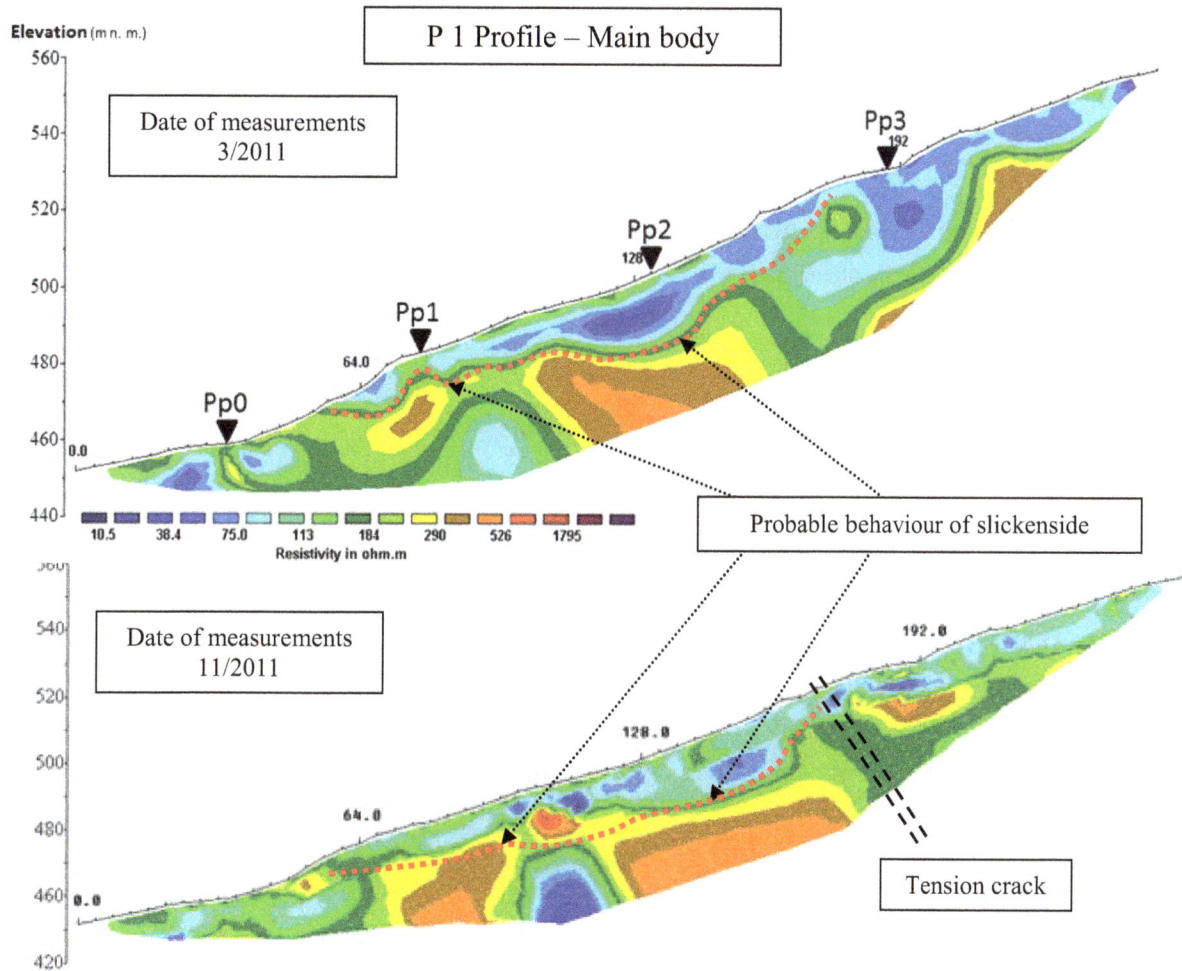

Fig. 8 Interpreted vertical resistivity cuts in the accumulation area of the slope deformation of the P 1 profile. (▼ – intersections with cross profiles)

6 CONCLUSIONS

➢ The ERT measurement results allow a more precise definition of structural, lithological and tectonic interfaces, identification of quasi-homogeneous blocks depending on the landslide construction. On the basis of the performed geophysical monitoring of the slope deformation in Lidečko, it was managed to define and by repeated measurements to confirm the existence of a significant slickenside, which occurs in the depth range of 8 to 15 m. The mentioned slickenside may be the main factor that allows designing optimal prevention and phase-to-phase remediation works that are often very expensive.

➢ By comparing the results obtained by the application of the ERT method with the methods of GPR and shallow refraction seismic (Ryšávka, Skopal 2008), it was showed that the ERT method is at least fully comparable with these methods, has lower economic exigency and allows more detailed division of measured environment into physically different blocks (a notable difference in resistivity particularly for distinguishing sandstone blocks and claystone positions).

> ➤ Optimal utilization of the ERT method is subjected to such a ground surface, which enables high-quality grounding of electrodes, in particular to sandy-clayey environment. The implementation of the measurement itself in the locality of Lidečko, see Figs 3 and 4, was considerably complicated by the nature of the landslide surface. Still it was managed to get the results comparable with the geological conditions of the area.

REFERENCES

[1] BAROŇ I. Hluboká svahová deformace na kopcích u Lidečka: Výsledky inventarizačního a geofyzikálního průzkumu. Geol. výzk. Mor. Slez. 2004, Brno. p. 82 – 87.

[2] CERVANTES B., POLÁČEK A. Metoda ERT (elektrická rezistivitní tomografie) jako prostředek k významnému zlepšení informace o fyzikální nehomogenně sesuvu včetně vymezení smykových ploch. Závěrečná zpráva SGS SP2011/113. 2011 VŠB – TU Ostrava.

[3] KAROUS M. Geofyzikální metody v inženýrské geologii a geotechnice. Geonika, s. r. o. 1998. Máchova 23, Praha.

[4] LOKE M.H., BARKER R.D. Rapid least-squares inversion of apparent resistivity pseudosections by a quasi-Newton method. European association of Geoscientists and Engineers.1996, Vienna, Austria. Geophysical prospecting, 44. p. 131 – 152.

[5] LOKE M.H. Constrained time lapse resistivity imaging inversion. The Environmental and Engineering Geophysical Society SAGEEP. 2001, Symposium Program. Denver: 34

[6] MACKA Z. Analýza vlivu 1. kroku stabilizace sesuvu Lidečko – nad vodovodním přivaděčem a predikce možného vývoje sesuvu. Diplomová práce. 2008, VŠB – TU Ostrava.

[7] MERTA P. Geotechnický průzkum – vodovodní přivaděč Lidečko, Unigeo a.s. 2066, Ostrava.

[8] RYŠÁVKA J., SKOPAL R. Lidečko-I Etapa sanačních prací. UNIGEO a.s. 2008, Ostrava. Divize SANEKO.

[9] RYŠÁVKA J., POLÁČEK A., CERVANTES B. Přínos elektrické rezistivitní tomografie (ERT) pro stanovení homogenity sesuvného tělesa Lidečko. Konference „Svahové deformace a Pseudokras", Ústav Geotechniky VUT FAST, Brno. 2011.

RESUMÉ

V článku jsou uvedeny výsledky získané metodou elektrické rezistivitní tomografie na svahové deformaci Lidečko v roce 2011. Zejména jsou diskutovány výsledky zjištěné ve svahové deformaci dlouhé 250 m, která z hlediska stavby sesuvu představuje část transportní, ale zejména část akumulační. Čelo sesuvu se v současné době nachází cca 17 m před vodovodním přivaděčem. Sesuv tak představuje značné riziko ohrožení jeho funkce.

Výsledky získané metodou ERT doplňují a upřesňují geofyzikální měření prováděné cca před třemi lety, které jsou součásti zprávy (Ryšávka, Skopal 2008). Jedná se zejména o vymezení smykové plochy, která byla interpretována v březnu 2011 a po té ověřena po pohybu sesuvu o 1 m blíže k vodovodnímu přivaděči. Geofyzikální měření bylo do značné míry limitováno nevhodným stavem povrchu terénu. Povrch sesuvu se vyznačoval v důsledku samotného horninového složení a sesouváním nesourodých hmot značnými nerovnostmi a hustým náletovým porostem, který znesnadňoval samotné měření, uzemňování elektrod a do určité míry i optimální volbu geofyzikálních profilů.

Publikace je součástí řešení grantového projektu SGS SP2011/113 " Metoda ERT (elektrická rezistivitní tomografie) jako prostředek k významnému zlepšení informace o fyzikální nehomogenitě sesuvu včetně vymezení smykových ploch".

19

APPLICATION OF DISCRIMINATE ANALYSIS TO PREDICTION OF COMPANY FUTURE ECONOMIC DEVELOPMENT

Radmila SOUSEDÍKOVÁ [1], Jaroslav DVOŘÁČEK [2], Igor SAVIČ [3]

[1] *RNDr. Ph.D., Institute of Economics and Control Systems, Faculty of Mining and Geology,*
VŠB–Technical University of Ostrava
17. listopadu 15, Ostrava-Poruba
e-mail: radmila.sousedikova@vsb.cz

[2] *prof. Ing. CSc., Institute of Economics and Control Systems, Faculty of Mining and Geology,*
VŠB–Technical University of Ostrava
17. listopadu 15, Ostrava-Poruba
e-mail: jaroslav.dvoracek@vsb.cz

[3] *Mgr., Cybex Industrial LTD, C/O Columbus Trading-Partners*
GMBH Riedinger str. 18, 95448 Bayreuth, German
e-mail: igor.savic@cybex-online.com]

Abstract

The paper takes into account applications of discriminate analysis as regards prediction of future economic development of companies. An assumption of multivariate normality of discriminators has been tested and outliers identified. An outlier reduction of original data files brings data distribution closer to multivariate normality, and substantially improves discriminate function classification abilities.

Abstrakt

Článek se zabývá využitím diskriminační analýzy pro predikci budoucího vývoje firmy. Je testován předpoklad vícerozměrné normality diskriminátorů a jsou identifikovány vybočující hodnoty. Redukce původních datových souborů o vybočující data přispívá k přiblížení rozložení dat vícerozměrné normalitě a vede k podstatnému zlepšení klasifikační schopnosti diskriminační funkce.

Key words: Default prediction, discriminate analysis, multivariate normality, outliers

1 INTRODUCTION

The initial comprehensive studies of modelling future development of companies date back to the thirties of the twentieth century. These works take into account analyses of financial ratios as regards successful and failed companies. Ramser and Foster (1931) compared ratios for successful and failed businesses and could prove that corporate finance of successful firms demonstrate better ratio figures than it would be the case with those in distress or bankrupt. What more, Fitzpatrick (1932) tried to identify those corporate finance ratios that might serve the purpose of predicting future economic developments. Smith and Winakor (1935) analyzed book keeping records of bankrupt companies, identifying indices of imminent default.

A study of cardinal importance is that of Beaver (1966) who analyzed the utilisation of financial ratios for predicting future economic development of companies. Beaver designed his study to be a benchmark for future investigations into alternative predictors of failure. He succeeded in providing an empirical verification of the predictive ability of financial statement accounting data, namely the financial ratios, for the prediction of corporate business failure.

All the studies that have been mentioned above demonstrate the ability of financial ratios to predict bankruptcy, and they try to identify a single ratio or a group of such ratios whose values would indicate imminence of default. Nevertheless, conclusions of these studies are not unambiguous, because each study exemplifies different financial ratios as indicators of default imminence. The period from the thirties to the end of the sixties of the 20th century can be qualified as the period of one-dimensional analyses of financial ratios, when the individual ratios are analyzed in their isolation and their interaction is disregarded.

Altman (1968) built on the existing results and could overcome limitations of one-dimensional approaches by combining several financial ratios for his developments of prediction models. He solved problems of the capability of specific ratios to predict default. He could provide for objective quantification of each specific ratio by defining weights that should be attached to these selected ratios and how these weights should be objectively established. Altman is innovative in his utilization of the multiple discriminate analyses by developing a multi-dimensional model of a corporate business default. In his later studies, Altman adapted his model several times and along with Halderman and Naryanan (1977) developed a second generation of prediction models for corporate business failure. Altman is aware of deficiencies of models that utilize linear discriminate analysis (Altman et. al., 1981), especially as regards non-conformity to the assumption multivariate normal distribution of discriminators in each class, and equality of class covariance matrices.

The impossible fulfilment of multidimensional normality resulted in the utilization of other distribution functions that would describe the input data distribution. Ohlson (1980) was the first one who used a logistic regression analysis for creating prediction models of corporate business defaults.

Since the beginning of the nineties, studies have been performed that utilize neuron networks for modelling of corporate failures (Odom and Sharda, 1990; Tam, 1991).

2 DISCRIMINATE ANALYSIS

The discriminate analysis enables the evaluation of differences between two or more subject groups that are characterised by a certain number of features. Such evaluation provides a base of classification that builds on it. If the discriminate analysis is applied for predicting future economic footing of companies, the subjects are particular firms that are structured into two groups of prosperous firms and those in jeopardy of default. Each firm is characterised by a definite number of quantitative variables called discriminators.

The choice of discriminators plays a major role as regards prediction abilities of final modelling. The discriminator calculation data can be drawn from common accounting documentation – Balance Sheet, Profit and Loss Statement – and it provides for calculations of ratios or indexes.

There are two ways for choosing the discriminators. The first way takes advantage of a mathematical approach to the problem, and takes a lot of various ratios and indexes into account, and so we are not sure which of them are effective for classifying a particular firm as prosperous or threatened by default. We can distinguish particular discriminator efficiencies by assessing the changing values of the Mahalanobis distance, D_M^2, between mean values of both classes, that have been effected by adding or removing a definite discriminator. The individual steps of discriminator addition or removal can be directed by various decision-making criteria (for example the Wilks' Criterion, λ).

The second way has been based on experience, knowledge, and intuition of researchers, when the choice of discriminators is supported by a theoretical model for solving the given task. This paper's investigation has opted for this second way and the discriminator selection has been based on the assumption that bankruptcy is caused by disrupted circulation of capital. That is why, 8 financial ratios were chosen, namely,

1. Quick liabilities/Total assets
2. Current assets index
3. Current liquid assets/Current assets
4. Sales index
5. Total assets index
6. Receivables/Current assets
7. Index, Loan capital/Total assets
8. Equity capital/Total assets

2.1. Model

The discriminate analysis model consists in a linear combination of variables, so called discriminators that best distinguish between prosperous and default companies.

This is the linear discriminate function formula:

$$D_i = d_1 X_{i1} + d_2 X_{i2} + \cdots + d_m X_{im},$$

(1)

where

n – Number of firms in the class,

m – Number of discriminators,

D_i – Discriminate score for firm, i,

X_{ij} – Discriminator value for firm, i, ($j = 1,...,m$),

d_j – Linear discriminate coefficient for discriminator, j, (for $j = 1,...,m$).

This formula combines several firm's characteristics (discriminators) into a single multivariate score, D_i, whose value is between $-\infty$ and ∞, and indicates financial health of the firm. The discriminate score low value, D_i, marks bad financial health of the firm.

A correct application of the linear discriminate analysis asks for observation of the following requirements:
- Discriminators evidence multivariate normality of distribution,
- Classes of prosperous firms and those in jeopardy of default have the same covariance matrices.

2.2. Testing normality of discriminators

In the following, we concentrate on multivariate normality of discriminators. The multivariate normality of independent variables (discriminators) should be verified by an appropriate statistical test. The test of multivariate normality is difficult. Although the random vector particular constituents of all discriminators evince one-dimensional normality, the associated density of probability does not necessarily have multivariate normal distribution. As such, the one-dimensional normality of particular discriminators is a necessary but insufficient prerequisite of vector multivariate normality of all discriminators. This fact can serve as a tool of practical verification of multivariate normality of discriminators. The one-dimensional normality of particular discriminators is verified first. If, at least, one discriminator does not evidence one-dimensional normality, it is obvious that no multivariate normality exists, and the testing is terminated. If all features evidence normality, it is necessary to continue in testing of multivariate normality.

For example, the normality of particular discriminators can be verified by tests of skewness and kurtosis.

The testing criterion is defined by the following formula [6]:

$$C_1 = \frac{\hat{g}_1^{\,2}}{D(\hat{g}_1)} + \frac{[\hat{g}_2 - E(\hat{g}_2)]^2}{D(\hat{g}_2)}, \tag{2}$$

where

n - Sample size,

x_i - Value of the discriminators tested for firm, i,

$\bar{x}$ - Mean value of the discriminator tested,

$$\hat{g}_1 = \frac{\sqrt{n}\sum_{i=1}^{n}(x_i - \bar{x})^3}{\left[\sum_{i=1}^{n}(x_i - \bar{x})^2\right]^{3/2}} \qquad \text{- Sample skewness,} \tag{3}$$

$$D(\hat{g}_1) = \frac{n-2}{(n+1)(n+3)} \qquad \text{- Variance of sample skewness,} \tag{4}$$

$$\hat{g}_2 = \frac{n\sum_{i=1}^{n}(x_i - \bar{x})^4}{\left[\sum_{i=1}^{n}(x_i - \bar{x})^2\right]^2} \qquad \text{- Sample kurtosis} \tag{5}$$

$$E(\hat{g}_2) = 3 - \frac{6}{n+1} \qquad \text{- Expected value of sample kurtosis,} \tag{6}$$

$$D(\hat{g}_2) = \frac{24\,n\,(n-2)(n-3)}{(n+1)^2(n+3)(n+5)} \qquad \text{- Variance of sample kurtosis.} \tag{7}$$

If normality is evidenced, the value, C_1, has an asymptotic distribution, χ^2, with 2 degrees of freedom. If it is verified that $C_1 > \chi_{1-\alpha}^2(2)$, it is necessary to reject hypotheses that the sample is normally distributed [6].

The normality of particular discriminators can be also assessed graphically by means of rankit plots, which enable comparison of each discriminator distribution with normal distribution. The rankit plots are executed by plotting the quantiles of the standard normal distribution, u_{p_i}, on the plot's horizontal axis, and the order statistics, $x_{(i)}$ (discriminator values structured hierarchically in ascending order), on its vertical one. If a discriminator's distribution equals normal distribution, we can observe a linear dependence, $x_{(i)}$, on u_{p_i}.

For the class of prosperous firms, as well as for those in jeopardy of default, the tests of one-dimensional normality for all eight discriminators were performed. The value of the testing criterion, C_l, for particular discriminators (see Tab. 1), and their comparisons with the quantile, $\chi^2_{0,95}(2) = 5,99$, make it obvious that none of the discriminator distributions is normal.

Tab. 1 Test criterion values for the normality verification of particular discriminators

Discriminator	Test criterion value, C_l	
	Prosperous firms	Firms in jeopardy of default
Quick liabilities/Total assets	206.39	4941.70
Current assets index	10146.11	138.12
Current liquid assets/Current assets	61.94	869.67
Sales index	10661.24	4890.87
Total assets index	2563.03	6.47
Receivables/Current assets	2325.0	24.37
Index, Loan capital/Total assets	104.05	12035.42
Equity capital/Total assets	11.64	4019.53

Also rankit plots of all discriminators for both classes of firms evidence the fact that no discriminator distribution is normal. The following Figs. 1, 2, provide for a comparison of rankit plots of sample discriminators for the class of profitable businesses. The distribution of the discriminator, Equity capital/Total assets (see Fig. 1) approximates the normal distribution most closely. In contrast to this, the distribution of the discriminator, Sales index (see Fig. 2) evidences the worst result as regards the distribution normality.

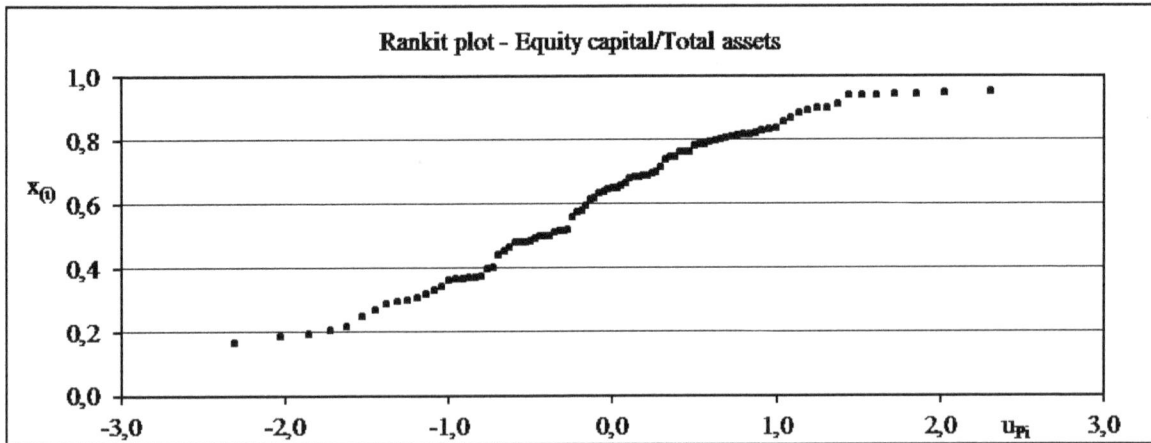

Fig. 1 Rankit plot for the Equity capital/Total assets (prosperous firms).

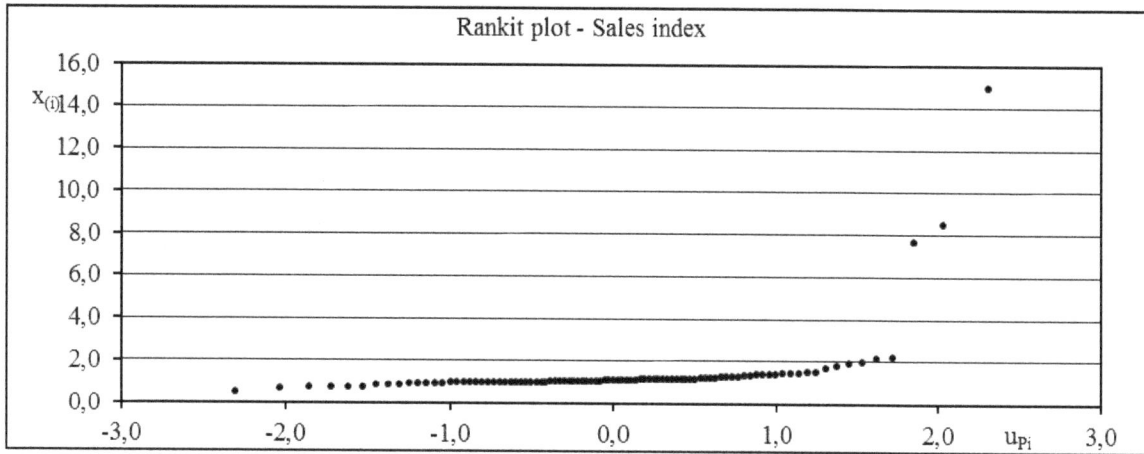

Fig. 2 Rankit plot for the Sales index (prosperous firms).

Analogically, for the class of firms in jeopardy of default, Fig. 3 provides for the rankit plot of the discriminators, Total assets index, which shows almost a linear dependence close to normality. In contrast to this, Fig. 4 testifies that the rankit plot for the discriminator, Index, Loan capital/Total assets, evidences a pronounced deviation from the normal distribution

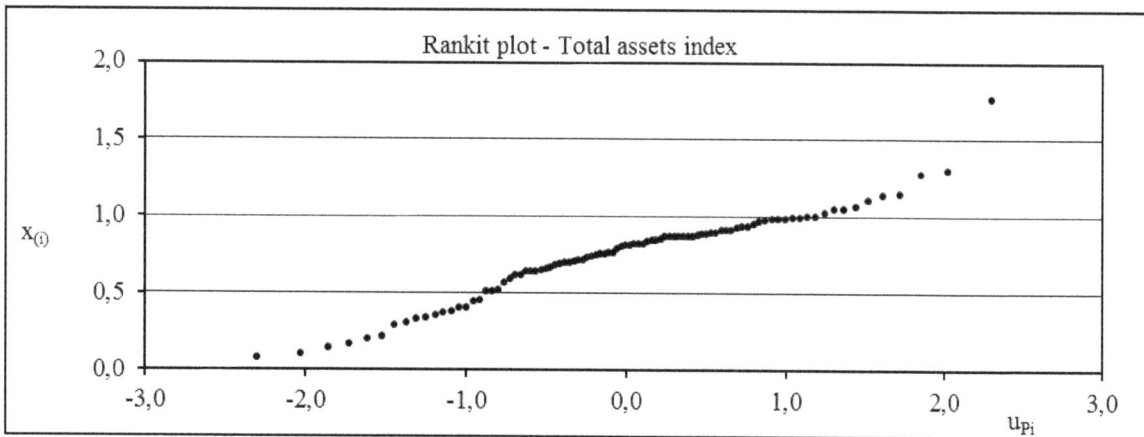

Fig. 3 Rankit plot for the Total assets index (default firms).

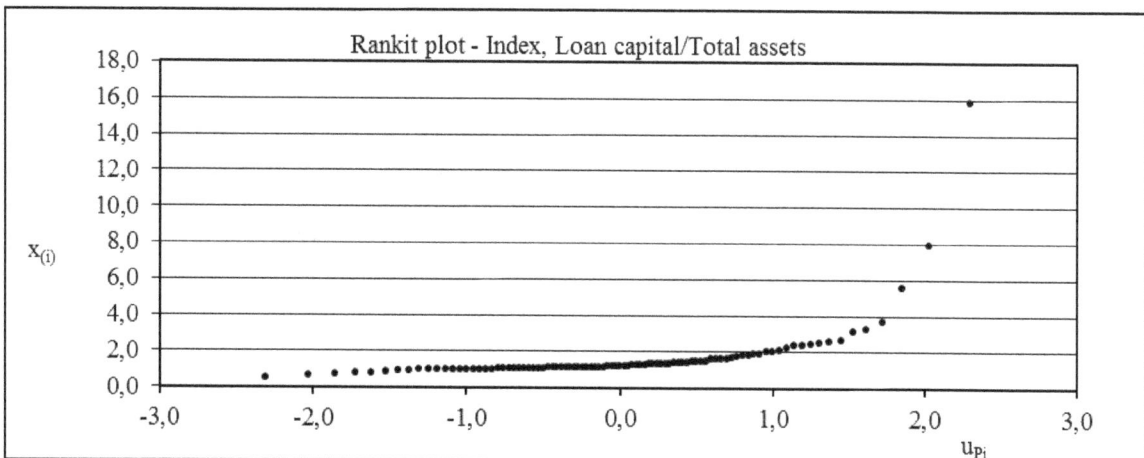

Fig. 4 Rankit plot for the Index, Loan capital/Total assets (default firms).

The above given data are not surprising. For the economic data, a normality deviation is rather the rule than the exception. The majority of research works (inclusive Altman) do not test multivariate normality and assume that models are sufficiently robust for providing rational approximations even without meeting the prerequisite of input data normal distribution.

Having this in mind, let us try to improve the data distribution by exclusion of outliers, and let us compare the results of the discriminate analysis before and after this modification of the data.

2.3. Identification of outliers

Two approaches will be taken for the outlier identification:

1. One-dimensional approach: Particular discriminator outliers are identified by so called inner fences.
2. Multivariate approach: Outliers are identified by their Mahalanobis distance from the mean value data.

Outlier identification by inner fences:

The outliers are all discriminator values that lie outside the interval,

$$\left(B_D^*; B_H^*\right),\tag{8}$$

$$B_D^* = \widetilde{x}_{25} - K\left(\widetilde{x}_{75} - \widetilde{x}_{25}\right),\tag{9}$$

$$B_D^* = \widetilde{x}_{75} - K\left(\widetilde{x}_{75} - \widetilde{x}_{25}\right),\tag{10}$$

$$K = 2{,}25 - \frac{3{,}6}{n},\tag{11}$$

where

n — Sample size,

$\widetilde{x}_{25}$ — Lower quartile,

$\widetilde{x}_{75}$ — Upper quartile [6].

The following Tab. 2 gives numbers of discriminator outliers for both classes of firms.

Tab. 2 Number of discriminator outliers

Discriminator	Number of outliers	
	Prosperous firms	Default firms
Quick liabilities/Total assets	2	5
Current assets index	8	1
Current liquid assets/Current assets	0	9
Sales index	7	3
Total assets index	1	1
Receivables/Current assets	1	0
Index, Loan capital/Total assets	4	6
Equity capital/Total assets	0	7

Firms that have one or more discriminator outliers were excluded from the investigation. The original samples of 93 prosperous and 93 default companies were reduced to 76 prosperous and 74 default companies.

The outlier identification by the Mahalanobis distance:

The outliers are all multivariate data of the formula,

$$d_i^2 = \left(\mathbf{x_i} - \bar{\mathbf{x}}\right)^T \mathbf{S}^{-1}\left(\mathbf{x_j} - \bar{\mathbf{x}}\right) > \chi^2_{1-\frac{\alpha}{n}}(m),\tag{12}$$

$\mathbf{x_j}$ — Discriminator vector of particular firms $(j = 1,...,n)$,

$\bar{\mathbf{x}}$ — Sample mean vector of particular discriminators,

$\mathbf{S}^{-1}$ — Inverse co-variance matrix,

$\chi^2_{1-\frac{\alpha}{n}}(m)$ - Quantile of distribution, χ^2, with degrees of freedom, m

where

m - Number of discriminators,
α - Significance level,
n - Number of firms of particular class [6].

The Mahalanobis distance identified five prosperous and eight default firms as having outlying discriminator values. As such, the original samples were reduced to 88 prosperous and 85 default companies.

The data distribution improvement can be demonstrated by a Q-Q plot. The plot illustrates dependence of order statistics, $C_{(i)}$ (C_i values structured hierarchically in ascending order) on values, C_i^*, where

$$C_i = \frac{n}{(n-1)^2}(\mathbf{x_i} - \overline{\mathbf{x}})^T \mathbf{S}^{-1}(\mathbf{x_j} - \overline{\mathbf{x}}), \tag{13}$$

$$C_i^* = \frac{i-a}{n-a-b+1}, \tag{14}$$

where

$$a = \frac{0,5m-1}{m},$$

$$b = \frac{0,5(n-m-1)-1}{n-m-1}.$$

In case of multivariate normality, the dependence of $C_{(i)}$ on C_i^* should be a linear one.

The Fig. 5 provides for the Q-Q plot of 93 prosperous firms

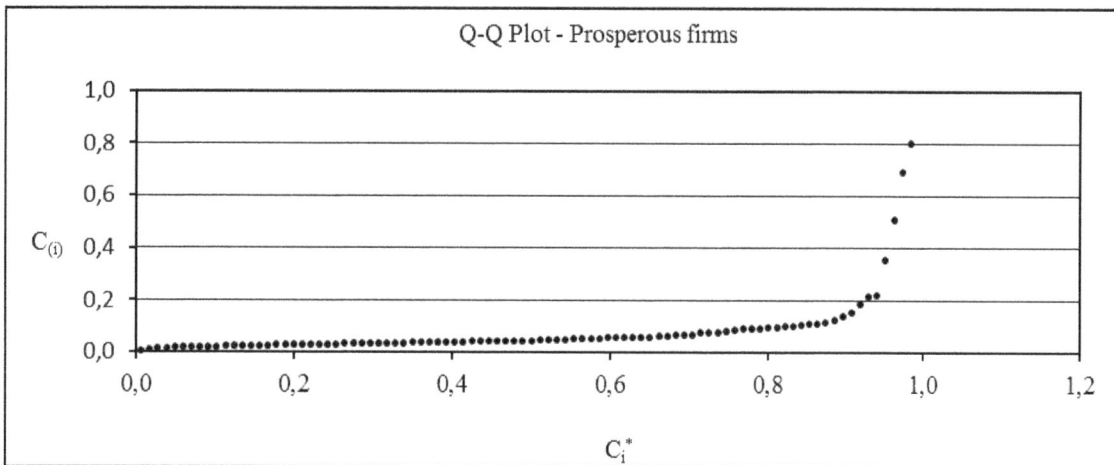

Fig. 5 Q-Q Plot of prosperous firms – Original sample of 93 firms

The following Fig. 6 gives the Q-Q Plot for the reduced sample of 76. The reduction was made by removing firms that had some discriminator values identified by inner fences as outliers.

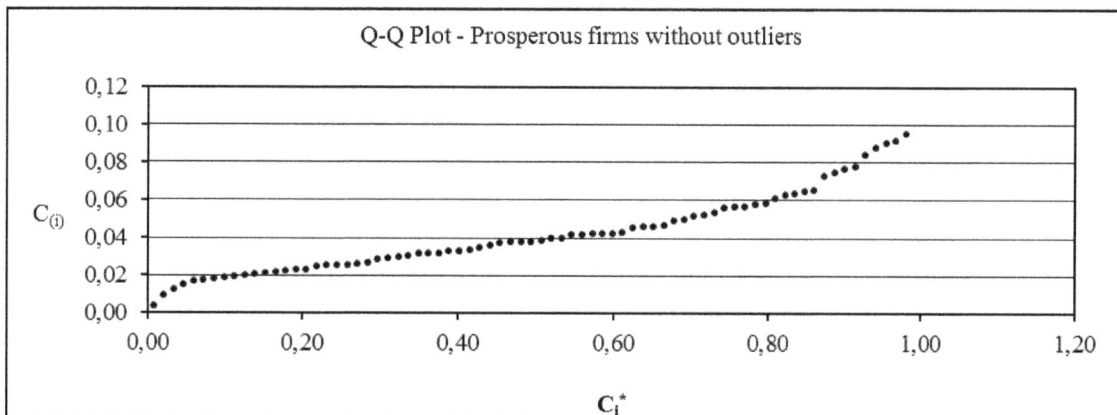

Fig. 6 Q-Q Plot of prosperous firms – Reduced sample of 76 firms

The comparison of these two plots makes it obvious that removal of outliers improved considerably the data distribution. The almost linear dependence of the plot in Fig. 6 provides for the possible conclusion that the reduced sample of prosperous firms closely approximates normal distribution.

3 DISCRIMINATE ANALYSIS OUTCOME COMPARISON

For the samples reduced by outlier elimination, the linear discriminate functions were calculated. These provided for the classification of firms as prosperous or in jeopardy of default. The results of this classification were compared with those, which had been the outcome of working with original non-reduced samples.

The comparison of results gives Tab. 3.

Tab. 3 Outcome comparison of working with original and reduced samples

	Classification success [%]		
	Prosperous firms	Bankrupt firms	Firms total
Original sample (93 prosperous firms + 93 in jeopardy of failure)	87.10	89.25	88.17
Sample reduced by inner fences (76 prosperous firms + 74 in jeopardy of failure)	100.00	90.54	95.33
Sample reduced by Mahalanobis distance (88 prosperous firms + 75 in jeopardy of failure)	95.45	94.12	94.80

It is obvious from the aforementioned results that approximation to multivariate normality improves the discriminate function classification abilities considerably. Concerning the approximation method specificity, i.e. the discriminate outlier elimination, the reduction by inner fences led to better results. Regarding the multivariate normality, the samples of which the outliers were eliminated by inner fences demonstrated better results which were corroborated by the Q-Q Plots.

4 CONCLUSION

The linear discriminate analysis represents a method of predicting future economic development of firms that is often applied. The choice of discriminators has a considerable impact on the prediction abilities of the final model. The selection of particular discriminators was based on the idea that bankruptcy is caused by disrupted circulation of capital. The correct application of discriminate analysis asks, apart from other requirements, for fulfilment of the prerequisite of the discriminator normal distribution. The majority of research that has been conducted in the field does not test the multivariate normality and assumes that models are sufficiently robust, providing for realistic results even without meeting the distribution normality requirements. The testing of the multivariate normality of discriminators was performed in the way, which first tested one-dimensional normality of discriminators by tests of skewness and kurtosis. The one-dimensional normality was also assessed by rankit plots. The testing of the aforementioned discriminators led to the conclusion that none of the discriminators tested had normal distribution, which was the reason why the requirement of multivariate normal distribution of all discriminators could not be met. Such results are not surprising because economic data often deviate from normal distribution. That was the reason why the data distribution was modified. The samples of prosperous and default firms were adapted in two ways. The one-dimensional way consisted in analysing particular discriminators and excluding the outlying discriminator values by inner fences. The multivariate way started with the elimination of outliers by the Mahalanobis distance. The improvement of the data distribution is testified by Q-Q Plots. The reduced samples were subject to a discriminate analysis which provided for the classification of firms as successful or failed. It is obvious from the results obtained that the approximation to multivariate normality improves considerably the classification abilities of the discriminate function.

REFERENCES

[1]   ALTMAN, E. I. Financial ratios, discriminate analysis and prediction of corporate bankruptcy. *The Journal of Finance*. 1968, XXIII. Nr. 4, pp. 589-609.

[2]   ALTMAN, E., HALDERMAN, R. & NARAYANAN, P. ZETA Analysis: A New Model to Identify Bankruptcy Risk of Corporations. *Journal of Banking and Finance.* 1977, I. Nr. 1, pp. 29-54.

[3]   ALTMAN, E. I. et al. *Application of Classification Techniques in Business, Banking and Finance. (Contemporary Studies in Economic and Financial Analysis).* Greenwich, Conn. : Jai Press, 1981. 418 pp.

[4]   BEAVER, W. Financial Ratios as Predictors of Failure. *Journal of Accounting Research. 1966,* IV. Empirical Research in Accounting: Selected Studies 1966, pp. 71-111.

[5]   FITZPATRICK, P. J. A Comparison of ratios of successful industrial enterprises with those of failed firms. *Certified Public Accountant,* 1932. October, November, December, pp. 598-605, pp. 656-662, pp. 727-731.

[6]   Meloun, M. & Militký, J. *Statistická analýza experimentálních dat (Statistical Analysis of Experimental Data).* Praha : Academia, 2004. ISBN 80-200-1254-0.

[7]   ODOM, M. D. a R. SHARDA. A Neural network model for bankruptcy prediction. *IJCNN International Joint Conference on Neural Networks.* 1990, I-III. pp. B163-B168.

[8]   OHLSON, J. Financial ratios and the probabilistic prediction of bankruptcy. *Journal of Accounting Research.* 1980, XVIII. Nr. 1, pp. 109-131.

[9]   RAMSER, J. R. & FOSTER, L. O. A Demonstration of Ratio Analysis. *Bulletin No. 40.* 1931, Bureau of Business Research, University of Illinois, Urbana.

[10] SMITH, R. F. and A. H. WINAKOR. *Changes in the financial structure of unsuccessful industrial corporations.* Urbana: University of Illinois, 1935.

[11] TAM, K. Y. Neural network models and prediction of bank bankruptcy. *Omega-International Journal of Management Science.* 1991, IXX. Nr. 5, pp. 429-445.

RESUMÉ

Článek se zabývá využitím diskriminační analýzy pro predikci budoucího vývoje firmy. Predikční schopnost výsledného modelu je významně ovlivněna volbou vstupních proměnných – diskriminátorů. Výběr použitých diskriminátorů byl založen na názoru, že příčinou bankrotu je porušení koloběhu kapitálu. Korektní aplikace diskriminační analýzy vyžaduje mimo jiné splnění předpokladu vícerozměrné normality rozdělení diskriminátorů. Většina výzkumných prací však vícerozměrnou normalitu netestuje a předpokládá, že modely jsou dostatečně robustní a dávají rozumné aproximace i bez splnění tohoto předpokladu. Negativní výsledky testování jednorozměrné normality jednotlivých diskriminátorů pomocí testu kombinace šikmosti a špičatosti vedly k závěru, že nemůže být splněna ani vícerozměrná normalita rozložení všech diskriminátorů. Proto bylo rozložení dat upraveno pomocí vyloučení vybočujících hodnot. Datové soubory zdravých firem i firem ohrožených bankrotem byly redukovány jednak pomocí vnitřních hradeb, jednak pomocí Mahalanobisovy vzdálenosti od středních hodnot. Na redukované soubory pak byla aplikována diskriminační analýza a provedena klasifikace firem na úspěšné a neúspěšné. Z výsledků klasifikace je zřejmé, že snaha o přiblížení rozložení dat vícerozměrné normalitě vede k podstatnému zlepšení klasifikační schopnosti diskriminační funkce.

20

MONITORING GNSS TEST BASE STABILITY

Marie SUBIKOVÁ [1], *Rostislav DANDOŠ* [2]

[1] *Ing., Institute of Geodesy and Mine Surveying, Faculty of Mining and Geology, VSB-Technical University of Ostrava,*
17. listopadu 15, 708 33 Ostrava Poruba
e-mail: marie.subikova.st@vsb.cz

[2] *Ing., Institute of Geodesy and Mine Surveying, Faculty of Mining and Geology, VSB-Technical University of Ostrava ,*
17. listopadu 15, 708 33 Ostrava Poruba
e-mail: rostislav.dandos.st@vsb.cz

Abstract

The article deals with monitoring the stability of the geodetic base Skalka. The introduction of the article briefly describes the history and purpose of the geodetic base Skalka, gradually resulting in the current characteristics of the base. The main part of the article deals with monitoring the stability of points of the inner and partly outer part of the base. The result is the evaluation of performed geodetic works with the assessment of the testing base stability. Identified vertical and horizontal shifts show that the base can be considered stable in terms of both geological and geodetic points of view.

Abstrakt

Článek se zabývá sledováním stability základny Skalka. V úvodu článku je stručně popsána historie a účel geodetické základny Skalka, která postupně navazuje na současnou charakteristiku základny. Hlavní část článku se zabývá sledováním stability bodů vnitřní a částečně i vnější části základny. Výsledkem je zhodnocení geodetických prací s posouzením stability testovací základny. Zjištěné vertikální a horizontální posuny prokazují, že lze testovanou základnu považovat za stabilní a to z hlediska geologického i geodetického.

Key words: GNSS, test base Skalka, vertical and horizontal shifts

1 INTRODUCTION

The test base Skalka, originally a national satellite station, is located in the village of Kostelní Střimelice. Originally, this part of the Pecný geodetic observatory was used to observe artificial Earth satellites; since 2000, the base Skalka has been used to test the functionality of GNSS apparatus.

The Skalka base lies mainly on bedrock of the Skalka hill. From the geological point of view, it is a very stable area, but still the base stability must be inspected and surveyed on a regular basis. The base is also measured by the GNSS technology when the survey results are used to draw up calibration protocols. These measurement results are not included in the stability assessment as they are the result of the testing complex "meter + software + GNSS equipment".

When testing the complex, the survey method (the accuracy of centration over points of the outer base, the method of determining the antenna height) and the subsequent data processing are assessed. Based on the size of deviations between reference and determined coordinates, calibration protocols must be drawn up. As the reference coordinates, the coordinates and heights determined using classical geodetic methods with high precision are taken.

2 HISTORY AND PRIMARY PURPOSE OF GEODETIC BASE SKALKA

According to [1], the Geodetic Observatory Skalka, originally the National Satellite Station Skalka, was built as an out-station of the Geodetic Observatory Pecný from 1962 to 1966. The main idea of the station was the utilization of the National Satellite Station for observations of artificial Earth satellites. Originally, the station should be located in the area of the Geodetic Observatory Pecný, but due to filling the observatory with astronomical instruments and observational houses preventing a good view to the horizon, the satellite station was built on a bare hill of Skalka.

Fig. 1 Air chamber Rb- 75 prepared for monitoring artificial Earth satellites

The satellite observations at the Skalka station were performed mainly by means of a photogrammetric method using stationary chambers Rb-75, Fig. 1. In 1969, according to [1], the position of Šankovský Grúň in the eastern Slovakia in relation to a point on Skalka was experimentally determined, using simultaneous observations of artificial Earth satellites. In 1969, the satellite chamber SBG (producer Carl-Zeiss Jena (GDR)) was installed at the observatory. In 1970, the first Czechoslovak laser rangefinder to measure the distances to satellites was tested in this chamber, see Fig.1. The rangefinder was equipped with a pulsed laser; the optical system of the chamber was then adapted for receiving reflected signals.

Fig. 2 The SBG chamber for photographic observations of artificial Earth satellites

In the years 1984 - 1989, groups of surveyors from the USSR performed two surveying campaigns of Doppler observations of NNSS Transit navigation satellites on Skalka. The photographic methods for monitoring artificial Earth satellites were completed in 1990. In 1991, the first GPS receiver was obtained, which was installed at the Geodetic Observatory Pecný. The receiver was used for GPS measurements in networks and for experiments only, and thus the primary importance of the Skalka station, i.e. observations of artificial Earth satellites, fell off. Since 1995, permanent GPS observations have been made at the Geodetic Observatory Pecný, performed mainly for the International GNSS Service (IGS). The data from permanent observations is sent e.g. to the GDDIS data centre where the data is used to determine the parameters of rotation of Earth and orbits. In

1999-2000, the base for testing GNSS equipment was built at the Geodetic Observatory Skalka, and thus its importance for the field of satellite geodesy increased again.

3 SKALKA, TEST BASE FOR GNSS EQUIPMENT

Test test base Skalka, Fig. 3, was built in 1999 by the Research Institute of Geodesy, Topography and Cartography in the area of the then Astronomical and Geodetic Observatory. The reason for building the base was the need to authenticate the GPS apparatus functionality, user software as well as the way of measurement by means of the apparatus and the subsequent evaluation of the measured data.

The base is divided into 3 parts - an inner base, outer base and connecting base. Each part of the base has different monumentation and performs different functions as well.

The inner part of the base is formed by five pillars with a forced centring system. The distribution of the pillars network is adapted to suit the requirement to ensure a free horizon above the horizon of about 10° as well as mutual visibility between the pillars. The coordinates of this part of the base are designed with the greatest possible precision. The maximum distance between the points is 224 m and the elevation is about 21 m. Two raised pillars spaced about 3 m are a part of the inner base. These pillars are used to identify positions of phase centres.

The outer base consists of three points of photogrammetric testing field control, levelling point and trigonometric point. The points are monumented with granite beams and fitted with a protective stave. This part of the base is used for verifying practical use of GPS [6] by means of a tripod and also for measuring the antenna height above the point.

According to [2], the connecting base consists of the GOPE station of IGS and EUREF permanent services, and then the trigonometric point Pecný (which was used in the first international GPS campaign in Czechoslovakia EUREF-CS-H/91 in 1991). The GOPE point is the reference point of the International Terrestrial Reference System (ITRS), which enables accurate connecting test measurements to the current geocentric system. According to [3], the Pecný point is monumented with a granite prism sealed to the base of the surveying tower with concrete. On the upper gallery, there is a central pillar of the tower, which is used for weighing the trigonometric point. Six pillars are symmetrically deployed around the central pillar. The Pecný point is secured with four locking points.

Fig. 3 The distribution of points of the geodetic base Skalka

Legend:

●   Points of connecting base

●   Points of inner base

●   Points of outer "technical" base

4 MONITORING BASE STABILITY

The term of monitoring stability is meant to be the finding out of horizontal and vertical shifts of points between individual measured stages. Surveying the inner and partly the outer base was made in the years 2000 to 2012. The height point stability was measured in a trigonometric way and also using the precise levelling. The positional stability was verified by means of the radius bar method.

The instrumentation used for surveying was always the most precise one for the certain time and had calibration protocols.

The survey was performed with classical geodetic methods, and for a subsequent calculation and alignment of coordinates, the local coordinate system was used, Fig. 4. According to [4], the origin of the local coordinate system is in the point 15 ($y = 0$ m and $x = 0$ m) and the positive X-axis is inserted into the point 11 ($y=0$ m, $x=s_{11,15}$). The starting point for the trigonometric determination of heights is the point 13 ($z = 100$ m).

Fig. 4 *L*ocal coordinate system with recording positions of points

The surveying methodology was the same in all phases. For the spatial determination of points, directions were measured (horizontal and vertical ones) in two groups, and lengths were measured bidirectionally. The precise levelling was measured 2 times within the vertical indication field.

Fig. 4 indicates the orientation of axes in the local system as well as the situation with the deployment of surveyed points. Total measurements were performed at all points of the inner base and two closest points of the outer base.

4.1 Instrumentation and its accuracy

The instrumentation used for surveying was always the most precise one for the certain time, and had calibration protocols.

In 2000, according to [7], the base was spatially surveyed by Kateřina Plecháčková, a student of the Faculty of Civil Engineering of the Czech Technical University in Prague. The student used the theodolite Wild T3000 with angular accuracy of ± 5mgon for angular measurements and for distance measurements – the electro-optical rangefinder Wild DI2000 with the accuracy of length determination of 1 mm +1 ppm.

In 2001, according to [8], the base heights were verified by the precise levelling. The participants of measurements were Ing. Vojtech Pálinkáš and Ing. Jakub Kostelecký. To verify the heights, the levelling device Zeiss Koni 007 and 3 m long levelling rod graduated by 5mm, were used.

In 2005, [9], the base was spatially surveyed by Ing. Jiří Lechner, CSc., Ing. Ilya Umnov and Ing. Mark Krátký. The total station Leica TCA 2003 was used for measurements with an angular accuracy of ± 0.15 mgon and length precision of 1 mm +1 ppm. The base was subsequently measured by means of the precise levelling method using the levelling device Zeiss H05 with an invar levelling rod.

In 2007, according to [10], the base heights were verified by the method of precise levelling. The measurements were made by Ing. Jiří Lechner, CSc., Ing. Ladislav Červinka, Ing. Ilya Umnov and Ing. Jiří Kratochvíl. The levelling device Zeiss H05 was used for measurements. In 2007, the base was measured also by the spatial polar method using the Total Station Leica TCA 2003. The coordinates of 2007 were consistent with the coordinates of 2008, and therefore not listed in Tab. 1 and in Tab. 2.

In 2008, according to [4], a survey was performed by Ivan Majorník, a student of the Faculty of Civil Engineering of the Czech Technical University in Prague, within his bachelor's thesis. The student used the universal electrooptical theodolite Leica TCA2003 for surveying. The instrument measures lengths with an accuracy of 1 mm +1 ppm and angles with an accuracy of ± 0.15 mgon.

For the last time, the base was surveyed by the authors Ing. Marie Subiková and Ing. Rostislav Dandoš. The survey was performed using the spatial polar method and heights were also verified by the method of precise levelling. For the spatial surveying, the Leica TS30 total station was used, with an accuracy of measuring lengths of 1mm +1 ppm and angular accuracy of ± 0.05 mgon. To verify the heights, the Leica DNA03 Digital Levelling Device was used with a moderate mileage error of ± 0.2 mm and a code invar rod.

4.2 Monitoring position stability

According to [5], it is possible to perform a comparison of positional coordinates of individual stages within the previous measurements. The comparison was always carried out in relation to the first stage (2000). It is possible to carry out a mutual comparison of individual stage, but due to the fact that Tab. 2 shows millimetre horizontal shifts in comparison with the measurements performed between 2000 and 2012, there is no need for further comparisons.

Tab. 1 Coordinates of points in the local network using a classical geodesy method

Point number (j)	2000		2005		2008		2012	
	y[m]	x[m]	y[m]	x[m]	y[m]	x[m]	Y[m]	x[m]
11	0.000	223.328	-	-	0.000	223.330	0	223.330
12	2.959	222.100	2.959	222.101	2.959	222.101	2.959	222.101
13	37.779	74.548	37.781	74.549	37.781	74.549	37.781	74.549
14	17.632	42.058	17.633	42.058	17.633	42.058	17.633	42.058
15	0.000	0.000	0.000	0.000	0.000	0.000	0.000	0.000
31	-24.691	265.735	-24.691	265.735	-24.691	265.735	-24.691	265.735
32	35.200	335.022	35.200	335.022	35.200	335.022	35.200	335.022

The coordinates given in Tab. 1 are calculated based on the formula (1) and then aligned within the network.

$$y_j = y_{15} + s_{15,j} \cdot \sin \sigma_{15,j}$$
$$x_j = x_{15} + s_{15,j} \cdot \cos \sigma_{15,j}$$

(1)

σ - bearing from point 15 to point j [gon]

$s_{15,j}$ - horizontal distance between point 15 and point j [m]

i - marking of measurement phase – year of measurements (2000,2001,2005 etc.)

j - marking of point

Tab. 2 Comparison of coordinates in relation to the default stage in 2000

Number of point (j)	2000-2005			2000-2008			2000-2012		
	$\Delta y_{2000,2005}$ [mm]	$\Delta x_{2000,2005}$ [mm]	$\Delta p_{2000,2005}$ [mm]	$\Delta y_{2000,2007}$ [mm]	$\Delta x_{2000,2007}$ [mm]	$\Delta p_{2000,2007}$ [mm]	$\Delta y_{2000,2012}$ [mm]	$\Delta x_{2000,2012}$ [mm]	$\Delta p_{2000,2012}$ [mm]
11	-	-	-	0.0	1.6	1.6	0.0	1.8	1.8
12	-0.4	1.4	1.5	-0.4	1.4	1.5	-0.5	1.2	1.3
13	2.5	1.3	2.8	2.5	1.3	2.8	2.9	1.4	3.2
14	0.6	0.2	0.6	0.6	0.2	0.6	0.7	0.5	0.9
15	Fixed point of local system								
31	0.0	0.0	0.0	0.0	0.0	0.0	0.2	-0.1	0.2
32	0.0	0.0	0.0	0.0	0.0	0.0	0.2	-0.5	0.5

Differences of coordinates and subsequent total horizontal shift of points are calculated on the basis of the relation (2).

$$\Delta y_{2000i} = y_i - y_{2000}$$
$$\Delta x_{2000i} = x_i - x_{2000}$$
$$\Delta p_{2000i} = \pm\sqrt{\Delta y_{2000i}^2 + \Delta x_{2000i}^2}$$

(2)

4.3 Monitoring height stability

In the area of the base being surveyed, the Vertical Indication Field Pecný is located, which is a part of the levelling line of the second order Jac Nespeky- Oleška. The levelling line in the area of Skalka is led through individual points of the indication field, through the gravity point and also through levelling marks located on the tops of pillars. The point stability is assessed on the basis of differences in height between the point 13, which is taken as the starting point when determining trigonometric heights, and the other base points.

Tab. 3 Differences in elevation for individual points in relation to the point 13 determined in a trigonometric way

Point number (j)	$\Delta H_{13,j}^{i(2000)}$ [m]			$\Delta h_{13,j}^{2000i}$ [mm]	
	2000	2008	2012	2000-2008	2000-2012
13	Point height taken as starting point for comparison				
11	-20.583	-20.583	-20.582	0	1
12	-20.462	-20.465	-20.460	-3	2
14	-0.743	-0.740	-0.744	3	-1
15	-3.618	-3.617	-3.620	1	-2
31	-24.941	-	-24.949	-	-8
32	-27.362	-	-27.374	-	-12

The heights are determined in a trigonometric way based on the relationship (3). The formula (4 a, b), i.e. the calculation of the difference in elevation between the point 13 and j and the calculation of vertical shifts, is valid for heights determined in a trigonometric way, Tab. 3, or by the method of precise levelling, Tab. 4.

$$H_j^{i(2000)} = H_{13} + s_{13,j} \cdot \cot g \; z_{13,j} + v_p - v_S \tag{3}$$

$$\Delta H_{13,j}^{i(2000)} = H_j - H_{13} \tag{4a}$$

$$\Delta h_{13,j}^{2000i} = \Delta h_{13,j}^{\cdot i} - \Delta h_{13,j}^{2000} \tag{4b}$$

H_j - height of point j [m]

H_{13} - height of point 13 [m]

$s_{13,j}$ - distance between point 13 and determined point m j [m]

$z_{13,j}$ - zenith angle measured from point 13 to point j[m]

v_p - instrument height [m]

v_p - signal height [m]

Tab. 4 Differences in elevation of individual points in relation to the point 13, the point heights are determined by the method of precise levelling

Point number	$\Delta H_{13,j}^{i(2000)}$ [m]				$\Delta h_{13,j}^{2000i}$ [mm]		
(j)	2001	2005	2007	2012	2001-2005	2001-2007	2001-2012
13	Point height taken as starting point for comparison						
11	-20.584	-20.583	-20.584	-20.584	1.5	-0.1	-0.3
12	-20.462	-20.460	-20.462	-20.462	1.6	0	-0.1
14	-0.743	-0.742	-0.743	-0.741	0.8	0.3	2.2
15	-3.617	-	-	-3.617	-	-	0.1
31	-	-24.933	-	-24.936	-	-	-2,4*
32	-	-27.358	-27.360	-27.360	-	-2*	-1.6*

*Note: The declines marked with * are determined in relation to the stage 2005*

5 CONCLUSIONS

This paper describes the way in which the stability of points of the geodetic base Skalka is monitored. The base is part of the Geodetic Observatory Pecný and originally served to observations of artificial Earth satellites. At present, the base is used primarily to testing the GPS apparatus. The testing is performed on the basis of the distance between the determined coordinates and the reference coordinates.

The base was built in 1999-2000 and since then it has been being regularly measured. Based on these measurements, the stability of the base, horizontal and vertical shifts in relation to the initial stage measured in 2000, is determined. Measurements are always carried out with the latest and most accurate geodetic instruments. Coordinates of base points are calculated in a local system, Fig. 4, and heights are calculated in a trigonometric way in relation to the point 13 and, from the levelling point of view, are related to the vertical indication field which is built on the premises.

Horizontal shifts Δp_{2000i} and vertical shifts $\Delta h_{13,j}^{2000i}$ vary in the order of mm. A maximum horizontal shift is at the point No. 13 – 3,2mm and a maximum vertical shift is at the point No. 31 – 2,4mm.

Vertical shifts were also determined in a trigonometric way, which is a less accurate method, but even so the differences range from -3 to 3 mm. The exception is the points 31 and 32 at which the vertical shift is about 1cm. This variation may be due to the inaccurate determination of the machine height over the point monumentation of the outer base, as it is not exactly possible to assign the end of the band to the defined point. Given that the points 31 and 32 are monumented by the "classical" geodesic monumentation, the vertical shift could occur during the monitored period as well.

The points inside the base are located on hard bedrock (granite) and it can reasonably be expected that their stability is not compromised in any way. Monitoring the height as well as positional stability shows a max shift ± 3 mm in about 12 years, which are negligible shifts due to the size of the entire base.

REFERENCES

[1]   Historie observatoře Pecny, [cit. 16/07/2012]. Available at: http://oko.asu.cas.cz/pecny/histo.html

[2]   STAŇKOVÁ, H. & ČERNOTA, P.: Principle of Forming and Developing Geodetic Bases in the Czech Republic, *Geodesy and Cartography*, Vilnius Technica, 2010, Vol. 36, No. 3p. 103- 112, ISSN 1392-1541 print / ISSN 1648-3502 online

[3]   Testovací základna pro GPS [cit. 16/07/2012]. Available at: http://oko.asu.cas.cz/pecny/zgpspol.html

[4]   MAJORNÍK, I.: *Sledování stability GPS základny Skalka*. Bakalářská práce, 2008, [cit. 2012-16-7] Available at: http://gama.fsv.cvut.cz/~cepek/proj/bp/2008/ivan-majornik-bp-2008.pdf

[5]   VILLIM, A. & HODAS, S. & STAŇKOVÁ, H.: Spoločné spracovanie družicových a terestrických meraní v priestorovej sieti pre dopravnú infraštruktúru, *Civil and Enviromental Engineering*, CEE/ SEI SvF ŽU v Žiline, December 2011, vol 7th/ 7, issue 2/2011, str.126- 138, ISSN 1336-5835

[6]   VITÁSKOVÁ, J.& STAŇKOVÁ, H.:*Návody na měření s GPS*, MZLU v Beně, VŠB- TU Ostrava, Brno 2004, ISBN 80- 7157- 828- 2

[7]   KOSTELECKÝ, J. (jr) : *Zaměření testovací základny pro GPS pozemními metodami (shrnutí prací provedených v roce 2000)*, Technická zpráva 1008/2000, VÚGTK Zdiby

[8]   KOSTELECKÝ, J. (jr): *Ověření výšek testovací základny pro GPS pomocí nivelace*, Technická zpráva 1015/2001, VÚGTK Zdiby

[9]   LECHNER, J.:*Technická zpráva o polohovém a výškovém měření v síti geodetických bodů testovací základny pro GPS na Geodetické observatoři Pecný i Skalka- o metrologickém navázání v parametru úhel a délka*, Technická zpráva 1086/2005, VÚGTK Zdiby´

[10]   LECHNER, J. & ČERVINKA, L. & UMNOV, I. & KRATOCHVÍL, J.:*Délkové, výškové a polohové určení geodetické bodové sítě Výzkumného ústavu geodetického, topografického a kartografického v areálu Skalka*, Technická zpráva VÚGTK Zdiby, 2007

RESUMÉ

Ověření funkčnosti a správnosti klasických geodetických přístrojů i GNSS aparatur je v současné době považována za zcela běžnou. Klasické geodetické přístroje se ověřují v autorizovaných metrologických střediscích, případně v oprávněných laboratořích. Funkčnost GNSS aparatur se v České republice ověřuje pouze na geodetické základně Skalka.

The article deals with monitoring the stability of the geodetic base Skalka. Základna se nachází v obci Kostelní Střimelice, na vrchu Skalka a je součástí Geodetické observatoře Pecný. Původně základna sloužila k fotografickým pozorováním umělých družic Země, ale v současné době je využívána hlavně k testování GNSS aparatur. Základna je rozdělena na tři části a každá část plní jinou funkci a má jiný druh stabilizace. Vnitřní část základny slouží k ověření funkčnosti GNSS aparatury a k stanovení polohy fázových center antény přijímače. Vnější, technická část základny ověřuje způsob horizontace a centrace GNSS aparatury nad bodem a také způsob určování výšky antény. Poslední část základny slouží k navázání testovacích měření na aktuální geocentrický systém.

Stěžejní částí článku je sledování stability bodů vnitřní a částečně i vnější části základny. Souřadnice bodů základny jsou určovány s vysokou přesností polární prostorové metody a kontrolně jsou výšky proměřeny přesnou nivelací. Základna byla proměřena v letech 2000 až 2012, kdy pro zaměření byla použita nejmodernější technika na tehdejší dobu. Stabilita základny se posuzuje na základě horizontálních a vertikálních posunů. Vertikální a horizontální posuny, zjištěné při porovnání měření z let 2000 až 2012 vykazují maximální posun ±3mm. Lze tedy prohlásit, že geodetická základna je stabilní a zjištěné posuny jsou vzhledem k rozloze celé základny zanedbatelné.

Permissions

All chapters in this book were first published in GE, by De Gruyter; hereby published with permission under the Creative Commons Attribution License or equivalent. Every chapter published in this book has been scrutinized by our experts. Their significance has been extensively debated. The topics covered herein carry significant findings which will fuel the growth of the discipline. They may even be implemented as practical applications or may be referred to as a beginning point for another development.

The contributors of this book come from diverse backgrounds, making this book a truly international effort. This book will bring forth new frontiers with its revolutionizing research information and detailed analysis of the nascent developments around the world.

We would like to thank all the contributing authors for lending their expertise to make the book truly unique. They have played a crucial role in the development of this book. Without their invaluable contributions this book wouldn't have been possible. They have made vital efforts to compile up to date information on the varied aspects of this subject to make this book a valuable addition to the collection of many professionals and students.

This book was conceptualized with the vision of imparting up-to-date information and advanced data in this field. To ensure the same, a matchless editorial board was set up. Every individual on the board went through rigorous rounds of assessment to prove their worth. After which they invested a large part of their time researching and compiling the most relevant data for our readers.

The editorial board has been involved in producing this book since its inception. They have spent rigorous hours researching and exploring the diverse topics which have resulted in the successful publishing of this book. They have passed on their knowledge of decades through this book. To expedite this challenging task, the publisher supported the team at every step. A small team of assistant editors was also appointed to further simplify the editing procedure and attain best results for the readers.

Apart from the editorial board, the designing team has also invested a significant amount of their time in understanding the subject and creating the most relevant covers. They scrutinized every image to scout for the most suitable representation of the subject and create an appropriate cover for the book.

The publishing team has been an ardent support to the editorial, designing and production team. Their endless efforts to recruit the best for this project, has resulted in the accomplishment of this book. They are a veteran in the field of academics and their pool of knowledge is as vast as their experience in printing. Their expertise and guidance has proved useful at every step. Their uncompromising quality standards have made this book an exceptional effort. Their encouragement from time to time has been an inspiration for everyone.

The publisher and the editorial board hope that this book will prove to be a valuable piece of knowledge for researchers, students, practitioners and scholars across the globe.

List of Contributors

Juraj GAŠINEC
Institute of Geodesy, Cartography and Geographic Information Systems, Faculty of Mining, Ecology, Process Control and Geotechnologies, Technical University of Košice, Park Komenského 19, 043 84 Košice, Slovak Republic

Silvia GAŠINCOVÁ
Institute of Geodesy, Cartography and Geographic Information Systems, Faculty of Mining, Ecology, Process Control and Geotechnologies, Technical University of Košice, Park Komenského 19, 043 84 Košice, Slovak Republic

Vladislava ZELIZŇAKOVÁ
Institute of Geodesy, Cartography and Geographic Information Systems, Faculty of Mining, Ecology, Process Control and Geotechnologies, Technical University of Košice, Park Komenského 19, 043 84 Košice, Slovak Republic

Jana PALKOVÁ
Institute of Geodesy, Cartography and Geographic Information Systems, Faculty of Mining, Ecology, Process Control and Geotechnologies, Technical University of Košice, Park Komenského 19, 043 84 Košice, Slovak Republic

Žofia KUZEVIČOVÁ
Institute of Geodesy, Cartography and Geographic Information Systems, Faculty of Mining, Ecology, Process Control and Geotechnologies, Technical University of Košice Park Komenského 19, 043 84 Košice, Slovak Republic

Eva JIRÁNKOVÁ
Institute of Geodesy and Mine Surveying, Faculty of Mining and Geology VŠB - Technical University of Ostrava, 17. listopadu 15, Ostrava

Slavomír LABANT
Institute of Geodesy, Cartography and Geographic Information Systems, Faculty of Mining, Ecology, Process Control and Geotechnologies, Technical University of Košice, Park Komenského 19, 043 84 Košice, Slovak Republic

Hana STAŇKOVÁ
Institute of Geodesy and Mine Surveying, Faculty of Mining and Geology, VSB-Technical University of Ostrava, 17. listopadu 15, 708 33 Ostrava - Poruba, Czech Republic

Roland WEISS
Institute of Geotourism, Faculty of Mining, Ecology, Process Control and Geotechnologies, Technical University of Košice, Letná 9, 042 00 Košice, Slovak Republic

Michal VANĚK
Institute of Economics and Control Systems, Faculty of Mining and Geology, VŠB - Technical University of Ostrava 17. listopadu 15/2172, Ostrava

Milan MIKOLÁŠ
Institute of Mining Engineering and Safety, Faculty of Mining and Geology, VŠB - Technical University of Ostrava 17. listopadu 15/2172, Ostrava

Kateřina ŽVÁKOVÁ
Hyundai Motor Manufacturing Czech. Průmyslová zóna Nošovice, 739 51 Dobrá

Silvia GAŠINCOVÁ
Institute of Geodesy, Cartography and Geographic Information Systems, Faculty of Mining, Ecology, Process Control and Geotechnologies, Technical University of Košice Park Komenského 19, 043 84 Košice, Slovak Republic

Juraj GAŠINEC
Institute of Geodesy, Cartography and Geographic Information Systems, Faculty of Mining, Ecology, Process Control and Geotechnologies, Technical University of Košice, Park Komenského 19, 043 84 Košice, Slovak Republic

Jiří PÁNEK
Mgr., Department of Development Studies, Faculty of Science, Palacky University in Olomouc, 17. listopadu 12, Olomouc

Ivan MUDRON
Institute of Geoinformatics, Faculty of Mining and Geology, VSB-TU OSTRAVA, 17.listopadu 15/2172, 70833, Ostrava, Czech Republic

Michal PODHORANY
Inovation Centre of Excellence VSB-TU OSTRAVA, 17.listopadu 15/2172, 708 33, Ostrava, Czech Republic Department of Physical Geography and Geology, Faculty of Science, University of Ostrva, Chittussiho 10, 710 00, Ostrava, Czech Republic

Juraj CIRBUS
Institute of Geoinformatics, Faculty of Mining and Geology, VSB-TU OSTRAVA, 17.listopadu 15/2172, 70833, Ostrava, Czech Republic

Branislav DEVEČKA
Institute of Geoinformatics, Faculty of Mining and Geology, VSB-TU OSTRAVA, 17.listopadu 15/2172, 70833, Ostrava, Czech Republic

Ladislav BAKAY
Department of Garden and Landscape Design, Slovak university of Agriculture, Trieda A. Hlinku 2, 949 76, Nitra, Slovak Republic

Martin BARTÍK
Department of Natural Environment, Faculty of Forestry, Technical university in Zvolen
T. G. Masaryka 2117/24, Zvolen

Matúš HRÍBIK
Department of Natural Environment, Faculty of Forestry, Technical university in Zvolen
T. G. Masaryka 2117/24, Zvolen

Miriam HANZELOVÁ
Department of Natural Environment, Faculty of Ecology and Environmental Sciences, Technical university in Zvolen, T. G. Masaryka 2117/24, Zvolen

Jaroslav ŠKVARENINA
Department of Natural Environment, Faculty of Forestry, Technical university in Zvolen,T. G. Masaryka 2117/24, Zvolen

Luděk KOVÁŘ
Institute of Geological Engineering, Faculty of Mining and Geology, VŠB – Technical University of Ostrava, tř. 17. listopadu 15/2172, 708 33 Ostrava-Poruba

Pavel POSPÍŠIL
Institute of Geological Engineering, Faculty of Mining and Geology, VŠB – Technical University of Ostrava, tř. 17. listopadu 15/2172, 708 33 Ostrava-Poruba,

Radoslav CHUDÝ
Department of Cartography, Geoinformatics and Remote Sensing, Faculty of Natural Sciences, Comenius University Mlynská dolina, 842 15, Bratislava, Slovenská republika

Martin IRING
Department of Cartography, Geoinformatics and Remote Sensing, Faculty of Natural Sciences, Comenius University Mlynská dolina, 842 15, Bratislava, Slovenská republika

Richard FECISKANIN
Department of Cartography, Geoinformatics and Remote Sensing, Faculty of Natural Sciences, Comenius University Mlynská dolina, 842 15, Bratislava, Slovenská republika

Jan THOMAS
Institute of Environmental Engineering , Faculty of Mining and Geology, VŠB-Technical University of Ostrava

Miroslav KYNCL
Institute of Environmental Engineering, Faculty of Mining and Geology, VŠB-Technical University of Ostrava

Silvie LANGAROVÁ
SmVaK Ostrava a.s., 28. října 169, Ostrava

Martin KLEMPA
Institute of Geological Engineering, Faculty of Mining and Geology, VSB – Technical University of Ostrava 17. listopadu 15, Ostrava Poruba

Michal PORZER
Institute of Geological Engineering, Faculty of Mining and Geology, VSB – Technical University of Ostrava 17. listopadu 15, Ostrava Poruba

Petr BUJOK
Institute of Geological Engineering, Faculty of Mining and Geology, VSB – Technical University of Ostrava 17. listopadu 15, Ostrava Poruba

Ján PAVLUŠ
Institute of Geological Engineering, Faculty of Mining and Geology, VSB – Technical University of Ostrava 17. listopadu 15, Ostrava Poruba

Ľudovít Kovanič
Institute of Geodesy, Cartography and GIS, BERG Faculty, Technical University of Košice, Letná 9, Košice, Slovak Republic

Michal VANĚK
Institute of Economics and Control Systems, Faculty of Mining and Geology, VŠB – Technical University of Ostrava, 17. listopadu 15/2172, Ostrava

Yveta TOMÁŠKOVÁ
Institute of Combined Studies in Most, Faculty of Mining and Geology, VŠB – Technical University of Ostrava, Dělnická 21, Most

Alena STRAKOVÁ
Institute of Combined Studies in Most, Faculty of Mining and Geology, VŠB – Technical University of Ostrava, Dělnická 21, Most

Kateřina ŠPAKOVSKÁ
Institute of Economics and Control Systems, Faculty of Mining and Geology, VŠB – Technical University of Ostrava, 17. listopadu 15/2172, Ostrava

Petr BORA
Institute of Economics and Control Systems, Faculty of Mining and Geology, VŠB – Technical University of Ostrava, 17. listopadu 15/2172, Ostrava

Lucie KRČMARSKÁ
Institute of Economics and Control Systems, Faculty of Mining and Geology, VŠB – Technical University of Ostrava 17. Listopadu 15, Ostrava

Igor ČERNÝ
Institute of Economics and Control Systems, Faculty of Mining and Geology, VŠB – Technical University of Ostrava 17. Listopadu 15, Ostrava

Michal VANĚK
Institute of Economics and Control Systems, Faculty of Mining and Geology, VŠB – Technical University of Ostrava 17. Listopadu 15, Ostrava

Michal Lesňák
VŠB - TU Ostrava, Institute of Physics, Faculty of Mining and Geology, VŠB –Technical University of Ostrava, 17. listopadu 15/2171, 708 33 Ostrava – Poruba

František Staněk
VŠB - TU Ostrava, Institute of Physics, Faculty of Mining and Geology, VŠB –Technical University of Ostrava, 17. listopadu 15/2171, 708 33 Ostrava – Poruba

Jaromír Pištora
Nanotechnology Centre, VŠB –Technical University of Ostrava, 17. listopadu 15/2171, 708 33 Ostrava – Poruba

Jan Procházka
VŠB - TU Ostrava, Institute of Physics, Faculty of Mining and Geology, VŠB –Technical University of Ostrava, 17. listopadu 15/2171, 708 33 Ostrava – Poruba

Vojtech DIRNER
Institute of Environmental Engineering, Faculty of Mining and Geology, VŠBTechnical University of Ostrava, 17.listopadu 15, Ostrava

Jozef KRNÁČ
Department of Environmental Management, Faculty of Natural Sciences, Matej Bel University, Tajovského 52, 974 01 Banská Bystrica, Slovensko

Lenka ČMIELOVÁ
Institute of Environmental Engineering, Faculty of Mining and Geology, VŠB-Technical University of Ostrava, 17.listopadu 15, Ostrava

Eva LACKOVÁ
Institute of Environmental Engineering, Faculty of Mining and Geology, VŠB-Technical University of Ostrava, 17.listopadu 15, Ostrava

Peter ANDRÁŠ
Geological Institute, Slovak Academy of Sciences, Ďumbierska 1, 974 01 Banská Bystrica, Slovensko

Bladimir CERVANTES
Institute of Geological Engineering. Faculty of Mining and Geology, VŠB – Technical University of Ostrava, tř. 17. listopadu 15/2172, 708 33 Ostrava- Poruba

Aleš POLÁČEK
Institute of Geological Engineering. Faculty of Mining and Geology, VŠB – Technical University of Ostrava, tř. 17. listopadu 15/2172, 708 33 Ostrava- Poruba

Jaroslav RYŠÁVKA
Unigeo a.s., Místecká 258, Ostrava- Hrabová,

Radmila SOUSEDÍKOVÁ
Institute of Economics and Control Systems, Faculty of Mining and Geology, ŠB–Technical University of Ostrava 17. listopadu 15, Ostrava-Poruba

Jaroslav DVOŘÁČEK
Institute of Economics and Control Systems, Faculty of Mining and Geology, VŠB–Technical University of Ostrava 17. listopadu 15, Ostrava-Poruba

Igor SAVIČ
Cybex Industrial LTD, C/O Columbus Trading-Partners GMBH Riedinger str. 18, 95448 Bayreuth, German

Marie SUBIKOVÁ
Institute of Geodesy and Mine Surveying, Faculty of Mining and Geology, VSB-Technical University of Ostrava, 17. listopadu 15, 708 33 Ostrava Poruba

Rostislav DANDOŠ
Institute of Geodesy and Mine Surveying, Faculty of Mining and Geology, VSB-Technical University of Ostrava, 17. listopadu 15, 708 33 Ostrava Poruba